L.H. Greene
March 1990

To Laura,
Good Luck in
all your life's work —

SUPERCONDUCTING DEVICES

SUPERCONDUCTING DEVICES

Edited by

Steven T. Ruggiero

Department of Physics
University of Notre Dame
Notre Dame, Indiana

David A. Rudman

Department of Materials Science and Engineering
Massachusetts Institute of Technology
Cambridge, Massachusetts

ACADEMIC PRESS, INC.
Harcourt Brace Jovanovich, Publishers
Boston   San Diego   New York
Berkeley   London   Sydney
Tokyo   Toronto

This book is printed on acid-free paper. ∞

Copyright © 1990 by Academic Press, Inc.
All rights reserved.
No part of this publication may be reproduced or
transmitted in any form or by any means, electronic
or mechanical, including photocopy, recording, or
any information storage and retrieval system, without
permission in writing from the publisher.

ACADEMIC PRESS, INC.
1250 Sixth Avenue, San Diego, CA 92101

United Kingdom Edition published by
ACADEMIC PRESS LIMITED
24–28 Oval Road, London NW1 7DX

Library of Congress Cataloging-in-Publication Data

Superconducting devices/edited by Steven T. Ruggiero, David A.
  Rudman.
    p. cm.
    Includes bibliographical references.
    ISBN 0-12-601715-8 (alk. paper)
    1. Superconductors.   I. Ruggiero, Steven T.   II. Rudman, David
  Albert.
    TK7872.S8S79   1990
    621.381'52—dc20   89-17902
                                                             CIP

Printed in the United States of America
90 91 92 93   9 8 7 6 5 4 3 2 1

CONTENTS

Contributors		ix
Preface		xi
1.	**New Possibilities for Superconductor Devices**	1
	K.K. Likharev, V.K. Semenov, and A.B. Zorin	
	1. Introduction	2
	2. High-T_c Superconductors	3
	3. Operating at Nitrogen Temperatures	9
	4. Components and Simple Devices	12
	5. Analog Devices	14
	6. Digital Integrated Circuits	30
	7. Conclusion	41
	Acknowledgments	42
2.	**SQUIDS: Principles, Noise and Applications**	51
	John Clarke	
	1. Introduction	52
	2. The Resistively Shunted Junction	53
	3. The dc SQUID	56
	4. The rf SQUID	66
	5. SQUID-Based Instruments	73
	6. The Impact of High-Temperature Superconductivity	86
	7. Concluding Remarks	93
	Acknowledgments	94
3.	**Computing**	101
	Hisao Hayakawa	
	1. Introduction	101
	2. Advantages of Josephson Switching Devices	103

	3.	Josephson Switching Gates	105
	4.	Logic Gates	107
	5.	Memory Cells	118
	6.	Circuits	123
4.	**Josephson Arrays as High Frequency Sources**		135

James Lukens

	1.	Introduction	135
	2.	Single-Junction Sources	137
	3.	Arrays	142
	4.	Phase-Locking	146
	5.	Phase-Locking in Arrays	149
	6.	Distributed Arrays	159
	7.	Prospects for the Future	163
		Acknowledgments	165
5.	**Quasiparticle Mixers and Detectors**		169

Qing Hu and P.L. Richards

	1.	Introduction	169
	2.	Photon-Assisted Tunneling	172
	3.	Quasiparticle Mixing	174
	4.	Mixer Noise	177
	5.	Imbedding Admittance and Computer Modeling	179
	6.	LO Power and Saturation	181
	7.	Series Arrays of Junctions	182
	8.	Types of Junctions	183
	9.	Quasiparticle Mixer Measurements	184
	10.	Quasiparticle Direct Detector	190
		Acknowledgments	193
6.	**Digital Signal Processing**		197

Theodore Van Duzer and Gregory Lee

	1.	Introduction	197
	2.	Analog-to-Digital Converters	199
	3.	Shift Registers	219
7.	**Wideband Analog Signal Processing**		227

Richard S. Withers

	1.	Functional Overview of Signal Processing	228
	2.	Mature Signal-Processing Device Technologies	236

	3. Superconductive Devices	238
	4. Fabrication Technology	255
	5. System Integration	261
	6. Comparisons and Conclusions	264
	Acknowledgments	269

8. "MBE" Growth of Superconducting Materials — 273

A. I. Braginski and J. Talvacchio

1. Introduction — 273
2. "MBE" Apparatus for Superconductor Growth — 277
3. Materials Systems — 285
4. Conclusions — 317

9. Three-Terminal Devices — 325

A.W. Kleinsasser and W.J. Gallagher

1. Introduction — 325
2. Hybrid Superconductor-Semiconductor Devices — 328
3. Nonequilibrium Superconducting Devices — 353
4. Magnetic and Other Devices — 364
5. Conclusions — 366
   Acknowledgments — 367

10. Artificial Tunnel Barriers — 373

S.T. Ruggiero

1. Introduction — 373
2. Early Work — 374
3. Semiconductor and Surface-Layer Barriers — 377
4. Deposited-Oxide Barriers — 378
5. Conclusions — 386

Index — 391

CONTRIBUTORS

Numbers in parentheses refer to the pages on which the authors' contributions begin.

A.I. Braginski (273), *Institut fur Schicht-und Ionentechnik (ISI), Kernforschungsanlage Juelich GmbH, Postfach 1913, Juelich D-5170, Federal Republic of Germany*

John Clarke (51), *Materials and Chemical Sciences Division, Lawrence Berkeley Laboratory, One Cyclotron Road, Berkeley, California 94720*

W.J. Gallagher (325), *IBM Research Division, T.J. Watson Research Center, P.O. Box 218, Yorktown Heights, New York 10598*

Hisao Hayakawa (101), *Department of Electronics Engineering, Nagoya University, Furo-cho, Chikusa-ku, Nagoya, Japan*

Qing Hu (169), *Materials and Chemical Sciences Division, Lawrence Berkeley Laboratory, One Cyclotron Road, Berkeley, California 94720*

A.W. Kleinsasser (325), *IBM Research Division, T.J. Watson Research Center, P.O. Box 218, Yorktown Heights, New York 10598*

Gregory Lee (197), *TRW Space and Technology Group, Redondo Beach, California 90278*

K.K. Likharev (1), *Department of Physics, Moscow State University, Moscow 119899 GSP, U.S.S.R.*

James Lukens (135), *Department of Physics, State University of New York, Stony Brook, New York 11794*

P.L. Richards (169), *Materials and Chemical Sciences Division, Lawrence Berkeley Laboratory, One Cyclotron Road, Berkeley, California 94720*

S.T. Ruggiero (373), *Department of Physics, University of Notre Dame, Notre Dame, Indiana 46556*

V.K. Semenov (1), *Department of Physics, Moscow State University, Moscow 119899 GSP, U.S.S.R.*

CONTRIBUTORS

J. Talvacchio (273), *Westinghouse Science & Technology Center, Pittsburgh, Pennsylvania 15235*

Theodore Van Duzer (197), *Department of Electrical Engineering, and the Electronics Research Laboratory, University of California, Berkeley, California 94720*

Richard S. Withers (227), *Analog Device Technology Group, MIT Lincoln Laboratory, Lexington, Massachusetts 02173*

A.B. Zorin (1), *Department of Physics, Moscow State University, Moscow 119899 GSP, U.S.S.R.*

PREFACE

Superconductivity has certainly come of age with the discovery of high-temperature materials. The field now enjoys a broad level of awareness among physicists, materials scientists, chemists, electrical engineers, and even the general public. The timeliness of the current volume of articles on superconducting devices with respect to these new discoveries has placed the material in a new light.

This collection presents an up-to-date discussion of the theory, fabrication, and qualification of superconducting device elements and integrated circuitry. The chapters present an in-depth look at issues key to the development of practical superconducting devices and systems. Included in the work are discussions of new ideas for superconducting device structures (including three-terminal devices) along with detailed discussions of device fabrication, with an emphasis on tunnel junctions and tunneling barriers. Integrated systems, including the fabrication and application of SQUIDs, Josephson arrays, microwave detectors, digital signal processors and computers, and analog signal processors are all discussed.

At present, superconducting devices are employed in a number of areas where their speed, sensitivity, or other characteristics stemming from the unique nature of superconductivity make them the device of choice. Superconducting devices fall into two basic classes: SQUID systems, which are designed to measure directly magnetic flux; and Josephson devices, which take advantage of the unique electrical characteristics of Josephson junctions—most often tunnel junctions—to perform traditional electronic functions.

Because of their quantum-level sensitivity to magnetic flux, SQUIDs remain without peer as practical ultra-sensitive magnetic-field detectors. SQUID systems are employed in passive medical diagnostics, in mineral surveying, in submarine detection, in relative motion detection, and in a variety of scientific instruments. SQUIDs can also be employed as ultra-sensitive dc voltage amplifiers. As active electronic components, superconductor devices presently serve as low-noise, high-frequency mixers, as front-end elements in high speed oscilloscopes, and in the more prosaic role of setting the magnitude of the standard volt.

We note that more and more microwave circuit designers are taking a hard look at passive superconducting elements such as resonators and filters because of their superior performance with respect to normal metals. Therefore, with the availability of superconducting phase shifters, local oscillators, and mixers, fully integrated all-superconducting receiver front ends and other sophisticated ultra-high performance all-superconducting systems are appearing on the horizon. Similar trends also seem in store for signal processing and computing systems.

SQUIDs and other active elements are employed at present only when there is no other device with comparable performance. However, with the above-noted microwave applications in mind and with the increasing number of systems applications that require cryogenic cooling (e.g., for noise–temperature considerations) it may well be that superconducting elements will soon begin to make their way into a much broader spectrum of commercial, military, and space applications.

In the face of this promise and success, however, an open challenge to the field remains: the development of a true three-terminal device. While this circuit element function can be duplicated with (often SQUID-based) combinations of Josephson device elements, a heavy price in circuit real estate, engineering flexibility, speed, and other key design considerations must be paid. This important issue is addressed in this work, both in the context of competing semiconducting systems and in light of imaginative new possibilities for addressing this important issue.

In closing, we note that contributions to this book were being written just as the new high-temperature superconductors were being discovered. It is exciting to see that these wonderful new materials are already being successfully explored as devices, due in large part to the tremendous background of knowledge and experience gained to date with more traditional superconducting systems.

We wish to thank the authors of this work for their professional and timely contributions, especially considering the unusual circumstances the high T_c revolution imposed on us all. One of us (S.T.R.) would also like to thank friends and family in South Bend for their help and support, those at Superconducting Technologies, Inc. (Santa Barbara) where some work on this volume was performed, and where a much broadened perspective on superconducting devices was acquired, and the National Science Foundation (D.M.R.) for support during the preparation of this volume. Thanks also go (from D.A.R.) to Terry Orlando at M.I.T., for support during our mutual efforts to complete our books.

S.T. Ruggiero
D.A. Rudman
December, 1989

CHAPTER 1

New Possibilities for Superconductor Devices

K.K. LIKHAREV, V.K. SEMENOV, AND A.B. ZORIN

Department of Physics
Moscow State University
Moscow, U.S.S.R.

1. Introduction 2
2. High-T_c Superconductors 3
   2.1. Basic Properties 3
   2.2. Josephson Junctions 7
3. Operating at Nitrogen Temperatures 9
   3.1. Refrigeration 9
   3.2. Thermal Fluctuations 9
4. Components and Simple Devices 12
   4.1. EMF Shields 12
   4.2. Transmission Lines and Resonators 13
   4.3. Modulators and Commutators 13
5. Analog Devices 14
   5.1. SQUIDs 14
   5.2. Samplers 20
   5.3. Receivers 22
   5.4. DC Voltage Standards and Multipliers 24
   5.5. A/D and D/A Conversion 27
6. Digital Integrated Circuits 30
   6.1. Superconductor Connections for Semiconductor ICs 31
   6.2. Josephson Junction ICs 32
   6.3. Single Electron Tunneling and Superconductivity 39
7. Conclusion 41
   Acknowledgments 42
   Notes 42
   References 43

1. Introduction

The discovery of high-T_c superconductivity of the copper oxides [1] and of new materials with $T_c \simeq 100$ K [2,3,4] have opened new prospects for applications of superconductors in many areas including electronics. In fact, the new materials allow one to increase the operation temperatures of superconductor devices from the helium range ($T \simeq 1$–20 K) to the nitrogen range ($T \simeq 60$–200 K) where refrigeration costs can be decreased by several orders of magnitude. This factor can lead to much wider use of various superconductor electronic devices in various fields of science and technology.

The transition to the nitrogen temperatures creates, however, a number of problems due to

(i) the specific properties of the high-T_c superconductors, and
(ii) the increase of the thermal fluctuations with T.

Solution of these problems would require a considerable effort combined with an essential use of experience accumulated in superconducting electronics since the late 1960s and of the new ideas and inventions that appeared in the field recently.

The only published survey of these problems we know is that by Malozemoff et al. [5]. This paper, however, deals with both large-scale and electronic applications of the high-T_c superconductors, so that the analysis of the latter is very brief and far from being complete. The purpose of the present chapter is to give a more detailed analysis of the problems faced by the superconductor electronics and possible ways of their solution.

We will start in Sections 2 and 3 with the analysis of the impact of the two factors listed above on the performance of the basic elements of superconductor electronic circuits, including Josephson junctions. The remaining sections of this chapter present a review of prospects arising in various directions of superconductor electronics. In Section 4, we will discuss superconductor components and the simplest devices. Section 5 is devoted to more complex analog devices employing the Josephson junctions, while Section 6 deals with the most complex devices based on superconductor digital integrated circuits. Finally, in Section 7 the main conclusions of our analysis are summarized.

The length of this chapter did not allow us to give comprehensive descriptions of traditional superconductor electronic devices. For those, the reader is referred to other chapters of this book.

2. High-T_c Superconductors

2.1. Basic Properties

At the time that this chapter was written, three types of single-phase superconductors with $T_c > 77$ K have been discovered and identified:

(i) "1–2–3" (or "yttrium") phases $M_1Ba_2Cu_3O_{7-\delta}$ with M = Y, Nd, La, Sm, Eu, Gd, Dy, Ho, Er, Tm, Yb, Lu [7,8] with $T_c \simeq 85$–95 K.

(ii) "Bismuth" compound $BiSrCaCu_2O_x$ [3] with $T_c \simeq 105$ K.

(iii) "Thallium" compound $Tl_2Ca_{1.5}BaCu_3O_{8.5+x}$ [4] with $T_c \simeq 120$ K.

We will concentrate on the more studied 1–2–3 structure because properties of the other high-T_c superconductors seem to be almost similar.

The high-T_c superconductors were first synthesized in the form of ceramics, i.e., disordered structures of randomly sized and oriented crystallites. The normal-state resistivity ρ (typically, from 500 to 10,000 μohm-cm) and critical supercurrent density j_c (typically, from 1 to 10^3 A/cm^2) of the ceramics are determined by weak contacts between the crystallites rather than their intrinsic properties. The low values of j_c make the ceramics virtually unsuitable for practical applications including those in electronics (rare exceptions will be noted below).

Fortunately, both bulk and thin-film monocrystals of the high-T_c superconductors can be readily fabricated. Structure of the crystals is highly anisotropic (Fig. 1a). The copper–oxygen planes ab provide metallic conductivity along the planes, while the conductivity σ_1 along the axis $c \perp ab$ is rather low (Fig. 1b) and is apparently due to the rare hopping of electrons between the ab planes (see, for example, [9,10,11]).

The layered structure also leads to high anisotropy of superconducting properties of the new materials. For example, their critical magnetic fields are approximately related as follows:

$$\frac{(H_{c1})_\|}{(H_{c1})_\perp} \simeq \frac{(H_{c2})_\perp}{(H_{c2})_\|} \equiv \varepsilon, \qquad (2.1)$$

where $\varepsilon \simeq 0.2$ for the 1–2–3 phase, at least for $T \simeq T_c$. The critical current density j_c is also much higher in the direction parallel to the ab planes. The data available presently imply that the high-T_c superconductors can be reasonably well described by the well-known simple model of layered superconductors [12,13] (see also [14,15]). According to the model, within the Ginzburg–Landau approach, the superconductor is characterized by two

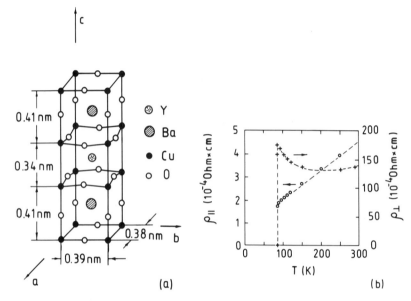

Fig. 1. The $Y_1Ba_2Cu_3O_7$ monocrystal: (a) atomic structure; (b) resistivity vs. temperature for two directions [10].

values of both the magnetic field penetration depth $\lambda(T)$ and the coherence length $\xi(T)$, related as follows:

$$\frac{\lambda_\parallel}{\lambda_\perp} = \frac{\xi_\perp}{\xi_\parallel} = \varepsilon. \tag{2.2}$$

Absolute values of these parameters can readily be restored from those of the critical fields if $\xi_\perp \gg d/\sqrt{2}$ [13]. Table 1 gives the resulting estimates of $\lambda(T)$ and $\xi(T)$ for the 1–2–3 phase [16] reduced to the most probable operation temperature $T = 77$ K.

Although these values should be considered as preliminary ones, they enable one to estimate other important parameters, including the Ginzburg–Landau critical current density

$$j_{GL}(T) = \frac{\phi_0}{3\sqrt{3}\,\pi\mu_0\lambda^2(T)\xi(T)}, \quad \phi_0 \equiv \frac{h}{2e}. \tag{2.3}$$

One can see from Table 1 that $j_{GL}(0)$ is quite high for any direction of the current flow. Experimentally achieved critical currents are considerably lower (for thin films, maximum values $(j_c)_\parallel \simeq 2 \times 10^6 \, \text{A/cm}^2$ at $T = 77$ K were

1. NEW POSSIBILITIES FOR SUPERCONDUCTOR DEVICES

Table 1. Estimates of basic parameters of a typical high-T_c superconductor ($YBa_2Cu_3O_{7-\delta}$ at $T = 77$ K, after [16]) versus those of the most important low-T_c superconductor (Nb at $T = 4.2$ K, after [17])

Material and direction	T_c (K)	$\xi(T)$ (nm)	$\lambda(T)$ (nm)	$j_{GL}(T)^a$ (A/cm^2)	$J_{c1}(T)^b$ (mA/μm)
Y-Ba-Cu-O ($\parallel ab$)	95	7	50	5×10^8	250
Y-Ba-Cu-O ($\perp ab$)	95	1.5	250	1×10^8	50
Nb (almost isotropic)	9.25	40	40	5×10^7	100

[a] See Eq. (2.2).
[b] See Eq. (2.3).

reported by Enomoto et al. [18], presumably because of nonideality of the sample surface and, as a consequence, a suppressed surface energy barrier for the Abrikosov vortices (see, for example [19]). In this situation, a more relevant parameter is the first critical current density J_{c1} [20]:

$$(J_{c1})_\parallel = (H_{c1})_\perp = \frac{\phi_0}{4\pi\mu_0 \lambda_\parallel^2} \ln\left(\frac{\lambda_\parallel}{\xi_\parallel}\right),$$

$$(J_{c1})_\perp = (H_{c1})_\parallel = \frac{\phi_0}{4\pi\mu_0 \lambda_\perp \lambda_\parallel} \ln\left(\frac{\lambda_\perp \lambda_\parallel}{\xi_\perp \xi_\parallel}\right)^{1/2}$$

(2.4)

This parameter can be used, in particular, to evaluate (as $I_c = wJ_{c1}$) the minimum critical current of a thin-film strip of width w placed close to the superconducting ground plane, i.e., of a microstrip line (Fig. 2). According to

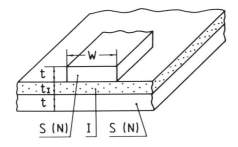

Fig. 2. A schematic view of a typical microstrip line.

Table 1, the critical currents of the new superconductor monocrystalline thin films at $T = 77$ K can be close to those of classical low-T_c superconductors like Nb at $T = 4.2$ K and thus be quite sufficient for most applications in electronics.

Another parameter of a superconductor, which is of importance for applications, is its surface impedance

$$Z_S(\omega, T) = i\omega\mu_0 \delta(\omega, T). \quad (2.5)$$

At $T \simeq T_c$, the effective (complex) penetration depth δ can be crudely estimated [21,22] as

$$\delta^{-2} = \lambda^{-2}(T) + i\delta_{sc}^{-2}(\omega), \quad \delta_{sc}(\omega) = (\mu_0\omega\sigma)^{-1/2}. \quad (2.6)$$

Figure 3 shows estimates of the energy loss factor Re $Z_S(\omega)$ and also the real penetration depth Re δ for a 1–2–3 phase monocrystal with the surface parallel to the ab planes. One can see that in this (most promising) case the surface losses in the new superconductors are somewhat lower than those in the best normal metals; for $T = 77$ K and $T_c = 95$ K this difference, however, is not very impressive (for lower T's and/or larger T_c's the difference should grow rapidly). Perhaps, a more important advantage of superconductors is the virtual absence of a frequency dispersion of Re δ (see Sections 4.2 and 6.1 for more discussion of this point).

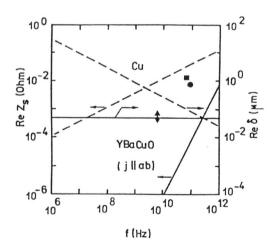

Fig. 3. Frequency dependencies of the real parts of the surface impedance Z_s and effective penetration depth δ at $T = 77$ K for: the 1–2–3 phase with ab planes parallel to the surface and $\rho = 50$ μOhm-cm (solid lines) and pure copper. Experimental data for the 1–2–3 phase at $T = 77$ K are taken from Klein et al. (1989) [112] (circle), Rubin et al. (1988) [116] (triangle), Vendik et al. (1988) [118] (square), and Sridhar et al. (1989) [117] (rhombus).

2.2. *Josephson Junctions*

Josephson junctions (see, for example, [23,24,25]) are the basic active elements of the superconductor electronics. Their frequency range (operation speed) can be evaluated approximately as

$$\Delta\omega \simeq \min[\omega_c, \omega_p]. \quad (2.7)$$

Here the "characteristic" frequency ω_c is determined by the product of the critical current I_c and normal resistance R_N of the junction:

$$\omega_c = \frac{2\pi}{\phi_0} V_c, \quad V_c = I_c R_N, \quad (2.8)$$

while the "plasma" frequency ω_p depends on I_c and the capacitance C of the junction:

$$\omega_p = (L_J C)^{-1/2}, \quad L_J = \frac{\phi_0}{2\pi I_c}. \quad (2.9)$$

The standard theory of superconductivity establishes the maximum value of ω_c

$$(\omega_c)_{\max} = \begin{cases} \dfrac{\pi}{\hbar}\Delta(0) \simeq 5\dfrac{k_B}{\hbar} T_c & \text{for } T \ll T_c, \\[2mm] \dfrac{\pi}{2}\dfrac{\Delta^2(T)}{\hbar k_B T_c} \simeq 15\dfrac{k_B}{\hbar}[T_c(T_c - T)]^{1/2} & \text{for } T \lesssim T_c, \end{cases} \quad (2.10)$$

but does not limit ω_p. Hence, for the 1-2-3 phase with $T_c \simeq 95$ K, one would expect a very high value $(\Delta\omega)_{\max} \simeq (\omega_c)_{\max} \simeq 3 \times 10^{13}$ s^{-1} ($V_c \simeq 12$ mV) for $T = 77$ K. Let us now discuss whether this value can be approached in various types of the Josephson junctions.

2.2.1. S–N–S Junctions

According to standard theory [24], the value (2.10) could be approached in S–N–S sandwich[1] structures if the following conditions were satisfied

$$t_N \lesssim 3\xi_N(0), \quad \frac{\sigma_N}{\xi_N(0)} \ll \frac{\sigma_S}{\xi_S(0)}. \quad (2.11)$$

Table 1 and Fig. 1b show that choosing a possible interlayer (N) material that would satisfy Eq. (2.11) when used with the high-T_c superconductor electrodes (S) is rather difficult. If one takes into account the natural requirements of sufficient R_N of the junction (ranging from 1 to 10^2 Ohm for various applications, see [25] and of technological compatibility of the materials, it becomes very unlikely that a suitable interlayer material could be found.

Note, however, that experiments (see for example [26,27,28,61]) indicate the presence of natural S–N–S Josephson junctions between crystallites ("grains") of the oxide superconductor ceramics. Their values of V_c are typically much lower than those given by Eq. (2.10) (for the 1–2–3 phase at $T = 77$ K, $V_c \lesssim 1$ mV) and are rather irreproducible. Nevertheless, this allows very simple fabrication of primitive Josephson junctions: it is enough to make a constriction in a sample of either bulk or thin-film ceramics, with the size of order that of the grains. and you have a fair chance to single out an intergrain Josephson junction with a reasonable V_c. Such junctions have proved to be quite useful at the first stage of development of simple nitrogen-cooled devices like SQUIDs [29,30]. However, both commercial fabrication of the simplest devices and development of more complex electronic circuits would require reproducible Josephson junctions.

2.2.2. Tunnel (S–I–S) Junctions

Tunnel (S–I–S) junctions seem to be the most probable candidates for such reproducible structures. Their advantages include high values of V_c approaching the fundamental limit (2.10) and less stringent requirements to the barrier material. The tunnel junctions, however, suffer from typically low values of the plasma frequency ω_p. For example, a typical junction with $j_c \equiv I_c/S \simeq 3 \times 10^3$ A/cm^2 and specific capacitance $c \equiv C/S \simeq 10^{-5}$ F/cm^2 has $\omega_p \simeq 10^{12}$ s^{-1} ($V_p \equiv \hbar\omega_p/2e \simeq 0.3$ mV). A further considerable increase of ω_p could be achieved only via that of the critical current density, accompanied as a rule by a simultaneous decrease of the junction area S in order to keep the critical current $I_c = j_c S$ at a certain level required by the device dynamics. For example, in order to make ω_p approach $(\omega_c)_{max}$, one should have values of $j_c \sim 3 \times 10^6$ A/cm^2, which is unrealistic from several points of view (submicron junction areas; monoatom-layer-thick barriers; electrode current densities exceeding J_{c1}).

Thus, it seems presently that the frequency ranges $\Delta\omega$ of the high-T_c superconductor tunnel junctions will be limited solely by their plasma frequencies ω_p, i.e., by the maximum critical current densities determined by minimum reproducible tunnel barrier thicknesses achievable with various barrier materials and fabrication techniques. It is hard to predict presently what values of j_c and ω_p will be approached even in the nearest future; it seems

1. NEW POSSIBILITIES FOR SUPERCONDUCTOR DEVICES

reasonable to believe that at least the value $\omega_p \simeq 10^{12}$ s^{-1} can be achieved for $T = 77$ K because even natural intergrain Josephson junctions exhibit such values while retaining nonhysteretic dc $I-V$ curves peculiar for $\omega_p \gtrsim \omega_c$ (see, for example, [31]).

Thus, in what follows, we will use the figure

$$\omega_p \simeq \omega_c \simeq 10^{12} \text{ s}^{-1} \qquad (2.12)$$

as a (conservative) estimate of possible parameters of the high-T_c Josephson junctions. Note that the same set of values is also typical for state-of-the-art low-T_c Josephson junctions externally shunted in order to avoid the dc $I-V$ curve hysteresis.

3. Operating at Nitrogen Temperatures

3.1. Refrigeration

The most important advantage of the transition from helium to nitrogen operation temperatures is a drastic reduction in the refrigeration costs. In fact, at normal pressure the specific heat of evaporation of liquid nitrogen is as large as 160 J/cm^3, compared with 2.6 J/cm^3 for liquid helium. Moreover, the cost per liter of liquid nitrogen is a factor of $\sim 10^2$ lower than that of helium (~ 10 dollars vs. ~ 10 cents). Hence, the cost of the cooling agent evaporated by a fixed power flow from a cooled object can be reduced by a factor of $\sim 10^4$ by the transition from 4.2 K to 77 K. Of course, the ratio of total refrigeration costs (including those of the cryostats, etc.) can be considerably lower than the above factor (see [5]); however, one can confidently describe the cost reduction as radical.

Another positive effect of the transition on superconductor electronics is a larger power flow that can be removed from a substrate by its simple placement into the cooling liquid without an excessive overheating resulting in the liquid boiling. For helium, Hatano *et al.* [32] report the figure 0.3 W/cm^2, with the critical overheating $\Delta T \simeq 0.3$ K, while for nitrogen, the figure is close to 10 W/cm^2 with $\Delta T \simeq 10$ K [33].

3.2. Thermal Fluctuations

A negative effect of the 4 K $\to$ 77 K transition is a 20-fold increase in the thermal fluctuations. This factor is especially important for devices using Josephson junctions. In these cases, the fluctuations can be characterized by a dimensionless factor (see, for example, [25], Chapter 1)

$$\gamma \equiv \frac{2\pi k_B T}{\phi_0 I_c} \equiv \frac{I_T}{I_c}, \qquad I_T \equiv \frac{2\pi k_B T}{\phi_0}. \tag{3.1}$$

In various devices, the factor should not exceed certain limits (from 10^{-3} to 10^{-1}, see below), so that the critical current I_c of the junction will always be considerably larger than the "thermal current" I_T. For $T = 4.2$ K, I_T is rather small ($\simeq 0.2$ μA) thus the condition $I_c > I_T/\gamma_{max}$ can be readily met. For $T = 77$ K, however, I_T approaches 3.2 μA, so the necessary values of I_c can be as large as 3 mA. Two major problems arise because of these high values of the current.

(i) Power dissipated by a junction scales as

$$P = \bar{I}\bar{V} \propto I_c \bar{V}, \tag{3.2}$$

and hence grows proportionally to T at fixed γ and $\bar{V}$. Moreover, some superconductor devices are based on the sharp increase of the quasiparticle current at the voltage

$$\bar{V} \simeq V_g \equiv \frac{2\Delta(T)}{e}, \tag{3.3}$$

where $\Delta(T)$ is the energy gap of the superconductor employed. If the new superconductors follow the standard BCS theory, then $\Delta(T)$ scales as T_c, and the power scales as T^2. This increase can have a forbidding effect on important classes of LSI superconductor circuits (see Section 5).

(ii) Most Josephson junction devices use superconducting quantum interferometers ([25], Chapters 6–8). Their basic parameter

$$l = \frac{2\pi L I_c}{\phi_0} \tag{3.4}$$

in most cases should be of order 3, so that the loop inductance L of the interferometer scales as T^{-1} for $\gamma = $ constant. For example, at $T = 77$ K and $\gamma = 10^{-3}$, we obtain $I_c \simeq 3$ mA and $L \simeq 0.3$ pH. The only way to provide inductances so small is to use superconducting microstrip line structures (Fig. 4), where L can be presented as follows:

$$L = L_\square N_\square, \qquad L_\square = \mu_0 t_M, \qquad t_M = t_I + 2\lambda(T). \tag{3.5}$$

Here $N_\square$ is a geometrical factor equal crudely to the number of thin film "squares" connected in series to form the loop of concern (see Fig. 4 for some

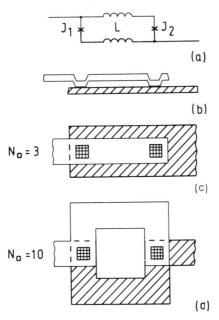

Fig. 4. Two-junction superconducting quantum interferometers: (a) the equivalent circuit; (b) and (c) the simplest thin-film configuration with relatively small parameter $N_\square$; (d) a configuration more suitable for application in dc SQUIDs (the indicated value of $N_\square$ suggests that the interferometer is placed close to a superconducting plane; in particular, this role can be played by the primary coil L_1 of the superconducting dc transformer fed from a signal coil $L_i \ll L_1$—see Fig. 5).

examples). Inductance "per square" $L_\square$ is a minimum for the case of planes ab parallel to the thin film surface ($\lambda \to \lambda_\parallel$). For a small but realistic value $t_I = 50$ nm of the insulator thickness, one obtains $t_M \simeq 120$ nm and $L_\square \simeq 0.15$ pH. One can see that for small γ, there exists a very strict limitation ($N_\square \lesssim 2$ for $\gamma \simeq 10^{-3}$) on the geometry of the interferometers.

Some other circuits can impose even stricter conditions. For example, external resistive shunting of the Josephson junction necessary to make it overdamped ($\omega_c \lesssim \omega_p$) requires the parameter l of the shunt loop to be as low as ~ 0.3 in order to prevent undesirable resonances. Taking into account the fact that values $N_\square \lesssim 1$ are hardly achievable in practice (cf. Fig. 4) one arrives at the conclusion that a good external shunting for junctions with $I_c \gtrsim 1$ mA ($\gamma \lesssim 3 \times 10^{-3}$ at $T = 77$ K) can constitute quite a problem.

Moreover, plane dimensions of the Josephson junction itself should not exceed the Josephson penetration depth λ_J ([25], Chapters 8 and 9) in order to avoid nonuniform distribution of the supercurrent through its area. This condition can be rewritten as follows:

$$I_c \lesssim I_J \equiv j_c \lambda_J^2 = \frac{\phi_0}{2\pi\mu_0 t_M}, \qquad t_M = t_1 + 2\lambda(T). \tag{3.6}$$

For $t_M \simeq 2\lambda(T) \simeq 100$ nm, one obtains $I_c \lesssim 3$ mA, i.e., $\gamma \gtrsim 10^{-3}$ at $T = 77$ K.

One can conclude that proper operation of the Josephson junction and superconducting quantum interferometers at nitrogen temperatures in most circuits is possible but requires their very careful design.

4. Components and Simple Devices

4.1. EMF Shields

The drastic decrease of the refrigeration costs can justify some applications of superconductors, which were well recognized but rarely used earlier. A very characteristic example of these is the shielding of closed volumes from external electromagnetic field (EMF) interferences.

Due to the Meissner effect, a virtually perfect shielding at all frequencies from dc to at least ultraviolet can be provided by even a thin superconducting shell, with thickness d of the order of several penetration depths $\lambda(T)$. According to Table 1, a monocrystalline thin film shield of proper orientation ($ab \parallel$ surface) can be as thin as a few tenths of a micron. Moreover, polycrystalline (ceramic) thin layers are quite suitable for this particular application, with an appropriate increase of their thickness: $d \gtrsim 3\lambda(T) \propto \rho^{1/2}$.

The shielding effect can be suppressed by the EMF field with power density $p = P/S$ exceeding some critical value p_c. At relatively low frequencies where $\delta_S \gtrsim \lambda(T)$ (Section 2.1), p_c can be estimated as follows:

$$p_c = \frac{\rho_0}{8} H_{c1}^2, \qquad \rho_0 \equiv \left(\frac{\mu_0}{\varepsilon_0}\right)^{1/2} = 120\pi \text{ Ohm}, \tag{4.1}$$

(at larger frequencises, p_c is even larger and scales as $\sim \omega^2$). For monocrystalline 1–2–3 phase with $S \parallel ab$, H_{c1} is of order 10^4 A/m, so that $p_c \sim 10^6$ W/cm². This figure is more than enough for most cases we know.

The superconducting shields can be most valuable for defence against low frequency interferences ($f \lesssim 10^3$ Hz), which can be alternatively filtered out only by relatively thick shells made of either normal metals or soft magnetic materials. In such cases, the lower weight of a superconducting shield can compensate the necessity of its nitrogen cooling.

4.2. Transmission Lines and Resonators

Figure 3 implies that monocrystalline high-T_c superconductors enable one to decrease power loss in microwave transmission lines and resonators at all frequencies below $\sim 10^{12}$ Hz. The resulting gain, however, is relatively minor and can hardly justify wide use of the new materials.

This is why the high-T_c superconductors seem to be promising only for decreasing frequency dispersion of microstrip lines (Fig. 2). In fact, the propagation constant γ' of such lines (in the usual case when $w \gg t_I$, $t \gg \delta$) can be expressed as follows (see, for example, [34]):

$$\gamma' = \frac{\omega}{c} \mathrm{Re}\left(1 + \frac{2\delta}{t_I}\right)^{1/2}. \tag{4.2}$$

Thus, when the insulator thickness t_I is comparable or less than the effective penetration depth δ (see Eq. (2.6)), γ' closely reproduces the frequency dispersion of δ. For normal metals, $\delta \propto \omega^{-1/2}$ for most frequencies of interest (Fig. 3), and such dispersion practically forbids transfer of pulses shorter than ~ 100 ps along microstrip lines applicable in VLSI circuits ($t_I \simeq 0.1$–0.3 μm, $w \simeq 1$–3 μm, length $\simeq 1$–10 μm). For superconducting microstrip lines of the same dimensions, the dispersion is virtually vanishing for pulses longer than ~ 1 ps, so that the lines can be quite valuable for digital microelectronics—see Section 6.

4.3. Modulators and Commutators

Switching the microstrip from superconducting to normal state and back, one can effectively change the attenuation of the line

$$\gamma'' = \frac{\omega}{c} \mathrm{Im}\left(1 + \frac{2\delta}{t_I}\right)^{1/2} \tag{4.3}$$

and/or its input impedance. This fact enables one to use the lines in modulators and commutators of both narrowband and pulse signals.

The switching can be provided by several means, including

(i) dc current, and
(ii) laser irradiation

(magnetic control is also possible but hardly practical, because of relatively large critical magnetic fields of the high-T_c superconductors). Required intensities of the control signal can be quite reasonable. For example, the typical value $j_c \simeq 10^6$ A/cm² yields $I_c \simeq 10$ mA for a microstrip with $d \simeq 0.3$ μm and $w = 3$ μm.

Maximum modulation speed is determined by the rate τ_Δ^{-1} of the gap relaxation (see, for example, [35]), which can hardly be lower than $\sim 10^{10}$ s^{-1}; such speed is more than sufficient for most applications.

5. Analog Devices

5.1. SQUIDs

Superconducting quantum interference devices, the celebrated SQUIDs, have become the main achievement of superconductor electronics. A SQUID is essentially a supersensitive detector of magnetic flux ϕ_x applied to the loop of its sensor, the superconducting quantum interferometer (for details, see Chapter 2 of the present collection). Let us analyze whether it is possible to retain this sensitivity at nitrogen operation temperatures.

5.1.1. RF SQUIDs

Sensitivity of the rf SQUID operating in hysteretic mode ($l \gtrsim \pi$) is usually determined by the noise of the rf amplifier rather than that of the sensor, because of relatively low transfer factor

$$H = \left|\frac{\partial V}{\partial \phi_x}\right|, \tag{5.1}$$

where V is output voltage of the sensor (in the present case, the rf voltage across the rf tank circuit). For a typical rf SQUID, H is of order 10 μV/ϕ_0 and the amplifier r.m.s. noise V_N is of order 10^{-9} V/Hz$^{1/2}$, so that the flux sensitivity $\delta\phi_x = V_N/H$ is of order 10^{-4} ϕ_0/Hz$^{1/2}$ (until the signal frequency is so low that the $1/f$ noise becomes essential—see Chapter 2 of this book).

An increase of operation temperature T does not directly affect H and V_N. One should, however, take into account that the "classical" hysteretic mode requires the following relations ([25], Chapter 14):

$$1 \lesssim \frac{l}{\pi} \ll \left(\frac{2}{\gamma}\right)^{2/3}. \tag{5.2}$$

At helium temperatures, these relations can be readily satisfied within a wide range of I_c and L. Increasing T, however, narrows the range considerably. The minimum value of γ that leaves the range acceptably large can be (somewhat empirically) established as 0.1, which yields $I_c \simeq 30$ μA at $T = 77$ K.

With $l \simeq \pi$, one obtains the maximum inductance $L \simeq 3 \times 10^{-11}$ H. Unfortunately, this value can hardly be approached in the standard point-contact version of the rf SQUID, because its holes would become too small to contain dc-transformer and rf-tank-circuit coils. In the first versions of the nitrogen-cooled SQUIDs (see, for example, [30]), the inductance was an order of magnitude larger; this is apparently why the SQUID noise was relatively high ($\delta\phi_x \simeq 5 \times 10^{-4} \phi_0/\text{Hz}^{1/2}$). If the problem of the reduction of L were solved by one way or another, one could presumably hope for reduction of the noise to values typical for the helium-cooled rf SQUIDs.

5.1.2. DC SQUIDs

Much higher sensitivity can be obtained using dc SQUIDs employing interferometers with two overdamped ($\omega_c \lesssim \omega_p$) Josephson junctions. The transfer factor H of such SQUID is relatively large:

$$H_{\max} \sim V_c/\phi_0 \qquad (5.3)$$

and reaches $\sim 1\,\text{mV}/\phi_0$ for the typical values (2.12) of ω_c and ω_p, so that the amplifier noise contribution can be made negligible in comparison with that of the interferometer. Beyond the $1/f$ noise range, the thermal fluctuations of the junctions limit the SQUID energy resolution $\varepsilon_v \equiv (\delta\phi_x)^2/2L\,\Delta f$ at the level [36,37]

$$(\varepsilon_v)_{\min} \simeq 9\frac{k_B T}{\omega_p} \qquad \text{for} \quad \omega_c \simeq \omega_p, \quad l \simeq 3, \quad \gamma \lesssim 0.1. \qquad (5.4)$$

For nitrogen temperatures, this formula yields the figure $\varepsilon_v \sim 2 \times 10^{-32}$ J/Hz, which is quite sufficient for virtually all applications of SQUIDs. In order to achieve this performance, one should provide relatively high critical currents I_c ($\gtrsim 30\,\mu\text{A}$) and low inductances L ($\lesssim 30$ pH). In this context, we should remind the reader about one possibility to reduce the effective value of L, which was rarely used in practice.

Figure 5a shows the equivalent circuit of a measurement by a SQUID using the superconducting dc transformer. If the coupling factor $k \equiv M/\sqrt{L_1 L_2}$ were much less than unity, the inductance L seen by the SQUID junctions (i.e., reduced to AA' port) equals just L_2. For the integrated dc SQUID, another situation, $k \simeq 1$, is more typical; in this case,

$$L \simeq L_2 \frac{L_i}{L_1 + L_i} \to L_i \frac{L_2}{L_1} = \frac{L_i}{N^2} \qquad (5.5)$$

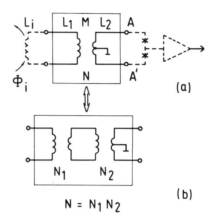

Fig. 5. Equivalent circuits of (a) a typical arrangement for measuring an external magnetic flux ϕ_i with the help of a single-stage superconducting dc transformer, and (b) the two-stage transformer.

(the latter limit concerns the case $L_1 \gg L_i$ of optimum transfer of the input signal). One can see that at the fixed ratio N of the coil turns, the effective SQUID inductance is determined by the source inductance L_i rather than L_2. Vice versa, if the signal source is fixed (as it usually is), one can (and should) decrease L via increase of the ratio N of turns. For a low-T_c SQUID with its relatively large L ($\sim 10^{-11}$ H), N's in the range between 30 and 100 are quite sufficient. For high-T_c SQUIDs, however, optimum values of N can become too large to be practical. In this case, a series connection of two transformers (Fig. 5b) with moderate $N_{1,2}$ can be an easier way to achieve the required large $N = N_1 N_2$.

Another problem becoming serious at nitrogen temperatures is that of external resistive shunting of the Josephson junctions. Such shunting is necessary for tunnel junctions in order to bring their effective characteristic frequency

$$\omega'_c = \frac{2\pi}{\phi_0} I_c \left(\frac{1}{R_{qp}} + \frac{1}{R_s} \right)^{-1} \tag{5.6}$$

down to their plasma frequency ω_p (here R_{qp} is effective quasiparticle resistance of the junctions at the operation voltage, and R_s is that of the shunt). The problem concerns not only R_s but the shunt loop inductance L_s as well, which should not impede the junction shunting at frequencies of the Josephson oscillations $\omega_J \sim \omega'_c$; this condition can be expressed as $\omega'_c L_s \lesssim 0.3 R_s$. For $I_c \simeq 30$ μA, the inductance should be as low as ~ 3 pH, so that the shunt loop would require a very thoughtful design (Fig. 4c).

5.1.3. RO-Driven SQUIDs

One way to avoid the latter problem is just to keep the shunt inductance large! In this case, the whole dynamics of the dc SQUID changes drastically and can be described in terms of relaxation oscillations [38–41].

Consider the simplest version of the RO-driven SQUID (Fig. 6a). Its tunnel Josephson junctions have $\omega_c \gg \omega_p$, and thus exhibit the usual hysteretic I–V curve with nonstable negative-slope branch at $0 \leq \bar{V} \leq V_r$, where $V_r \simeq V_p = (\phi_0/2\pi)\omega_p \ll V_g$ (Fig. 7a). If the bias circuit with relatively low $R_s \lesssim V_p/I_c(\phi_x)$ and relatively large $L_s \gg \phi_0/I_c(\phi_x)$ tries to fix the dc bias point at this unstable branch, relaxation oscillations with frequency

$$f_r \simeq \frac{V_b}{L_s I_c(\phi_x)}, \quad V_b = I_b R_s \lesssim V_r, \tag{5.7}$$

arise in the circuit. During the longer part of the oscillation period, the superconducting interferometer remains in its superconducting (S) state with $V = 0$ and the current I growing linearly in time:

$$\dot{I} = V_b/L_s. \tag{5.8}$$

As soon as I reaches $I_c(\phi_x)$, the interferometer switches to its resistive (R) state with $V \simeq V_g \gg V_b$ and the current I decreases rapidly to a value I_r, where the

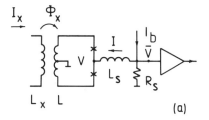

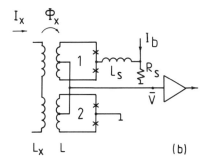

Fig. 6. Equivalent circuits of the relaxation-oscillation-driven dc SQUIDs: (a) the simplest RO-driven SQUID; (b) the balanced SQUID with better sensitivity [41]. © 1989 IEEE.

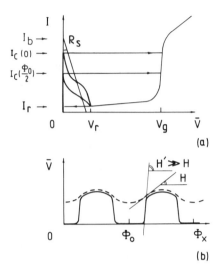

Fig. 7. RO-driven dc SQUIDs: (a) a scheme of operation for two values of the critical current $I_c(\phi_x)$ of the interferometer; (b) output voltage as a function of the measured flux ϕ_x for (1) the simplest SQUID (dashed line) and (2) the balanced SQUID (solid line), schematically.

reverse switching to the superconducting state occurs. The mean (dc) voltage

$$\bar{V} = V_b - R_s(I_b - \bar{I}) \simeq V_b - \tfrac{1}{2}R_s I_c(\phi_x) \tag{5.9}$$

developed across the interferometer transfers information about its critical current $I_c(\phi_x)$ to the circuit output. One can see that the transfer factor H of the RO-driven SQUID

$$H = \left|\frac{\partial \bar{V}}{\partial \phi_x}\right| = \frac{1}{2}R_s\left|\frac{dI_c}{d\phi_x}\right| \simeq \frac{1}{2}\frac{I_c R_s}{\phi_0} \lesssim \frac{1}{2}\frac{V_p}{\phi_0} \tag{5.10}$$

can be only slightly lower than that of the traditional dc SQUIDs; in practice, values of order 10^2 $\mu V/\phi_0$ were obtained even for relatively low ω_p tunnel junctions [40].

The noise of the RO-driven SQUID is fundamentally limited by thermal fluctuations in the Josephson junctions, providing randomness of the S → R and R → S switching times. For optimum values of parameters ($l \simeq 3$, $R_s \simeq (R_s)_{max} \simeq V_p/I_c$, $\omega_r \simeq (\omega_r)_{max} \simeq \omega_p^2/\omega_c$), this noise allows the following energy resolution [39]:

$$(\varepsilon_v)_{min} \simeq 10\frac{k_B T \omega_c^{7/20}}{\omega_p^{27/20}} \quad \text{for} \quad \omega_c \gg \omega_p. \tag{5.11}$$

1. NEW POSSIBILITIES FOR SUPERCONDUCTOR DEVICES 19

Comparison of Eqs. (5.4) and (5.11) shows that maximum sensitivity of RO-driven SQUIDs can be only slightly worse than that of their traditional counterparts. Experimentally, values $\varepsilon_v \simeq 4 \times 10^{-31}$ J/Hz have been achieved at $T = 4.2$ K [40] in spite of the use of low-ω_p junctions ($j_c \simeq 10^2$ A/cm^2, $S \simeq 10^2$ μm^2) and an apparently substantial contribution of amplifier noise. For nitrogen temperatures, tunnel junctions with $\omega_p \simeq 10^{12}$ s^{-1} and $\omega_c \simeq 10^{13}$ s^{-1} should allow quite decent sensitivities ε_v of order 3×10^{-32} J/Hz.

5.1.4. Balanced SQUIDs

Performance of the RO-driven SQUIDs can be apparently improved even further via utilization of their balanced version (Fig. 6b) suggested by Gudoshnikov et al. [41]. In contrast to the previous version, here the linear rise of the bias current described by Eq. (5.8) leads to the S → R switching of only one of the interferometers (with a smaller critical current I_c at a given ϕ_x). As a result, the output dc voltage equals either zero (if $I_{c1} < I_{c2}$) or a nonvanishing value $\bar{V}$ yielded by Eq. (5.9) in the opposite case (solid line in Fig. 7b).

Crossover between the regimes takes place at $I_{c1}(\phi_x) = I_{c2}(\phi_x)$, providing a voltage step at the resulting $V(\phi_x)$ dependence. Width of the step is fundamentally limited by thermal fluctuations as ([25], Chapter 3)

$$\delta(I_{c1} - I_{c2}) \simeq I_c \gamma^{2/3}. \tag{5.12}$$

As a result, the transfer factor can be much higher than that in the original version of the SQUID:

$$(H)'_{max} \simeq \left(\frac{\bar{V}}{\delta \phi_x}\right)_{max} \simeq \gamma^{-2/3} \frac{V_p}{\phi_0}; \tag{5.13}$$

for $\omega_p \simeq 10^{12}$ s^{-1} and $\gamma \sim 0.1$, one obtains an estimate $H_{max} \sim 1$ mV/ϕ_0.

Analysis [41] has shown that output noise of the balanced RO-driven dc SQUID differs from that of its predecessor and is given by very simple formula

$$\delta_V(0) \equiv \frac{\langle V^2 \rangle_\omega}{2 \Delta \omega} = \frac{\bar{V}^2}{\omega_r} p_1 p_2, \tag{5.14}$$

where p_i is probability of S → R switching of the ith interferometer ($p_1 + p_2 = 1$). Equations (5.13) and (5.14) yield the energy resolution of the device

$$(\varepsilon_v)_{min} \simeq 1.5 \, \gamma^{1/3} \frac{k_B T \omega_c}{\omega_p^2} \quad \text{for} \quad \omega_c \gtrsim \omega_p, \tag{5.15}$$

which can apparently be of order of that for the traditional dc SQUIDs. This conclusion still is to be confirmed experimentally.

To summarize, it seems that nitrogen-cooled dc SQUIDs of various types can have sensitivities quite acceptable for most applications. This is why the sharp reduction of the refrigeration costs (and hence of the general costs of these otherwise simple devices) can make them much more practical for widespread use in not only traditional areas like biomedicine, geophysics, and experimental instrumentation, but also in other areas of science and technology.

5.2. Samplers

A similar large effect due to the transition to nitrogen cooling can be expected for superconductor samplers, devices capable of measuring waveforms of short periodic pulses (see, for example, [42–46]). The main component of the device is a Josephson junction comparator with a well-defined current threshold I_t for switching from superconducting to resistive state. Figure 8a

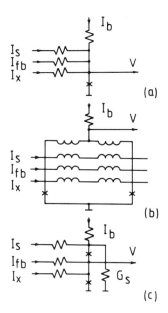

Fig. 8. Josephson junction comparators used in samplers and parallel-type A/D converters: (a) the simplest single-junction comparator; (b) the interferometer comparator allowing galvanic decoupling of the input circuits; (c) comparator with a "Goto pair" of similar junctions, which reduces the influence of technical fluctuations.

shows the simplest version[2] of the comparator comprised of a single underdamped Josephson junction fed by the measured current $I_x(t)$, dc bias current I_b, dc feedback current I_{fb}, and periodic short "sampling" pulses $I_s(t - nT)$. The junction performs S → R switching as soon as

$$I_\Sigma \equiv I_x + I_b + I_{fb} + I_s > I_t \simeq I_c. \qquad (5.16)$$

Thus, if the dc feedback loop is arranged to keep I_Σ just at the switching threshold, I_{fb} reflects the value of I_x at the time of maximum $I_s(t)$. Slowly changing the time delay τ between the measured pulses and the sampling pulses, one can record the whole signal waveform.

The extremely low intrinsic noise of the Josephson junctions enables one to obtain very high sensitivity of the samplers ($\delta I_x \lesssim 5 \mu A$ was reported by Fujimaki et al. [46]) together with few-picosecond time resolution. These features have made possible the recent commercial introduction of a helium-cooled version of the sampler by Hypres Inc. [44]. If this high performance could be reproduced with nitrogen cooling, the samplers would become much more attractive for users.

The time resolution δt is fundamentally limited by the inertia of the comparator. The inertia can be minimized by an optimum choice of the junction damping [47]:

$$(\delta t)_{min} \simeq \frac{1}{\omega_p \varepsilon^{1/2}} \quad \text{at} \quad \omega_c \gtrsim \omega_p, \qquad (5.17)$$

where the parameter $\varepsilon = 1 - I_t/I_c$ is limited by thermal fluctuations as $\varepsilon_{min} \simeq \gamma^{1/3}$. With a realistic value of $\gamma \simeq 10^{-2}$ ($I_c \simeq 300 \mu A$, $T = 77$ K) and using Eq. (2.12), one obtains the figure $\delta t \sim 2$ ps, which is close to that achieved with helium-cooled Josephson junction samplers [43] and is considerably less than those of the semiconductor tunnel diode samplers. (Better time resolution can be apparently obtained only in optoelectronic devices (see, for example, [48]) with their much worse sensitivity.) Note, however, that in order to approach the limit (5.17), one should make negligible the contribution δt introduced by the delay line employed in the sampler.

The sensitivity of the samplers is limited, firstly, by technical fluctuations of the dc bias currents and sampling pulse amplitude. These fluctuations can apparently be reduced by using comparators comprised of a "Goto pair" of either similar Josephson junctions (Fig. 8c) or similar interferometers (see Fig. 8b). The shunt conductance G_s provides a negative feedback enabling the sampling pulse to switch S → R only one junction/interferometer at a time (see preliminary experiments by Gudkov et al. [49]). One can readily get convinced that small fluctuations of I_s and I_b do not contribute to the statistics of the switching events, i.e., to noise of the output voltage V of this circuit[3].

Secondly, the sensitivity is fundamentally limited by thermal fluctuations in the junction(s) of the comparator. The existing semiquantitative theory [25] yields the following estimates of the associated resolution:

$$\delta I_x = N^{-1/2} \Delta I_x,$$

$$\Delta I_x \simeq \begin{cases} (I_t^2 I_c)^{1/3} & \text{for } \Delta t \gtrsim \dfrac{6\pi}{\omega_p}\left(\dfrac{3\gamma}{2}\right)^{-5/6}, \\ (I_t I_c)^{1/2} & \text{for } \Delta t \lesssim \dfrac{\pi}{\omega_p}, \end{cases} \quad (5.18)$$

where Δt is length of the sampling pulse and N is the number of signal periods averaged to obtain one point of the measured waveform. If the input signal can be reproduced with a large frequency f (say, 1 GHz), and one is satisfied with a reasonable number of the measured waveform points $m \simeq \tau_{\max}/\delta t$ (say, 10^3) and total measurement time T (say, 10^{-1} sec), very large averaging factors $N = Tf/m$ of order 10^5 can be achieved. This means that even the considerable thermal fluctuations $\Delta I_x \sim 30\ \mu A$ typical for nitrogen temperatures would limit the sampler sensitivity to the level $\delta I_x \sim 1$ nA (!), which is more than sufficient for any imaginable application of the device. The problem of whether it is possible to reduce the technical fluctuations to such low levels is still to be studied experimentally. In any case, it seems that Josephson junction samplers will be indispensable tools for studies of dynamics of future ultrafast devices, including superconductor digital circuits (see Section 6).

5.3. Receivers

Low-noise receivers of EMF radiation are another important achievement of low-temperature superconductor electronics (see, for example, Chapter 5 of this collection). Most promising results have been obtained in the millimeter waveband, where noise temperatures T_N of the SIS mixers have been lowered to approach closely the so-called quantum limit $\hbar\omega/2k_B^4$, and thus leave all other receivers behind.[4] (At lower (microwave) frequencies, the superconductor receivers meet strong competition from the cryogenic GaAs MESFET and HEMT amplifiers (see, for example, [54]), while at larger (infrared) frequencies the bulk mixers and bolometers can have lower T_N's (see, for example, [55,56,57]).)

Nevertheless, practical use of the superconductor receivers even in the millimeter waveband is restricted to radioastronomy. The main reason for this is their low saturation power P_s, typically $\sim 10^{-10}$ W or less and, as a consequence, their narrow dynamic range insufficient for radar and communication systems.

The advent of the high-T_c superconductors cannot improve the situation considerably. In fact, P_s can be estimated as [25]

$$P_s \sim \frac{V_{ef}^2}{R_N}, \quad V_{ef} \sim \frac{\phi_0}{2\pi}\min[\omega_p, \omega_c], \tag{5.19}$$

where the normal resistance R_N of the superconducting junction should be close to the typical impedance ρ of the waveguide employed (10 to 400 Ohm). One can see that even the possible larger ω_c's of the high-T_c junctions cannot increase V_{ef} and P_s. Thus, new broad areas of applications can hardly be open for the receivers. Moreover, the 20-fold increase in the operation temperature would increase the intrinsic noise of the receivers on approximately the same scale ([25], Chapters 10–13). Such an increase is hardly tolerable in radioastronomy where, on the other hand, helium cooling is quite acceptable.

One could conclude that the nitrogen-cooled high-T_c superconductor receivers have low chances to become popular. Note, however, that one can meet the conditions when the dynamic range is not so crucial and only nitrogen rather than helium cooling is available. In such a case, some types of Josephson-junction receivers (wideband and narrowband quadratic videodetectors and external-local-oscillator mixers—see [25]) can be quite competitive with the best semiconductor receivers like Schottky diode mixers through all the millimeter waveband, especially if coherent multijunction arrays are employed (see Chapter 5 of this book).

Note also one very simple structure (see, for example, [58,59]) shown in Fig. 9. It consists of merely a granular superconductor thin film patterned to form a zigzag-shaped strip of width $w \lesssim D$, where D is the average grain size. Analysis [60] shows that the structure should exhibit at least three different mechanisms of videoresponse to EMF radiation. At lower signal frequencies ($\omega/2\pi$ below $\sim 10^{12}$ Hz), the wideband detection of radiation by random intergrain Josephson junctions should prevail, with the responsivity $\eta \equiv V_{out}/P_{in}$ scaling as ω^{-2}. In the far infrared, photoexcitation of the Cooper pair across the superconductor energy gap $2\Delta(T)$ (see for example, [62]) with subsequent transfer of the resulting nonequilibrium single-electron quasiparticles through the junctions can constitute a

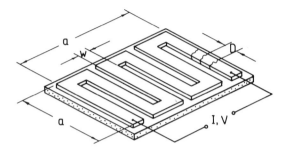

Fig. 9. The simplest structure formed from granular high-T_c superconducting thin film, which should allow effective quadratic videodetection of EMF radiation from microwave through infrared to optical bands.

more effective detection mechanism with $\eta \propto \omega^{-1}$. Finally, in the near infrared and optical frequency band, the simple bolometric response [55] with frequency-independent responsivity η should dominate. Estimates show that at $T = 77$ K one can expect maximum values of η from $\sim 10^5$ V/W in the millimeter waveband to $\sim 10^3$ V/W in the optical band, with NEPs of the order from $\sim 10^{-14}$ to $\sim 10^{-12}$ W/Hz$^{1/2}$, respectively. Somewhat better sensitivity can be obtained with uniform 1-D arrays of similar Josephson junctions, when a reproducible technology of their fabrication is developed.

5.4. DC Voltage Standards and Multipliers

The technology just mentioned becomes necessary when one comes to other multijunction Josephson-effect devices, for example, to the dc voltage standards—see, for example, a comprehensive recent review by Kautz et al. [63]. Recently, people from PTB and NBS produced a real breakthrough in this field by the fabrication of carefully designed arrays of several thousand (low-T_c) Josephson junctions dc-connected in series. Output dc voltage of such single-chip circuits can be increased to 1 Volt and beyond [64]. This factor allows one to use the voltage directly, avoiding complex systems of comparators necessary for the single-junction standards (see, for example, [65]), and thus allowing one to use the fascinating intrinsic stability of the standards, which approaches $\sim 10^{-17}$ [66].

In order to estimate the possibility of creating nitrogen cooled dc-voltage standards, one can readily use the thorough analysis by Kautz et al. [63], because our estimate (2.12) of ω_p for high-T_c junctions closely coincides with the figure used by those authors for the low-T_c case. Thus, one should merely

1. NEW POSSIBILITIES FOR SUPERCONDUCTOR DEVICES

insert the higher level of T in the well-known formula ([25], Chapter 3)

$$\tau \simeq \frac{2\pi}{\omega_p} \exp\left(\frac{4\sqrt{2}}{3} \frac{T_n}{T} \delta^{3/2}\right), \quad T_n = \frac{\phi_0 I_n}{2\pi k_B}, \quad (5.20)$$

for the lifetime of the (metastable) state of the array with the dc bias point fixed at the nth Josephson–Shapiro current step of each junction (parameter $\delta = 1 - (\bar{I} - I_{n0})/I_n$ describes deviation of the bias point from the center I_{n0} of the step with halfwidth I_n). According to Kautz et al. [63], for junctions of size $\sim 10 \times 30 \, \mu m^2$ with $j_c \simeq 10^2$ A/cm^2 irradiated by a wave of frequency $\omega/2\pi \simeq 100$ GHz and optimum power, the constant T_n can approach $\sim 0.7 \times 10^4$ K for a suitable value $n = 5$. For $T = 77$ K and $\delta_{min} \simeq 0.5$, this estimate yields an estimate $\tau_{min} \sim 10^{14}$ s, which is still sufficient for stable operation of all $N \simeq 10^3$ junctions during any realistic run period. One can thus conclude that the nitrogen-cooled dc voltage standard of the traditional type do look promising for development. Very good care, however, should be taken to ensure the junction uniformity which determines δ_{min}.

One should, nevertheless, note that these standards are complex in operation: they require extremely stable high-power millimeter-wave generators and very stable interference-free dc bias current sources. Recently, an alternative circuit, apparently free of these drawbacks, has been suggested [67,68].

The circuit consists of cells using six overdamped Josephson junctions arranged to provide dc voltage doubling (Fig. 10a). Input dc voltage $V_{12} \lesssim V_c$ induces Josephson oscillation of the left-most junction of the cell; in other words, it generates a continuous train of short ($\Delta t \simeq \pi/\omega_c$) voltage pulses with amplitude $\sim 2V_c$ and fixed area

$$\int V(t) \, dt = \phi_0 \simeq 2 \text{ mV ps.} \quad (5.21)$$

These "single-flux-quantum" (SFQ) pulses are reproduced with current gain in the following two junctions having larger critical currents ($I_{c1} < I_{c2} < I_{c3}$) and induce similar pulses in junction 4 with smaller I_c (the relation $I_{c4} \ll I_{c3}$ virtually excludes back-reaction of the cell load). These pulses are again current-amplified in junctions 5 and 6 ($I_{c4} < I_{c5} < I_{c6}$) and ensure that the equality $V_{34} = V_{12}$ holds despite the absence of a parallel dc connection of the input and output terminals. Relatively large superconducting inductance L does not affect the SFQ pulse propagation process but connects terminals 2 and 3 for dc current, so that $V_{14} = V_{12} + V_{34} = 2V_{12}$. Numerical simulation

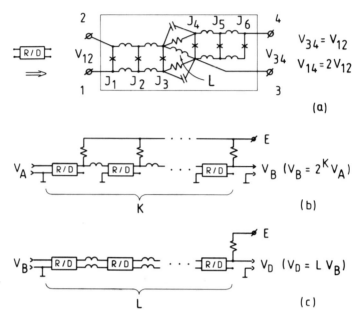

Fig. 10. DC voltage multipliers: (a) the basic stage of the multipliers, the dc voltage reproducer/doubler ($V_{34} = V_{12}$, $V_{14} = 2V_{12}$); (b) an efficient multiplier of small dc voltages ($V_B \lesssim V_c$); (c) a multiplier with higher output voltages [68] ©1989 IEEE.

[67] has shown that if inductances of each half of the doubler are chosen in the proper way (they should provide equal ϕ_i across all junctions at $V_{12} = 0$) and the coupling circuits RC satisfy the relations

$$|Z_{RC}(\omega_c)| \ll R_4, \qquad Z_{RC}(0) \gtrsim R_4, \tag{5.22}$$

the doubler should operate properly with wide parameter margins until V_{12} approaches $\sim V_c/2$.

Now, connecting the voltage doublers as Fig. 10b shows, one gets

$$V_B = 2^K V_A \tag{5.23a}$$

as long as $V_B \lesssim V_c \sim 300\ \mu\text{V}$. A further increase of the voltage can be achieved by series connection of the doublers:

$$V_D = LV_B, \quad \text{i.e.,} \quad V_D = L2^K V_A. \tag{5.23b}$$

Thus, in order to obtain $V_D = 1$ V, one would need $M = 6(K + L) \simeq 2 \times 10^4$ Josephson junctions, almost independently of V_A, i.e., of the frequency of the

1. NEW POSSIBILITIES FOR SUPERCONDUCTOR DEVICES

rf reference source which should phase lock the first junction of the first doubler alone.

This number of junctions required is an order of magnitude larger than that in the usual voltage standards. As a return from this hardware excess, one can obtain a number of advantages:

- virtually arbitrary frequency range of the rf reference source;
- large output currents (up to several milliamperes);
- much better stability with respect to thermal fluctuations and dc bias interferences;
- dc connection of the doublers is achieved by large-inductance leads, so that one can decompose the whole device into several (many) chips, each of a moderate integration scale.

Other possible applications of the voltage multipliers shown in Fig. 10 include absolute comparison of dc voltages [68] and D/A conversion (see the next subsection).

5.5. A/D and D/A Conversion

High operation speed and natural quantization of the magnetic flux, peculiar for the Josephson-junction circuits make them quite promising for analog-to-digital (A/D) and digital-to-analog (D/A) conversion of information. Although no commercially available devices of this class have been introduced so far, the last decade has witnessed a substantial progress in development of at least two of their types.

5.5.1. Parallel A/D Converters

Parallel A/D converters can be used for rapid signal quantization with not very high precision $\varepsilon \equiv 2^{-n}$ (where n is the number of correct output bits of the converter). In contrast to traditional semiconductor parallel converters requiring at least 2^n comparators, the Josephson junction devices developed by Hamilton et al. [69–72] can be restricted to only n comparators. The latter devices have, however, a grave disadvantage, a need for high-precision components: analog signal divider and two-junction interferometer comparators; relative accuracy of both components should be of order ε, i.e., of order 0.1% for a decent precision $n = 10$.

In subsequent works, several improvements of the basic concept have been suggested in order to remove this and other drawbacks. A special redundant coding [50,73] has been shown to increase the parameter margins of the comparator drastically (up to ten percent for any n), with only insignificant complication of the whole device. Utilization of balanced comparators (Section 5.2) and/or latching comparators [51] can decrease the aperture time τ_a of the converter and relax requirements to the sampling pulses. Finally, special multibit parametric-quantron-type comparators [74] can yield improved sensitivity.

Nevertheless, the need for a high-precision analog divider still remains a hurdle on the way to practical parallel superconductor A/D converters. Recent progress of their series counterparts seemingly make the latter devices preferable.

5.5.2. Series A/D Converters

Series A/D converters suggested originally by Hurrell and Silver [75] (see also [76]) are based on counting the number of SFQ pulses (5.20) generated by a superconducting quantum interferometer fed by the input analog signal. Figure 11 shows a more developed version of such a device, able to convert arbitrary signals (with arbitrary signs of their time derivatives). An analog signal current $I_x(t)$ applies a proportional flux $\phi_x(t)$ to the symmetrical interferometer formed mainly by the overdamped junctions J_1 and J'_1. When $I_x(t)$ increases ($\dot{I}_x > 0$), it induces switchings of the phase of J_1 each time ϕ_x reaches a level $n\Phi_0 + \text{const}$ (where n is integer). The SFQ pulse of voltage generated across the junction at the switching is current-amplified by the junction array $J_2, \ldots, J_m$ (cf. Fig. 10a and its discussion). At the same time, the negative SFQ pulses arising across J'_1 during this process are not reproduced

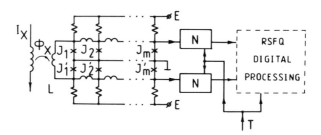

Fig. 11. A series analog-to-digital converter consisting of a delta A/D converter and a digital signal processor. The delta converter consists of the two-junction interferometer ($L - J_1 - J_2$) producing SFQ pulses, two Josephson-junction transmission lines ($J_2, J_3, \ldots, J_m$, and $J'_2, J'_3, \ldots, J'_m$) amplifying the pulses, and two "time normalizers" N (see text) timed by the clock SFQ pulse train T.

by the array $J'_2, \ldots, J'_m$. The latter array becomes operative only when $\dot{I}_x$ becomes negative; in this case, an SFQ pulse arrives to the lower output port (but not to the upper one) each time ϕ_x approaches $n'\Phi_0 + \text{const}'$.

The SQF pulses from both ports arrive at the "time normalizer" circuits N which delay them if they arrive during a certain small fraction τ' of each period τ between the periodic clock pulses T. This operation makes the pulses acceptable for the superfast RSFQ logic circuits (Section 6.2) performing digital processing of the information. The A/D conversion itself, however, is completed by the time normalizers, because the difference in the number of the SFQ pulses provided by them during each sampling period τ corresponds exactly to $\text{INT}\{[\phi_x(\tau) - \phi_x(0)]/\Phi_0\}$ (such devices are usually referred to as delta converters).

Performance of the series converter shown in Fig. 11 is determined mainly by the aperture time $\tau_a \simeq \tau'$ of the time normalizers. According to our numerical simulations, values of τ_a of order 3 ps are attainable with state-of-the-art 3 μm low-T_c superconductor technology. This means that the achievable accuracy

$$n' \simeq \log_2\left(\frac{\tau}{\tau_a}\right) \tag{5.24}$$

of the comparator can be quite good: $n' \simeq 10$ for $f_\tau \equiv \tau^{-1} = 100$ MHz or $n' \simeq 21$ for $f_\tau = 100$ kHz. Moreover, such samplers and their RSFQ signal processors admit very high sampling frequencies f_τ up to ~ 30 GHz. If the upper frequency f_s of the input signal is less than $1/2\tau$, additional information obtained from extra counts can be used for a further improvement of the accuracy by the "radiometric factor"

$$n'' \simeq \log_2\left(\frac{f_\tau}{f_s}\right)^{1/2} \tag{5.25}$$

via digital filtering of the output signal. In this way, fascinating overall precisions $n = n' + n''$ could be presumably achieved, up to $n \simeq 15$ for $f_s = 100$ MHz, or $n \simeq 30$ (i.e., ten correct decimal digits!) for $f_s = 100$ kHz.

Unfortunately, practical use of such converters at lower frequencies is restricted by a similarly large dynamic range of their input (analog) stage. This range is limited, on one hand, by r.m.s. thermal fluctuations I_T in the Josephson junctions and, on the other hand, by critical current I_c of the superconducting wires used to form the interferometer. If one accepts realistic figures $I_T \simeq 3$ μA ($T \simeq 77$ K) and $I_c \simeq 1$ A (cf. Table 1), one gets $m \equiv \log_2(I_c/I_T) \simeq 20$.

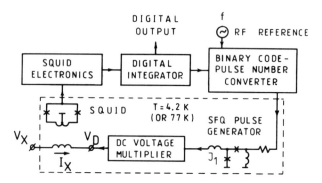

Fig. 12. A block diagram of a possible low-frequency A/D converter based on compensation of the input signal V_x by a signal V_D developed by the D/A converter consisting of a BC → PN converter, a SFQ pulse generator, and a dc voltage multiplier.

5.5.3. Low-frequency D/A and A/D Converters

The last problem can be presumably solved by design of the A/D converter shown in Fig. 12 [68]. It consists of a relatively high-speed ($f_r \gg f_s$) but low-bit A/D converter (for example, the SQUID with digital output), a digital integrator (providing the output signal of the whole device), and a negative feedback loop with a precise D/A converter with its analog output signal V_D compensating the input signal V_x. Note that one can compose this D/A converter from the dc voltage multiplier discussed above (Fig. 10) fed from the Josephson junction J_1. This junction is forced to make its 2π-switchings by a (presumably, room-temperature) circuit that converts the output digital code of the integrator to a number of output pulses during a certain time interval determined by a reference rf signal of a well-characterized frequency f. One can readily get convinced that static precision of the A/D converter as a whole is determined only by the frequency stability of the reference alone, while the dynamic range of the low-bit A/D converter (the SQUID in Fig. 12) limits only the frequency band of the whole device.

It seems presently that A/D converters of the type shown in Figs. 11 and 12 can become indispensable tools for metrology, even if helium cooling is required.

6. Digital Integrated Circuits

Digital LSI and VLSI circuits have always been the most desirable but also the least accessible application of the ideas and methods of superconducting electronics. In order to enter the extremely competitive market of microelectronics, superconductivity should provide integrated circuits with at least one substantial advantage, sufficient to overweigh not only refrigeration costs but also investments in new fabrication technology[5]. Presently, three major ways to ensure an advantage of some kind are discussed.

6.1. Superconductor Connections for Semiconductor ICs

The first possibility is to use superconducting thin-film interconnections in digital integrated circuits, with semiconductor transistors as active devices. In fact, a number of the transistor types improve their performance when cooled to $T = 77$ K, so that nitrogen-cooled integrated circuits are becoming reality even without any regard to superconductivity (see, for example, the special issue of *IEEE Trans. Electron. Dev.*, **34**, January 1987). Moreover, this family includes high-speed Si and GaAs transistors with extremely small intrinsic logic delays τ_0 of a few tens of picoseconds and even few picoseconds (see, for example, [77,78]). Due to interconnections, however, the real average logic delay τ per gate in the LSI circuits is much larger than τ_0 (see, for example, [79]), so that one could dream of reducing τ by making the connections superconducting.

Nevertheless, this dream does not correspond to present-day realities. The extra delay $\tau_i = \tau - \tau_0$ can be closely approximated as a time of recharge of the interconnection capacitance through the total resistance $R = R_o + R_i$, where R_o is the effective output impedance of the transistor and R_i is the resistance of the connection itself. For present-day high-speed transistors, R_o ranges between a few and a few tens kOhm.

On the other hand, even the longest interconnecting line (few millimeters) with cross-section area ~ 1 μm $\times$ 0.3 μm typical for VLSI circuits and made of either copper or aluminum would have R_i of order 10^2 Ohm at liquid nitrogen temperatures, so that $R_i \ll R_o$, and introduction of superconducting lines ($R_i \to 0$) would not result in a noticeable reduction of τ. (Malozemoff et al. [5] have arrived at a similar conclusion.)

Note, however, that superconducting lines can be useful in overcoming some other problems faced by VLSI circuits, namely electromigration and considerable voltage drops across dc supply lines. These possibilities are still to be studied, but their success depends crucially on the quality of contacts between semiconductors (like Si and GaAs) and the new superconductors (like 1–2–3 phase)[6]. According to the first experimental attempts, the prospects for good contacts are not very favorable.

6.2. Josephson Junction ICs

Completely superconductor circuits using Josephson junctions as active elements could provide not only a natural technological matching of the elements with interconnecting lines but also the impedance matching of these two major components of microelectronic circuits. In fact, the effective impedance of the Josephson junction R_o (which is of the order of its normal resistance) can readily be made close to the wave impedance ρ of the microstrip line (Fig. 2):

$$\rho\,[\text{Ohm}] \simeq 60 \frac{(t_I t_M)^{1/2}}{w \varepsilon^{1/2}}, \qquad t_M = t_I + 2\,\text{Re}\,\delta, \qquad (6.1)$$

with width ($w \simeq 1\ \mu$m) and insulation thickness ($t_I \simeq 0.3\ \mu$m) typical for the present-day technologies ($\rho \simeq 10$ Ohm). The ballistic (rather than diffusive) character of the signal propagation along these lines makes their contribution τ_i to the logic delay per gate quite small ($\tau_i \simeq 10$ ps for lines as long as ~ 1 mm). Moreover, the absence of a noticeable dispersion for frequencies up to $\sim 10^{12}$ Hz (see Sections 2.1 and 4.2 and Fig. 3) makes them the only means available for the transfer of picosecond pulses all along a chip.

What about the capabilities of Josephson junctions? They are, firstly, able to produce picosecond pulses: according to Eqs. (2.7) and (2.12) $\tau_{\min} \simeq \pi/\Delta\omega$ can be as short as ~ 3 ps even for a conservative value of ω_p. Secondly, power dissipation in the junction in its "open" (resistive) state is of the order

$$P_{\max} \sim \frac{V_{\max}^2}{R_N}. \qquad (6.2)$$

As we will see below, $V_{\max}$ can be of the order of $\max[V_c, V_p]$, so that with the above estimate of $R_N \simeq \rho$, one obtains $P_{\max}$ as low as $\sim 0.1\ \mu$W, i.e., two to three orders of magnitude less than the best semiconductor transistors.

Since the late 1960s these factors resulted in an eager interest in Josephson-junction digital technology, with the well-known IBM project (see, for example, the special issue of *IBM J. Res. Develop.*, **24**, March 1980) as a climax. By the end of 1983, however, the project failed due to several factors, including the two following ones:

(i) The project was based on relatively simple lead alloy technology, which proved to be insufficient to provide the necessary durable junctions, in particular to ensure their proper stability at repeated thermal cycling (300 K $\rightleftarrows$ 4.2 K);

1. NEW POSSIBILITIES FOR SUPERCONDUCTOR DEVICES 33

(ii) All logic families used in the IBM design were based on the "latching" circuits with underdamped Josephson junctions ($\omega_p \ll \omega_c$), with the dc voltage representation of the digital bits ($V = 0$ as binary "0" and $V \simeq V_g$ as binary "1"). These circuits exhibit relatively short "0" → "1" switching times of the order 10 ps, but their backward ("1" → "0") switching takes as long as ~ 1 ns because of the punch-through effect. Moreover, the latching logics suggested a need for the ac power supply also to play the role of global clock for the circuit, but the frequency of the supply was limited to a few hundred MHz due to numerous technical problems.

Note also that the latching logic makes power dissipation very large (on the Josephson junction scale, of course). In fact, for this logic $V_{max} \simeq V_g$, and the junctions remain in their open state for approximately half of the clock period. Even for low-T_c superconductors with $V_g \simeq 3$ mV, this leads to average power P of order 10 μW per gate. If such circuits were reproduced with high-T_c superconductors with their 20-fold larger energy gaps, P would reach several milliwatts (!) per gate. This figure is forbiddingly large for any LSI circuit because of the finite heat removal ability of liquid nitrogen (Section 3.1).

Fortunately, since 1983, ways have been found to overcome both major problems encountered by the IBM project. First of all, a family of Josephson junction fabrication technologies based on "rigid" (Nb and/or NbN) superconductors have been improved to make commercial production of LSI or even VLSI circuits quite feasible (see, for example, [83,32] and also Chapters 3 and 8 of this book).

Secondly, an alternative RSFQ (Resistive or Rapid Single Flux Quantum) logic family has been proposed[7] [84,85], and its basic components were successfully tested [86,87,88]. This family is based on a representation of information totally different from that used in the latching logics. The information is transferred generally through two lines, one carrying short (quasi)periodic clock pulses T, while the other, the bit pulses S. Binary "1" is denoted by a single S pulse arriving between two consequent T pulses (i.e., during the given clock period), while absence of the S pulse during this period denotes binary "0"—see Fig. 13. (Note that a small fraction τ' of the clock period τ around the T pulses should be forbidden for the S pulses.)

In principle, one can use such a representation in semiconductor transistor circuits as well, but in Josephson junction circuits it is very natural, especially if the single flux quantum (SFQ) pulses described by Eq. (5.20) are used in both the S and T lines. Such pulses can be readily generated and

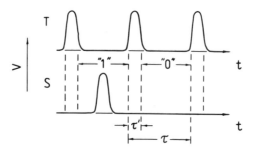

Fig. 13. Representation of binary units in the RSFQ logic circuits. The signal pulses S are allowed to arrive during the whole clock period τ except some close vicinity τ' of the clock pulses T.

reproduced/amplified either by single overdamped ($\omega_p \gtrsim \omega_c$) Josephson junctions [60,91] or by simple circuits consisting of such junctions.

Figures 14a,b,c show the simplest circuit of this kind, the buffer/amplifier stage, and the results of its numerical modeling [85] using the Josephson-junction-oriented COMPASS simulator [92].

If a short pulse arrives at the A terminal (Fig. 14b), it induces the 2π-switching of the Josephson phase ϕ of the junction J_2, which is dc-current biased below its critical current value I_{c2}. Such switching produces the standard SFQ pulse at the output terminal D. The input pulse, on the other hand, can be weaker than the standard one, so that the circuit can provide some gain. This property allows one to transfer the pulses between the logic gates through not only superconducting but also resistive circuits.

On the other hand, if the SFQ pulse arrives from the device output (terminal D), it 2π-switches junction 1 rather than 2 (Fig. 14c), because of the somewhat lower critical current of the former one. Thus, no SFQ pulse passes to the input of the circuit, so that it performs the function of a one-directional buffer as well.

Figure 14d shows another simple circuit, the R-S trigger, which illustrates methods to store and process information that are typical for the RSFQ logic family. The two-junction interferometer $L_1 - J_1 - J_2$ is dc biased to have two possible stable states that differ by $\Delta\phi \simeq \phi_0$ in trapped flux, i.e., differ by the direction of circulation of the persistent current $I_p \simeq \phi_0/2L$. Let us denote binary "0" by the state when the current is circulating counterclockwise. In this case, if an SFQ pulse arrives at the terminal S, it increases current through J_1 beyond the critical value, inducing its 2π-switching, i.e., "0" → "1" switching of the trigger (Fig. 14d). Now, the consequent S pulses switch junction 4 rather

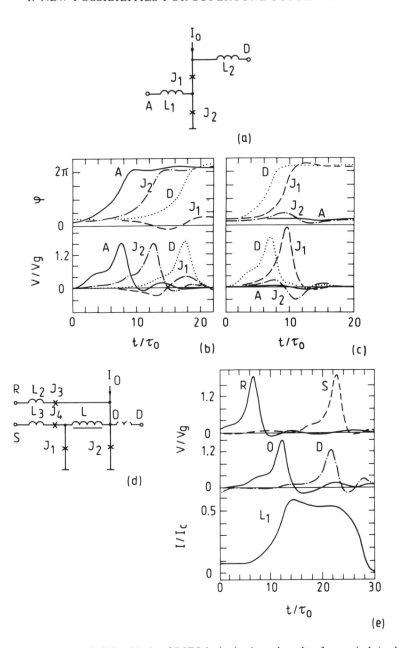

Fig. 14. The simplest building blocks of RSFQ logic circuits and results of numerical simulation of their dynamics: (a), (b), and (c) the buffer/amplifier stage; (d) and (e) the R-S trigger [85]. The constant $\tau_0 = \hbar/2\Delta(T)$ is close to 0.1 ps for 1-μm low-T_c technology ©1987 IEEE.

than 1, i.e., they do not change the state of the trigger. On the other hand, the first pulse arriving from the R terminal restores the trigger to its "0" state, simultaneously producing a similar SFQ pulse at the output terminal D.

One can see that within the RSFQ logic the normalized SFQ form of the transferred pulses is naturally matched with the information storage in the form of single flux quanta (in the IBM concept where similar storage was accepted for the main memory cells, its matching with the latching logic gates presented a major problem). Each member of the RSFQ logic family has an internal memory (i.e. is naturally "registered"), performs logic operations during a short time period following arrival of the input SFQ pulses, and produces the read-out SFQ pulse at its output under the action of the clock pulse T.

This general concept is illustrated by Fig. 15 which shows several logic gates of the RSFQ family [85]. Note the natural hardware-saving structure of the gates, including the NOT gate (Fig. 15a) crucial for any Josephson junction digital technology. One of the first versions of this gate has been verified to operate successfully at clock frequencies up to 30 GHz [86] despite relatively primitive 10-μm technology used in the experiments. Recently, a core circuit of the family, a T-trigger similar to the R-S trigger shown in Fig. 14c), fabricated using 5-μm technology, was demonstrated to operate at input pulse frequencies up to 70 GHz [88] with wide margins ($\pm 25\%$) for the dc bias current. Numerical simulation shows that using 3-μm technology one can increase the clock frequency of the RSFQ logic circuits to ~ 150 GHz, while 1-μm technology would make ~ 500 GHz operation possible [85]. These figures should be compared with record ~ 30 GHz speeds for performing similar logic functions achieved with GaAs transistors with much more complex 0.3-μm technology and much larger power consumption [78].

By the way, in the RSFQ logic circuits, the Josephson junctions are open only a small (typically, one tenth) part of the clock period, so that their average power consumption P is very low, typically only $\sim 10^{-7}$ Watt per gate (in other units, $P\tau \sim 10^{-18}$ Joule per bit) for liquid helium operation temperatures. This level of consumption (cf. Section 3.1) is quite acceptable for practical circuits with integration scales at least as large as $\sim 10^7$ gates per cm^2.

One more advantage of the RSQF logic circuitry is its local synchronization carried out by the pulses T, which can be readily generated, transferred, and (if necessary) processed at large speeds. Together with the natural intrinsic memory of each logic gate, this feature opens the way to very hardware-saving digital devices with bit-level conveyoring. An examples of this, a series

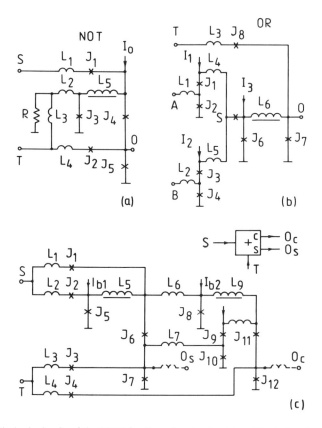

Fig. 15. The basic circuits of the RSFQ family performing "registered" logic functions: (a) NOT; (b) OR; (c) full single-bit addition [85] ©1987 IEEE.

multiplier, is shown in Fig. 16. It consists of just two types of single-bit elements: NDRO memory cells (denoted by circles) and full adders (squares). During the multiplication process, the bits of one operand (A) are constantly stored in the register of the memory cells, while those of another operand (B) are consequently fed into the device in RSFQ form. The latter bits act like read-out signals for the NDRO memory cells, so that the output signals of the cells present partial $A_i B_j$ products. The array of adders sums up those products to provide a continuous train of RSFQ bits of the full product C at the output, one bit per each clock period [91].

Estimates show that even with relatively primitive 3-μm technology such a device containing $\sim 10^3$ Josephson junctions could multiply 32-bit numbers in approximately 1 ns, i.e., as rapidly as a parallel multiplier containing $\sim 10^5$

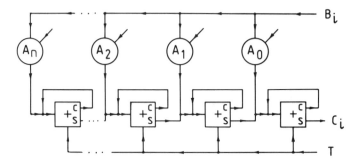

Fig. 16. A series n-bit number multiplier based on the RSFQ single-bit adders (squares) and NDRO memory cells performing single-bit multiplication (circles).

GaAs FETs with 0.3-μm-wide gates. The reader has seen that the deal was an exchange of speed for simplicity. Such an approach also simplifies interchip communications, because during the nanosecond the data can be readily transformed to the usual dc voltage form, passed to other chips, and transferred to the RSFQ form again if necessary. (Figure 17 shows a circuit capable of carrying out the former transformation, while the latter one can be made just by a single Josephson-junction interferometer [75].)

Another possible direction for RSFQ circuitry is the development of relatively complex devices for parallel digital processing with very large speeds. For example, a pipelined parallel multiplier (see, for example, [93,94]) containing $\sim 4 \times 10^4$ Josephson junctions designed with 3-μm rules could multiply 32-bit operands at a rate of above 5×10^{10} numbers per second! Such devices can be core components of extremely high-speed signal processors that can be fed, in particular, by the Josephson-junction A/D converters described in Section 5.5.

To summarize, we do not see obvious reasons why helium-cooled RSFQ circuits could not be used in the near future to fabricate either simple series devices with their performance comparable with that of much more complex semiconductor VLSI chips, or signal processors with very large scale integration and extremely high productivity. For the latter devices, the refrigeration costs are not the main concern, so that it is hardly urgent to reproduce them at the nitrogen level. Estimates show that such reproduction *is* possible, but would require very careful layout designs, mainly because of lower inductances (Section 3.2), not to mention the evident need of a well-developed technology for the fabrication of Josephson junctions with high-T_c superconductor electrodes.

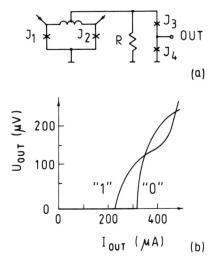

Fig. 17. A converter of binary information from the RSFQ form to the usual dc voltage form: (a) equivalent circuit; (b) experimental I–V curves recorded at the output port for two states of the input interferometer [87] ©1987 IEEE.

6.3. Single Electron Tunneling and Superconductivity

RSFQ logic gates may be considered as the ultimate in low power consumption[8], because their $P\tau$ products are limited only by unavoidable thermal fluctuations [97,25]

$$(P\tau)_{\min} = Ak_B T, \qquad A \simeq \ln\left(\frac{\Delta\omega\tau}{p}\right) \sim 10^3. \tag{6.3}$$

Here p is the permissible probability of a fluctuation-induced error per one gate (say, $\sim 10^{-30}$) during a typical operation period τ (say, $\sim 10^{-8}$ s). Computation speeds of RSFQ circuits are also close to the fundamental limits determined by light propagation velocity. Nevertheless, there is hardly anything fundamental in their third major parameter, the maximum integration scale.

In fact, the minimum feature size of all Josephson junction circuits (including the RSFQ logic family) is limited by the magnetic field penetration effects at the level of order $\lambda(T) \sim 0.1$ μm (just by chance this coincides with the limits typical for semiconductor transistors, in spite of the rather different physics of the limitations). This factor limits integration scales of both technologies by a maximum figure of order 10^7 gates per cm^2. This level is rapidly approached by the modern patterning technologies; thus, there is clearly a need for new physical ideas that will allow larger integration scales.

The recently predicted [100] and even more recently observed [101,102,103] effect of correlated single-electron tunneling is apparently one of such new ideas. Having no space here for a comprehensive description of the new effect (for reviews, see [25], Chapter 16, and [104]), we will only note that it allows externally-controlled transfer of single electrons[9] through tunnel junctions of very small area S with very small capacitance C and tunnel conductance G:

$$C \ll \frac{e^2}{2k_B T}, \qquad G \ll \frac{2e^2}{\pi \hbar}. \tag{6.4}$$

In particular, this effect allows one to design "Single Electron Logic" (SEL) circuits [108], which are close qualitative analogs of RSFQ circuits. In particular, presentation of binary information in SEL circuits corresponds to that shown in Fig. 13, with the only difference being that the single-flux-quantum pulses of voltage should be replaced by single-electron-charge pulses of current with the fixed area

$$\int I(t)\,dt = e \simeq 1.6 \text{ nA ps} \tag{6.5}$$

(in other words, transfer of single flux quanta is replaced by that of single electrons). Figure 18 shows an example of a universal SEL gate capable of performing either the OR or OR-NOT logic function, depending on what input is fed in by the S pulses.

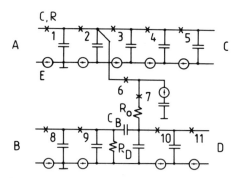

Fig. 18. A possible basic logic cell of single electron logic (SEL) circuits. Depending of what input (A,B) of the cell is fed by the signal (S) pulses, it performs either the OR or OR-NOT function [108].

1. NEW POSSIBILITIES FOR SUPERCONDUCTOR DEVICES   41

The major advantage of SEL circuits is virtually vanishing physical limitations imposed on their minimum feature size: it is determined by the tunnel barrier thickness plus two Debye screening depths of the metal electrodes, typically of order 3 nm. This means that the maximum integration scale can be as large as $\sim 10^{10}$ gates per cm^2.

It will by no means be easy, however, to approach this fabulous level! The basic problem is apparently fabrication of integrated circuits with thin-film electrodes of size $\sim 10 \times 10$ nm^2 interconnected by tunnel junctions of similar area (only such areas would allow acceptably low-error operation at liquid helium temperatures, and forget about nitrogen cooling). Such features are on the very frontier of the modern nanolithography—see, for example, the reviews by Owen [109] and Smith [110].

Another important problem is that of the gate interconnections. The picosecond (or even subpicosecond) pulses generated by SEL gates can be transferred the necessary distances (~ 1 cm) exclusively via superconducting microstrip lines (see Section 4.2). However, a line with a cross-section $S \simeq 3 \times 10$ nm^2 compatible with the junction size would have a wave impedance of order 10^4 Ohm (determined mainly by large specific kinetic inductance $L_0 \simeq \mu_0 \lambda^2(T)/S \sim 10^4$ nH/cm—see, for example, [111]), while the output impedance of the gates should be not less than $\sim 10^6$ Ohm in order to avoid computation errors resulting from quantum fluctuations [104]. This matching problem still should be studied in detail. More generally, real design and fabrication of these SuperLSI circuits would require a lot of thought, effort, and time (probably, not less than a decade).

7. Conclusion

The picture of the field presented in this chapter seems too complex to be summarized in a few words. One trend is, nevertheless, quite clear: recent "internal" development of low-T_c superconductor electronics has created a large potential for its intensive future progress, even without any regard to the advent of high-T_c materials. The latter event, on the other hand, has resulted in a drastic reduction of the refrigeration costs, rather than an improvement of the device performance (although one cannot exclude a possibility of sudden breakthroughs in the latter direction as well). The reduction will accelerate practical introduction of mostly the simplest superconductor components and devices, while for the most complex of them (e.g., digital VLSI circuits), the transfer from helium to liquid nitrogen operation temperatures would take more time and in some cases might be not advantageous at all.

There remains no doubt, however, that the future of superconductor electronics looks incomparably more bright now than it did just few years ago.

Acknowledgments

Fruitful discussions of the problems considered in this paper with Zh.I. Alferov, I.A. Devyatov, M.Yu. Kupriyanov, L.S. Kuzmin, V.F. Lukichev, O.A. Mukhanov, A.T. Rakhimov, S.V. Rylov, O.V. Snigirev, S.V. Vyshenskii, and other colleagues are gratefully acknowledged.

Notes

[1] The second requirement of Eq. (2.11) can be avoided in 3-D structures like variable-thickness microbridges (see, for example, [24]). Fabrication of such structures, however, requires definition of very small ($\lesssim 3\xi$) planar dimensions and hardly seems practical for the high-T_c superconductors with their very small coherence lengths.

[2] An alternative version of the comparator using two-junction interferometers (Fig. 7b) has similar properties but allows one to avoid galvanic coupling of the input circuits.

[3] Similar designs for the A/D converter comparators have been proposed earlier by Rylov and Semenov [50] and Petersen et al. [51]. Note also that the single junctions in the design by Faris [42] form a sort of Goto pair with the last junction of his sampling pulse generator. However, due to dissimilarity of the junctions, the technical fluctuations do not cancel in this case.

[4] In fact, this "limit" can presumably be overcome in coherent receivers using squeezed quantum states (see, for example, [52,53]).

[5] Note, however, that the technology of superconductor integrated circuits (at least of the low-T_c ones) is generally *simpler* than that of their semiconductor counterparts.

[6] The same remark can be made with respect to attempts to develop hot-electron transistors with superconducting bases [80,81,82].

[7] The RSFQ logic family is based on numerous ideas put forward during the 1970s and early 1980s (see, for example, [89,90,75,72]). Nevertheless, its basic concept illustrated by Fig. 13 was not formulated until 1985 [84].

[8] The limitation expressed by Eq. (6.3) *can*, in fact, be overcome in parametric-quantron-type circuits [95] performing reversible calculations [96–99]. These circuits are, however, rather hardware-wasting and only able to handle a limited class of algorithms. Nevertheless, this approach can be used as a last resort in the future if the dissipation level expressed by Eq. (6.3) becomes unacceptable (say, for development of genuine 3-D integration circuits).

[9] Almost simultaneously, similar effects of the correlated transfer of the Cooper pairs through small Josephson junctions have been predicted ([105,106]; see also [25], Chapter 16) and possibly observed experimentally [107]. Practical utilization of these effects would impose, however, additional conditions on parameters of the junction and is thus less feasible than that of the single-electron tunneling.

References

1. Bednortz, J.G., and Müller, K.A. Possible high-T_c superconductivity in the Ba-La-Cu-O system. *Z. Phys.* B **64**, 189–193 (1986).
2. Wu, M.K., Ashburn, J.R., Torng, C.J., Hor, P.H., Meng, R.L., Gao, L., Huang, Z.J., Wang, Y. Q., and Chu, C.W. Superconductivity at 93 K in a new mixedphase Y-Ba-Cu-O compound at ambient pressure. *Phys. Rev. Lett.* **58**, 908–910 (1987).
3. Maeda, H., Tanaka, Y., Fukutomi, M., and Toshihisa, A. A new high-T_c oxide superconductor without a rare earth element. *Jpn. J. Appl. Phys.* **27**, L209–L210 (1988).
4. Sheng, Z.Z., and Hermann, A.M. Bulk superconductivity at 120 K in the Tl-Ca/Ba-Cu-O system. *Nature* **332**, 138–139 (1988).
5. Malozemoff, A.P., Gallagher, W.J., and Schwall, R.E. Applications of high-temperature superconductivity. In *"Chemistry of High-Temperature Superconductors"* (D.L. Nelson, M.S. Whittingham, and T.F. George, eds.), pp. 280–306. ACS Symposium Series 351, Washington, D.C., 1987.
6. Cava, R.J., Batlogg, B., van Dover, R.B., Murphy, D.W., Sunshine, S., Siegrist, T., Remeika, J.P., Reitman, E.A., Zahurak, S., and Espinosa, G.P. Bulk superconductivity at 91 K in singlephase oxygen-deficient perovskite $Ba_2YCu_3O_{9-\delta}$. *Phys. Rev. Lett.* **58**, 1676–1679 (1987).
7. Hor, P.H., Meng, R.L., Wang, Y.Q., Gao, L., Huang, Z.J., Bechtold, J., Forster, K., and Chu, C.W. Superconductivity above 90 K in the square-planar compound system $ABa_2Cu_3O_{6+x}$ with A = Y, La Nd, Sm, Eu, Gd, Ho, Er, and Lu. *Phys. Rev. Lett.* **58**, 1891–1894 (1987).
8. Tamegai, T., Watanabe, A., Oguro, I., and Iye, Y. Structure and upper critical fields of high T_c superconductors $(RE)Ba_2Cu_3O_x$. *Jpn. J. Appl. Phys.* **26**, L1304–L1306 (1987).
9. Dinger, T.R., Worthington, T.K., Gallagher, W.J., and Sandstrom, R.L. Direct observation of electronic anisotropy in the single-crystal $Y_1Ba_2Cu_3O_{7-x}$. *Phys. Rev. Lett.* **58**, 2687–2690 (1987).
10. Tozer, S.W., Kleinsasser, A.W., Penney, T., Kaiser, D., and Holtzberg, F. Measurement of anisotropic resistivity and Hall constant for single-crystal $YBa_2Cu_3O_{7-x}$. *Phys. Rev. Lett.* **59**, 1768–1771 (1987).
11. Makarenko, I.N., Nikiforov, D.V., Bykov, A.B., Mel'nikov, O.K., Stishov, S.M. Analysis of electric resistance in single crystals of high-T_c superconductor. *Pis'ma Zh. Eksp. Teor. Fiz.* **47**, 52–56 (1988) [*JETP Lett.*].
12. Kats, E.I. On some properties of lamellar structures. *Zh. Eksp. Teor. Fiz.* **56**, 1673–1684 (1969) [*Sov. Phys. JETP*].
13. Lawrence, W.E., and Doniach, S. Theory of layer structure superconductors. In "Proceedings of the 12-th International Conference on Low Temperature Physics" (E. Kanda, ed.), pp. 361–362. Academic, Tokyo, 1971.
14. Bulaevskii, L.N. Superconductivity and electronic properties of layered compounds. *Usp. Fiz. Nauk* **116**, 447–483 (1975) [*Sov. Phys. Uspekhi*].
15. Buzdin, A.I., and Bulaevskii, L.N. Organic superconductors, *Usp. Fiz. Nauk.* **144**, 415–437 (1984) [*Sov. Phys. Uspekhi*].
16. Worthington, T.K., Gallaghar, W.J., and Dinger, T.R. Anisotropic nature of

high-temperature superconductivity in single-crystal $Y_1Ba_2Cu_3O_{7-x}$. *Phys. Rev. Lett.* **59**, 1160–1163 (1987).
17. Maxfield, B.W., and McLean, W.L. Superconducting penetration depth of niobium. *Phys. Rev.* **139**, A1515–A1522 (1965).
18. Enomoto, Y., Murakami, T., Suzuki, M., and Moriwaki, T. Largely anisotropic superconducting critical current in epitaxially grown $Ba_2YCu_3O_{7-y}$ thin films. *Jpn. J. Appl. Phys.* **26**, L1248–L1250. (1987).
19. de Gennes, P.G. "Superconductivity of Metals and Alloys," Chapter 3. Benjamin, New York, 1966.
20. Likharev, K.K. Formation of a mixed state in plane superconducting films. *Izvestia Vuzov—Radiofizika* **14**, 919–925 (1971) [*Radiophys. Quant. Electon.*].
21. Mattiss, D.C., and Bardeen, J. Theory of the anomalous skin effect in normal and superconducting metals. *Phys. Rev.* **111**, 412–417 (1958).
22. Likharev, K.K. Nonlinear electrodynamics of narrow superconducting films. *Izvestia Vuzov—Radiofizika* **14**, 1232–1241 (1971) [*Radiophys. Quant. Electron.*].
23. Barone, A., and Paterno, G. "Physics and Applications of the Josephson Effect." Wiley, New York, 1982.
24. Likharev, K.K. Superconducting weak links. *Rev. Mod. Phys.* **51**, 101–159 (1979).
25. Likharev, K.K. "Dynamics of Josephson Junctions and Circuits." Gordon and Breach, New York, 1986.
26. Esteve, D., Martinis, J.M., Urbina, C., Devoret, M.H., Collin, G., Monod, P., Ribault, M., and Revcolevschi, A. Observation of the ac Josephson effect inside copper-oxide-based superconductors. *Europhys. Lett.* **3**, 1237–1242 (1987).
27. Tsai, J.-S., Kubo, Y., and Tabuchi, J. All-ceramics Josephson junctions operative up to 90 K. *Jpn. J. Appl. Phys.* **26**, L701–L703 (1987).
28. Varlashkin, A.I., Vasiliev, A.L., Golovashkin, A.I., Ivanenko, O.M., Kuzmin, L.S., Likharev, K.K., Mitsen, K.V., Romanchikova, G.V., and Soldatov, E.S. Microscopic structure and contact properties of high-T_c superconducting ceramics Y-Ba-Cu-O. *Pis'ma Zh. Eksp. Teor. Fiz.* **46** (suppl.), 59–62 (1987) [*JETP Lett.*].
29. Koch, R.H., Umbach, C.P., Clark, G.J., Chaudhari, P., and Laibowitz, R.B. Quantum interference devices made from superconducting oxide thin films. *Appl. Phys. Lett.* **51**, 200–202 (1987).
30. Zimmerman, J.E., Beall, J.A., Cromar, M.W., and Ono, R.H. Operation of a Y-Ba-Cu-O RF SQUID at 81 K. *Appl. Phys. Lett.* **51**, 617–618 (1987).
31. Golovashkin, A.I., Gudkov, A. L., Krasnosvobodtsev, S.I., Kuzmin, L.S., Likharev, K.K., Maslennikov, Yu.V., Pashkin, Yu.A., Petchen, E.V., Snigirev, O.V. Josephson effect and macroscopic quantum interference in high-T_c superconducting thin film weak links at $T = 77$ K. Report at 1988 Applied Superconductivity Conference, and *IEEE Trans. Magn.*, **25**, 943–946 (1989).
32. Hatano, Y., Yano, S., Hirano, M., Tarutari, Y., and Kawabe, U. A subnanosecond Josephson data processor model. *In* "Extended Abstracts ISEC '87," pp. 239–242. Tokyo, 1987.

33. Lyon, D.N. Pool boiling of cryogenic liquids. *Advances in Cryogenic Heat Transfer* **64**, 82–85 (1968).
34. Kautz, R.L. Miniaturization of normal-state and superconducting striplines. *J. Res. of NBS* **84**, 247–259 (1979).
35. Kaplan, S.B., Chi, C.C., Langenberg, D.N., Chang, J.-J., Jafarey, S., and Scapalino, D.J. Quasiparticles and photon lifetimes in superconductors. *Phys. Rev. B* **14**, 4854–4873 (1976).
36. Danilov, V.V., Likharev, K.K., Snigirev, O.V., and Soldatov, E.S. Limit characteristics of two-junction magnetic flux detector. *IEEE Trans. Magn.* **13**, 240–241 (1977).
37. Tesche, C.D., and Clarke, J. DC SQUID: Noise and optimization. *J. Low Temp. Phys.* **29**, 301–331 (1977).
38. Gutmann, P. DC SQUID with high energy resolution. *Electron. Lett.* **15**, 372–373 (1979).
39. Snigirev, O.V. Ultimate sensitivity of the DC SQUIDs using unshunted tunnel junctions. *IEEE Trans. Magn.* **19**, 584–586 (1983).
40. Maslennikov, Yu.V., Snigirev, O.V., and Vasiliev, A.V. Integrated relaxation-oscillation-driven DC SQUIDs. *In* "Extended Abstracts ISEC '87," pp. 144–146. Tokyo. 1987.
41. Gudoshnikov, S.A., Maslennikov, Yu.V., Semenov, V.K., Snigirev, O.V., and Vasiliev, A.V. Integrated RO-driven DC SQUIDs. Report at 1988 Applied Superconductivity Conference, and *IEEE Trans. Magn.* **25**, 1178–1181 (1989).
42. Faris, S.M. Generation and measurement of ultrashort current pulses with Josephson devices. *Appl. Phys. Lett.* **36**, 1005–1007 (1980).
43. Wolf, P., Van Zeghbroek, B.J., and Deutsch, U. A Josephson sampler with 2.1 ps resolution. *IEEE Trans. Magn.* **21**, 226–229 (1985).
44. Weber, S. A practical way to turn out Josephson junction chips. *Electronics* **60**, 49–52 (1987).
45. Whiteley, S.R., Hohenwarter, G.K.G., and Faris, S.M. A Josephson junction time domain reflectometer with room temperature access. *IEEE Trans. Magn.* **23**, 899–902 (1987).
46. Fujimaki, A., Nakajima, K., and Sawada, Y. Direct measurement of the switching waveform in a DC-SQUID. *Jpn. J. Appl. Phys.* **26**, 74–80 (1987).
47. Kornev, V.K., and Semenov, V.K. The Josephson Goto pair as a basic element of high sensitive samplers. *In* "Extended Abstracts ISEC '87," pp. 131–134. Tokyo, 1987.
48. Gallagher, W.J., Chi, C.-C., Duling, I.N., Grischkowsky, D., Halas, N.J., Ketchen, M.B., and Kleinsasser, A.W. Subpicosecond optoelectronic study of resistive and superconductive transmission lines. *Appl. Phys. Lett.* **50**, 350–352 (1987).
49. Gudkov, A.L., Kornev, V.K., Makhov, V.I., Mushkov, S.I., Semenov, V.K., and Shchedrin, V.D. Josephson regenerative pulse triod as a high-sensitive comparator. *Pis'ma Zh. Tekhn. Fiz.* **14**, 888 (1988) [*JTP Lett.*].
50. Rylov, S.V., and Semenov, V.K. A wide-margin Josephson-junction A/D converter using the redundant coding. *Electron. Lett.* **21**, 829–830 (1985).
51. Petersen, D.A., Ko, H., and Van Duzer, T. Dynamic behavior of a Josephson

latching comparator for use in a high-speed analog-to-digital converter. *IEEE Trans. Magn.* **23**, 891–894 (1987).
52. Devyatov, I.A., Kuzmin, L.S., Likharev, K.K., Migulin, V.V., and Zorin, A.B. Quantum-statistical theory of microwave detection using superconducting tunnel junctions. *J. Appl. Phys.* **60**, 1808–1828 (1986).
53. Yurke, B., Kaminsky, P.G., Miller, R.E., Whittaker, E.A., Smith, A.D., Silver, A.H., and Simon, R.W. Observation of 4.2 K equilibrium-noise squeezing via a Josephson-parametric amplifier. *Phys. Rev. Lett.* **60**, 764–767 (1988).
54. Duh, K.H.G., Pospieszalski, M.W., Kopp, W.F., Ho, P., Jabra, A.A., Chao, P.-C., Smith, P.M., Lester, L.F., Ballingall, J.M., and Weinreb, S. Ultra-low-noise cryogenic HEMTs. *IEEE Trans. Electron. Devices* **35**, 249–256 (1988).
55. Clarke, J., Hoffer, G.I., Richards, P.L., and Yeh, N.-H. Superconductive bolometers for submillimeter wavelengths. *J. Appl. Phys.* **48**, 4865–4880 (1977).
56. Blaney, T.G. Detection techniques at short millimeter and submillimeter waves: An overview. *In* "Infrared and Millimeter Waves" (K.J. Button, ed.), Vol. 3, pp. 2–67. Academic Press, New York, 1980.
57. Haller, E.E. Physics and design of advanced IR bolometers and photoconductors. *Infrared Physics* **25**, 257–266 (1986).
58. Enomoto, Y., and Noda, J. Far infrared detector using boundary Josephson junction. *Jpn. J. Appl. Phys.* **26** (suppl. 26-3), 1145–1146 (1987).
59. Leung, M., Strom, U., Culberston, J.C., Claassen, J.H., Wolf, S.A., and Simon, R.W. NbN/BN granular films—A novel broadband bolometric detector for pulsed laser far infrared radiation. *IEEE Trans. Magn.* **23**, 714–716 (1987).
60. Likharev, K.K., Semenov, V.K., and Zorin, A.B. "Novye vosmozhnosti dlya sverkhprovodnikovoi electronik." VINITI, Moscow, 1988 (in Russian).
61. Wu, P.H., Cheng, Q.H., Yang, S.Z., Chen, J., Li, Y., Ji, Z.M., Song, J.M., Lu, H.X., Gao, X.K., Wu, J., and Zhang, X.Y. The Josephson effect in a ceramic bridge at liquid nitrogen temperature, *Jpn. J. Appl. Phys.* **26**, L1579–L1580 (1987).
62. Langenberg, D.N., and Larkin, A.I., eds. "Nonequilibrium Superconductivity." North-Holland, Amsterdam, 1986.
63. Kautz, R.L., Hamilton, C.A., and Lloyd, F.L. Series-array Josephson voltage standards, *IEEE Trans. Magn.* **23**, 883–890 (1987).
64. Lloyd, F.L., Hamilton, C.A., Beall, J.A., Go, D., Ono, R.H., and Harris, R.E. A Josephson array voltage standard at 10 V. *IEEE Electron. Device Lett.* **8**, 449–450 (1987).
65. Dziuba, R.F., Field, B.F., and Finnegan, T.F. Cryogenic voltage comparator system for $2e/h$ measurements. *IEEE Trans. Instr. Meas.* **23**, 264–267 (1974).
66. Kautz, R.L., and Lloyd, F.L. Precision series-array Josephson voltage standards. *Appl. Phys. Lett.* **51**, 2043–2045 (1987).
67. Semenov, V.K. Multi-Josephson-junction structures for DC voltage multiplication. *In* "Extended Abstracts ISEC '87," pp. 138–140. Tokyo, 1987.
68. Semenov, V.K., and Voronova, M.A. DC voltage multiplicator: A novel application of phase locking in Josephson junction arrays. Report at 1988 Applied Superconductivity Conference, and *IEEE Trans. Magn.* **25**, 1432–1435 (1989).
69. Hamilton, C.A., and Lloyd, F.L. A superconductivity 6-bit A/D converter with operation to 2×10^9 samples/second. *IEEE Trans. Electron. Devices Lett.* **1**, 92–94 (1980).

70. Hamilton, C.A., Lloyd, F.L., and Kautz, R.L. Analog measurements applications for high speed Josephson switches. *IEEE Trans. Magn.* **17**, 577–582 (1981).
71. Hamilton, C.A., and Lloyd, F.L. Design limitations for superconducting A/D converters. *IEEE Trans. Magn.* **17**, 3414–3419 (1981).
72. Hamilton, C.A., and Lloyd, F.L. 8-bit superconducting A/D converter. *IEEE Trans. Magn.* **19**, 1259–1261 (1983).
73. Rylov, S.V., and Semenov, V.K. Josephson-junction A/D converter using parametric quantrons and redundant coding. *In* "SQUID '85" 'H.-D. Hahlbonm and H. Lübbig, eds.), pp. 1109–1114. W. de Gruyter, Berlin, 1985.
74. Rylov, S.V., Semenov, V.K., and Likharev, K.K. Josephson junction A/D converters using differential coding. *IEEE Trans. Magn.* **23**, 735–738 (1987).
75. Hurrell, J.P., and Silver, A.H. SQUID digital electronics. *In* "Future Trends in Superconductive Electronics" (B.S. Deaver, Jr., C.M. Falko, J.H. Harris, and S.A. Wolf, eds.), pp. 437–447. AIP Conference Proceeding, Charlottesville, 1978.
76. Hurrell, J.P., Pridmore-Brown, D.C., and Silver, A.N. Analog-to-digital conversion with unlatched SQUIDs. *IEEE Trans. Electron. Devices* **27**, 1887–1896 (1980).
77. Heiblum, M., and Eastman, L.F. Ballistic electrons in semiconductors. *Sci. Amer.* **256**, 65–73 (1987).
78. Greiling, P.T. High-speed digital IC performance outlook. *IEEE Trans. Microwave Theor Techn.* **35**, 245–259 (1987).
79. Ito, H., Ishibashi, T., and Sugeta, T. Fabrication and characterization of AlGaAs/GaAs heterojunction bipolar transistors. *IEEE Trans. Electron. Devices* **34**, 224–229 (1987).
80. Frank, D.J., Brady, M.J., and Davidson, A. A new superconductivity-base-transistor. *IEEE Trans. Magn.* **21**, 721–725 (1985).
81. Tonouchi, M., Sakai, H., Kabayashi, T., and Fujisawa, K. A novel hot-electron transistor employing superconducting base. *IEEE Trans. Magn.* **23**, 1674–1677 (1977).
82. Davidson, A., Brady, M.J., Frank, D.J., and Woodall, J.M. Transport properties of a superconductor ohmic contact. *IEEE Trans. Magn.* **23**, 727–730 (1987).
83. Kotani, S., Fujimaki, N., Morohashi, S., Ohara, S., and Hasuo, S. Feasibility of an ultra-high-speed Josephson multiplier. *IEEE J. Sol. St. Circ.* **22**, 98–103 (1987).
84. Likharev, K.K. Mukhanov, O.A., and Semenov, V.K. Resistive single flux quantum logic for the Josephson-junction digital technology. *In* "SQUID '85" (H.-D. Hahlbohm and H. Lübbig, eds.), pp. 1103–1108. W. de Gruyer, Berlin, 1985.
85. Mukhanov, O.A., Semenov, V.K., and Likharev, K.K. Ultimate performance of RSFQ logic circuits. IEEE Trans. Magn. **23**, 759–762 (1987).
86. Koshelets, V.P., Likharev, K.K., Migulin, V.V., Mukhanov, O.A., Ovsyannikov, G.A., Semenov, V.K., Serpuchenko, I.L., and Vystavkin, A.N. Experimental realization of a resistive single flux quantum logic circuit. *IEEE Trans. Magn.* **23**, 755–758 (1987).
87. Kaplunenko, V.K., Koshelets, V.P., Likharev, K.K., Migulin, V.V., Mukhahov,

O.A., Ovsyannikov, G.A., Semenov, V.K., Serpuchenko, I.L., and Vystavkin, A.N. Experimental study of the RSFQ logic circuits. *In* "Extended Abstracts ISEC '87," pp. 127–130. Tokyo, 1987.

88. Kaplunenko, V.K., Khabipov, M.I., Koshelets, V.P., Likharev, K.K., Mukhanov, O.A., Semenov, V.K. Serpuchenko, I.L., and Vystavkin, A.N. Experimental study of the RSFQ logic elements. Report at 1988 Applied Superconductivity Conference, and *IEEE Trans. Magn.* **25**, 861–864 (1989).

89. Likharev, K.K. Properties of superconducting loop closed by a Josephson junction, as a multistable element. *Radiotekhnika i Elektronika* **19**, 1494–1502 (1974) [*Sov. Eng. Electron. Phys.*].

90. Nakajima, K., and Onodera, Y. Logic design of Josephson network. *J. Appl. Phys.* **47**, 1620–1627 (1976).

91. Mukhanov, O.A., Rylov, S.V., Semenov, V.K., and Vyshenskii, S.V. RSFQ logic arithmetic. Report at 1988 Applied Superconductivity Conference, and *IEEE Trans. Magn.* **25**, 857–860 (1989).

92. Odintsov, A.A., Semenov, V.K., and Zorin, A.B. Specific problems of numerical analysis of the Josephson junction circuits. *IEEE Trans. Magn.* **23**, 763–766 (1987).

93. Hatamian, M., and Cash, G.L. A 70-MHz 8-bit × 8-bit parallel pipelined multiplier in 2.5-μm CMOS. *IEEE J. Sol. St. Circ.* **21**, 505–513 (1986).

94. Hatamian, M., and Cash, G.L. Parallel bit-level pipelined VLSI design for high-speed signal processing. *Proc. IEEE* **75**, 1192–1202 (1987).

95. Likharev, K.K. Dynamics of some single flux quantum devices. 1. Parametric quantron. *IEEE Trans. Magn.* **13**, 242–244 (1977).

96. Bennett, C. Logic reversibility of computation. *IBM J. Res. Develop.* **17**, 525–532 (1973).

97. Likharev, K.K. Classical and quantum limitations on energy consumption in computation. *Int. J. Theor. Phys.* **21**, 311–326 (1983).

98. Likharev, K.K., Rylov, S.V., and Semenov, V.K. Reversible conveyor computation in array of parametric quantrons. *IEEE Trans. Magn.* **21**, 947–950 (1985).

99. Landauer, R. Reversible computation. *In* "Der Informations begriff im Technik und Wissenschaft" (O.G. Folberth, and C. Haskl, eds.), pp. 139–158. R. Oldenboug, Munchen, 1986.

100. Averin, D.V., and Likharev, K.K. Coulomb blockade of the single electron tunneling and coherent oscillations in small tunnel junction. *J. Low Temp. Phys.* **62**, 345–372 (1986).

101. Kuzmin, L.S., and Likharev, K.K. Direct experimental observation of discrete correlated single electron tunneling. *Pis'ma v Zh. Eksp. Teor. Fiz.* **45**, 389–390 (1987) [*Sov. Phys. JETP Lett.*].

102. Fulton, T.A., and Dolan, G.J. Observation of single electron charging effects in small tunnel junction. *Phys. Rev. Lett.* **59**, 109–112 (1987).

103. Barner, J.B., and Ruggiero, S.T. Observation of the incremental charging of Ag particles by single electrons. *Phys. Rev. Lett.* **59**, 807–810 (1987).

104. Likharev, K.K. Correlated discrete transfer of single electrons in ultrasmall tunnel junctions. *IBM J. Res. Develop.* **32**, 144–158 (1988).

105. Likharev, K.K., and Zorin, A.B. Theory of the Bloch-wave oscillations in small Josephson junctions. *J. Low Temp. Phys.* **59**, 347–382 (1985).
106. Likharev, K.K., and Zoring, A.B., Simultaneous Bloch and Josephson oscillations, and resistance quantization in small superconducting tunnel junctions. *Jpn. J. Appl. Phys.* **26** (suppl. 26-3), 1407–1408 (1987).
107. Yoshihiro, K., Kinoshita, J., Inagaki, K., Yamanouchi, C., Kobayashi, S., and Karasawa, T. Observation of the Bloch oscillations in granular films of oxidized tin particles. *Jpn. J. Appl. Phys.* **26** (suppl. 26-3), 1379–1380 (1987).
108. Likharev, K.K., and Semenov, V.K. Possible logic circuits based on the correlated single-electron tunneling in ultra-small junctions. *In* "Extended Abstracts ISEC '87," pp. 182–185. Tokyo, 1987.
109. Owen, G. Electron lithography for fabrication of microelectronic devices. *Repts. Progr. Phys.* **48**, 795–851 (1985).
110. Smith, H.I. A review of submicron lithography. *Superlat. and Microstruct.* **2**, 129–142 (1986).
111. Likharev, K.K. Linear electrodynamics of finite-width superconducting films. *Izvestia Vuzov—Radiofizika* **14**, 909–918 (1971) [*Radiophys. Quant. Electron.*].
112. Klein, N. Müller, G., Piel, H., Roas, B., Schultz, L., Klein, U., and Peiniger, M. Millimeter wave surface resistance of epitaxially grown $YBa_2Cu_3O_{7-x}$ thin films. *Appl. Phys. Lett.* **54**, 757–759 (1989).
113. Padamsee, H. Cornell University, CLNS 88/844 (1988).
114. Vendik, O.G., Gaidukov, M.M., Golovashkin, A.I., Karpuk, A., Kovalevich, L., Kozyrev, A.B., Krasnoslobodtsev, S.I., and Pechen, E.V. Surface resistance of single crystalline films $Ho_1Ba_2Cu_3O_{7-\delta}$ at 60 GHz. *Pis'ma Zh. Tekhn. Fiz.* **14**, 2209–2210 (1988).
115. Klein, N., Muller, G., Piel, H., Roas, B., Schultz, L., Klein, U., and Peiniger, M. Millimeter wave surface resistance of epitaxially grown $YBa_2Cu_3O_{7-x}$ thin films. *Appl. Phys. Lett.* **54**, 757–758 (1989).
116. Rubin, D.L., Green, K., Gruschus, J., Kirchgessner, J., Moffat, D., Padamsee, H., Sears, J., Shu, Q.S., Schneemeyer, L.F., and Waszczak, J.V. Observation of a narrow superconducting transition at 6 GHz in crystal of $YBa_2Cu_3O_7$. *Phys. Rev. B.* **38**, 6538–6542 (1988).
117. Sridhar, S., Wu, D.H., Kennedy, W.L., and Zahopoulus, C. Microwave and rf properties of high quality $Y_1Ba_2Cu_3O_y$ single crystals. Subm. to *Physica C* (1989).
118. Vendik, O.G., Gaidukov, M.M., Golovashkin, A.I., Karpuk, A., Kovalevich, L., Kozyrev, A.B., Krasnoslobodtsev, S.I., and Pechen, E.V. Surface resistance of single crystalline films $Ho_1Ba_2Cu_3O_{7-\delta}$ at 60 GHz. *Pis'ma Zh. Tekhn. Fiz.* **14**, 2209–2210 (1988).

CHAPTER 2

SQUIDs: Principles, Noise, and Applications*

JOHN CLARKE

Department of Physics
University of California
and
Materials and Chemical Sciences Division
Lawrence Berkeley Laboratory
Berkeley, California

1. Introduction 52
2. The Resistively Shunted Junction 53
3. The dc SQUID 56
    3.1. Theory of Operation 56
    3.2. Practical dc SQUIDs 59
    3.3. Flux-Locked Loop 61
    3.4. Thermal Noise in the dc SQUID: Experiment 63
    3.5. $1/f$ Noise in dc SQUIDs 64
    3.6. Alternative Read-Out Schemes 65
4. The rf SQUID 66
    4.1. Principles of Operation 66
    4.2. Theory of Noise in the rf SQUID 71
    4.3. Practical rf SQUIDs 73
5. SQUID-Based Instruments 73
    5.1. Magnetometers and Gradiometers 75
    5.2. Susceptometers 79
    5.3. Voltmeters 81
    5.4. The dc SQUID as a Radiofrequency Amplifier 81
    5.5. Gravity Wave Antennas 83
    5.6. Gravity Gradiometers 86
6. The Impact of High-Temperature Superconductivity 86
    6.1. Predictions for White Noise 87
    6.2. Practical Devices 88
    6.3. Flux Noise in YBCO Films 89
    6.4. Future Prospects for High-T_c SQUIDs 91

* This chapter is a shortened version of a chapter in the proceedings of the NATO Advanced Study Institute on Superconductive Electronics, edited by M. Nisenoff and H. Weinstock.

7. Concluding Remarks	93
Acknowledgments.	94
References.	94

1. Introduction

Superconducting QUantum Interference Devices (SQUIDs) are the most sensitive detectors of magnetic flux available. A SQUID is, in essence, a flux-to-voltage transducer, providing an output voltage that is periodic in the applied flux with a period of one flux quantum, $\Phi_0 \equiv h/2e \approx 2.07 \times 10^{-15}$ Wb. One is generally able to detect an output signal corresponding to a flux change of much less than Φ_0. SQUIDs are amazingly versatile, being able to measure any physical quantity that can be converted to a flux, for example, magnetic field, magnetic field gradient, current, voltage, displacement, and magnetic susceptibility. As a result, their applications are wide ranging, from the detection of tiny magnetic fields produced by the human brain and the measurement of fluctuating magnetic fields in remote areas to the detection of gravity waves and the observation of spin noise in an ensemble of magnetic nuclei.

SQUIDs combine two physical phenomena, flux quantization—the fact that the flux Φ in a closed superconducting loop is quantized [1] in units of Φ_0—and Josephson tunneling [2]. There are two kinds of SQUIDs. The first [3], the dc SQUID, consists of two Josephson junctions connected in parallel on a superconducting loop and is so named because it operates with a steady current bias. Relatively crude devices were developed in the second half of the 1960s and used successfully by low-temperature physicists to measure a variety of phenomena occurring at liquid helium temperatures. At the end of the decade, the rf SQUID [4,5] appeared. This device involves a single Josephson junction interrupting the current flow around a superconducting loop and is operated with a radiofrequency flux bias.

Because it required only a single junction, at the time, the rf SQUID was simpler to manufacture and quickly became commercially available. It has remained the more widely used device ever since. However, in the mid-1970s, it was shown that the dc SQUID was the more sensitive device, and since then, there has been an ongoing development of thin-film dc SQUIDs and instruments based on them. By contrast, there has been little development of the rf SQUID in the last decade.

This chapter gives an overview of the current state of the SQUID art; we cannot hope to describe all of the SQUIDs that have been made or, even less, all of the applications in which they have been successfully used. This overview begins in Section 2 with a brief review of the resistively shunted Josephson junction, with particular emphasis on the effects of noise. Section 3 contains a description of the dc SQUID—how these devices are made and operated, and the limitations imposed by noise. Section 4 contains a similar description of the properties of rf SQUIDs, but because there has been little development of these devices in the 1980s, this section has been kept relatively brief. In Section 5, we describe a selection of instruments based on SQUIDs and mention some of their applications. Section 6 is a discussion of the impact of high-temperature superconductivity on SQUIDs and of future prospects in this area, and Section 7 contains a few concluding remarks.

2. The Resistively Shunted Junction

The Josephson junction [2] consists of two superconductors separated by a thin insulating barrier. Cooper pairs of electrons (or holes) are able to tunnel through the barrier, maintaining phase coherence in the process. The applied current I controls the difference $\delta = \phi_1 - \phi_2$ between the phases of the two superconductors according to the current–phase relation

$$I = I_0 \sin \delta, \tag{2.1}$$

where I_0 is the critical current, that is, the maximum supercurrent the junction can sustain. When the current is increased from zero, initially there is no voltage across the junction, but for $I > I_0$, a voltage V appears and δ evolves with time according to the voltage-frequency relation

$$\dot{\delta} = \frac{2eV}{\hbar} = \frac{2\pi V}{\Phi_0}. \tag{2.2}$$

A high-quality Josephson tunnel junction has a hysteretic current–voltage (I–V) characteristic. As the current is increased from zero, the voltage switches abruptly to a nonzero value when I approaches I_0, but returns to zero only when I is reduced to a value much less than I_0. This hysteresis must be eliminated for SQUIDs operated in the conventional manner, and one does so by shunting the junction with an external shunt resistance. The "resistively shunted junction" (RSJ) model [6,7] is shown in Fig. 1a. The junction has a critical current I_0 and is in parallel with its self-capacitance C and the shunt resistance R, which has a current noise source $I_N(t)$ associated with it. The equation of motion is

$$C\dot{V} + I_0 \sin \delta + \frac{V}{R} = I + I_N(t). \tag{2.3}$$

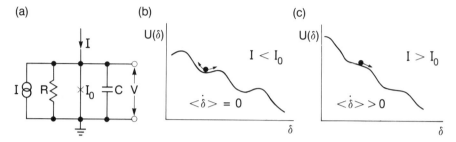

Fig. 1. (a) The resistively shunted Josephson junction; (b) and (c) show the tilted washboard model for $I < I_0$ and $I > I_0$.

Neglecting the noise term for the moment and setting $V = \hbar\dot{\delta}/2e$, we obtain

$$\frac{\hbar C}{2e}\ddot{\delta} + \frac{\hbar}{2eR}\dot{\delta} = I - I_0 \sin\delta = -\frac{2e}{\hbar}\frac{\partial U}{\partial \delta}, \quad (2.4)$$

where

$$U = -\frac{\Phi_0}{2\pi}(I\delta + I_0 \cos\delta). \quad (2.5)$$

One obtains considerable insight into the dynamics of the junction by realizing that Eq. (2.4) also describes the motion of a ball moving on the "tilted washboard" potential U. The term involving C represents the mass of the particle and $1/R$ the damping of the motion; the average "tilt" of the washboard is proportional to $-I$. For values of $I < I_0$, the particle is confined to one of the potential wells (Fig. 1b), where it oscillates back and forth at the plasma frequency [1]

$$\omega_p = \left(\frac{2\pi I_0}{\Phi_0 C}\right)^{1/2}\left[1 - \left(\frac{I}{I_0}\right)^2\right]^{1/4};$$

in this state, $\langle\dot{\delta}\rangle$ and hence the average voltage across the junction are zero ($\langle\ \rangle$ represents a time average). When the current is increased to I_0, the tilt increases so that the particle rolls down the washboard; in this state, $\langle\dot{\delta}\rangle$ is nonzero and a voltage appears across the junction (Fig. 1c). As the current is increased further, $\langle\dot{\delta}\rangle$ increases, as does V. For the nonhysteretic case, as soon as I is reduced below I_0, the particle becomes trapped in one of the wells, and V returns to zero. In this, the overdamped case, we require [6,7]

$$\beta_C \equiv \frac{2\pi I_0 R}{\Phi_0} RC = \omega_J RC \lesssim 1; \quad (2.6)$$

$\omega_J/2\pi$ is the Josephson frequency corresponding to the voltage $I_0 R$.

We introduce the effects of noise by restoring the noise term to Eq. (2.4) to obtain the Langevin equation

$$\frac{\hbar C}{2e}\ddot{\delta} + \frac{\hbar}{2eR}\dot{\delta} + I_0 \sin\delta = I + I_N(t). \tag{2.7}$$

In the thermal noise limit, the spectral density of $I_N(t)$ is given by the Nyquist formula

$$S_I(f) = \frac{4k_B T}{R}, \tag{2.8}$$

where f is the frequency. It is evident that $I_N(t)$ causes the tilt in the washboard to fluctuate with time. This fluctuation has two effects on the junction. First, when I is less than I_0, from time to time fluctuations cause the total current $I + I_N(t)$ to exceed I_0, enabling the particle to roll out of one potential minimum into the next. For the underdamped junction, this process produces a series of voltage pulses randomly spaced in time. Thus, the time average of the voltage is nonzero even though $I < I_0$, and the I–V characteristic is "noise-rounded" at low voltages [8]. Because this thermal activation process reduces the observed value of the critical current, there is a minimum value of I_0 for which the two sides of the junction remain coupled together. This condition may be written as

$$\frac{I_0 \Phi_0}{2\pi} \gtrsim 5k_B T, \tag{2.9}$$

where $I_0\Phi_0/2\pi$ is the coupling energy of the junction [2] and the factor of 5 is the result of a computer simulation [9].

The second consequence of thermal fluctuations is voltage noise. In the limit $\beta_C \ll 1$ and for $I > I_0$, the spectral density of this noise at a measurement frequency f_m, which we assume to be much less than the Josephson frequency f_J, is given by [10,11]

$$S_V(f_m) = \left[1 + \frac{1}{2}\left(\frac{I_0}{I}\right)^2\right]\frac{4k_B T R_d^2}{R}, \quad \begin{cases}\beta_C \ll 1 \\ I > I_0 \\ f_m \ll f_J\end{cases}. \tag{2.10}$$

The first term on the right-hand side of Eq. (2.10) represents the Nyquist noise current generated at the measurement frequency f_m flowing through the dynamic resistance $R_d \equiv dV/dI$ to produce a voltage noise. The second term, $\frac{1}{2}(I_0/I)^2(4k_B T/R)R_d^2$, represents Nyquist noise generated at frequencies $f_J \pm f_m$ mixed down to the measurement frequency by the Josephson oscillations and the inherent nonlinearity of the junction. The factor $\frac{1}{2}(I_0/I)^2$ is the mixing coefficient, and it vanishes for sufficiently large bias currents. The mixing coefficients for the Nyquist noise generated near harmonics of the Josephson frequencies, $2f_J, 3f_J, \ldots$, are negligible in the limit $f_m/f_J \ll 1$.

At sufficiently high bias current, the Josephson frequency f_J exceeds $k_B T/h$, and quantum corrections [12] to Eq. (2.10) become important provided the term $\frac{1}{2}(I_0/I)^2$ is not too small. It turns out that the requirement for observing significant quantum corrections is $eI_0 R/k_B T \gg 1$. The spectral density of the voltage noise becomes

$$S_V(f_m) = \left[\frac{4k_B T}{R} + \frac{2eV}{R}\left(\frac{I_0}{I}\right)^2 \coth\left(\frac{eV}{k_B T}\right)\right] R_d^2, \quad \begin{Bmatrix} \beta_C \ll 1 \\ I > I_0 \\ f_m \ll f_J \end{Bmatrix}, \quad (2.11)$$

where we have assumed that $hf_m/k_B T \ll 1$ so that the first term on the right-hand side of Eq. (2.11) remains in the thermal limit. In the limit $T \to 0$, the second term, $(2eV/R)(I_0/I)^2 R_d^2$, represents noise mixed down from zero point fluctuations near the Josephson frequency.

This concludes our review of the RSJ, and we now turn our attention to the dc SQUID.

3. The dc SQUID

3.1. Theory of Operation

The essence of the dc SQUID [3] is shown in Fig. 2a. Two Josephson junctions are connected in parallel on a superconducting loop of inductance L. Each junction is resistively shunted [6,7] to eliminate hysteresis on the current–voltage ($I-V$) characteristics, which are shown in Fig. 2b for $\Phi = n\Phi_0$ and $(n + \frac{1}{2})\Phi_0$; Φ is the external flux applied to the loop and n is an integer. If we bias the SQUID with a constant current ($>2I_0$), the voltage across the

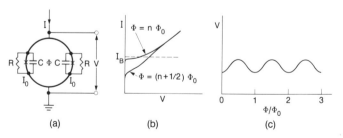

Fig. 2. (a) The dc SQUID; (b) $I-V$ characteristics; (c) V vs. Φ/Φ_0 at constant bias current I_B.

SQUID oscillates as we steadily increase Φ, as indicated in Fig. 2c. The period is Φ_0. The SQUID is generally operated on the steep part of V–Φ curve where the transfer coefficient, $V_\Phi \equiv |(\partial V/\partial \Phi)_I|$, is a maximum. Thus, the SQUID produces an output voltage in response to a small input flux $\delta\Phi$ ($\ll \Phi_0$) and is effectively a flux-to-voltage transducer. The resolution can be characterized by the equivalent flux noise $\Phi_N(t)$, which has a spectral density

$$S_\Phi(f) = \frac{S_V(f)}{V_\Phi^2} \tag{3.1}$$

at frequency f. Here, $S_V(f)$ is the spectral density of the voltage noise across the SQUID at fixed current bias. A more useful characterization, however, is in terms of the flux noise energy associated with $S_\Phi(f)$,

$$\varepsilon(f) = \frac{S_\Phi(f)}{2L}, \tag{3.2}$$

where L is the inductance of the loop.

Thermal noise imposes two constraints on the parameters of the SQUID. First, the coupling energy $I_0 \Phi_0/2\pi$ of each junction must be significantly greater than $k_B T$. This requirement is expressed by Eq. (2.9) which, at 4.2K, implies $I_0 \gtrsim 0.9$ μA. Second, the root mean square thermal noise flux in the loop, $\langle \Phi_N^2 \rangle^{1/2} = (k_B T L)^{1/2}$, must be significantly less than Φ_0; computer simulations [9] leads to

$$L \lesssim \frac{\Phi_0^2}{5 k_B T}. \tag{3.3}$$

This constraint implies $L \lesssim 15$ nH at 4.2 K.

We now outline the signal and noise properties of the SQUID. Each junction in Fig. 2a has a critical current I_0, a self-capacitance C, and a resistive shunt R chosen so that $\beta_C \equiv 2\pi I_0 R^2 C/\Phi_0 \lesssim 1$. The two resistors generate statistically independent Nyquist noise currents, $I_{N1}(t)$ and $I_{N2}(t)$, each with a spectral density $4 k_B T/R$ at temperature T. The differences in the phase across each junction, $\delta_1(t)$ and $\delta_2(t)$, obey the following equation [13,14,15]:

$$V = \frac{\hbar}{4e}(\dot\delta_1 + \dot\delta_2), \tag{3.4}$$

$$J = \frac{\Phi_0}{2\pi L}\left(\delta_1 - \delta_2 - \frac{2\pi \Phi}{\Phi_0}\right), \tag{3.5}$$

$$\frac{\hbar C}{2e}\ddot{\delta}_1 + \frac{\hbar}{2eR}\dot{\delta}_1 = \frac{I}{2} - J - I_0 \sin \delta_1 + I_{N1}, \quad (3.6)$$

$$\frac{\hbar C}{2e}\ddot{\delta}_2 + \frac{\hbar}{2eR}\dot{\delta}_2 = \frac{I}{2} + J - I_0 \sin \delta_2 + I_{N2}. \quad (3.7)$$

Equation (3.4) relates the voltage to the average rate of change of phase, Eq. (3.5) expresses fluxoid quantization, and Eqs. (3.6) and (3.7) are Langevin equations coupled via J. These equations have been solved numerically for a limited range of values of the noise parameter $\Gamma = 2\pi k_B T/I_0 \Phi_0$, reduced inductance $\beta = 2LI_0/\Phi_0$, and hysteresis parameter β_C. For typical SQUIDs in the ^{4}He temperature range, $\Gamma = 0.05$. One computes the time averaged voltage V vs. Φ, and hence finds V_Φ, which, for a given value of Φ, peaks smoothly as a function of bias current. The transfer coefficient exhibits a shallow maximum around $(2n + 1)\Phi_0/4$. One computes the noise voltage for a given value of Φ as a function of I, and finds that the spectral density is white at frequencies much less than the Josephson frequency. For each value of Φ, the noise voltage peaks smoothly at the value of I where V_Φ is a maximum. From these simulations, one finds that the noise energy has a minimum when $\beta \approx 1$. For $\beta = 1$, $\Gamma = 0.05$, and $\Phi = (2n + 1)\Phi_0/4$ and for the value of I at which V_Φ is a maximum, the results can be summarized as follows:

$$V_\Phi \approx \frac{R}{L}, \quad (3.8)$$

$$S_V(f) \approx 16 k_B T R, \quad (3.9)$$

$$\varepsilon(f) \approx \frac{9 k_B T L}{R}. \quad (3.10)$$

It is often convenient to eliminate R from Eq. (3.10) using the expression $R = (\beta_C \Phi_0/2\pi I_0 C)^{1/2}$. We find

$$\varepsilon(f) \approx 16 k_B T \left(\frac{LC}{\beta_C}\right)^{1/2} \quad (\beta_C \lesssim 1). \quad (3.11)$$

Equation (3.11) gives a clear prescription for improving the resolution: one should reduce T, L, and C. A large number of SQUIDs with a wide range of parameters have been tested and found to have noise energies generally in good agreement with the predicted values. It is common practice to quote the noise energy of SQUIDs in units of $\hbar$ ($\approx 10^{-34}$ J sec $= 10^{-34}$ J Hz^{-1}).

3.2. Practical dc SQUIDs

Modern dc SQUIDs are invariably made from thin films, with the aid of either photolithography or electron beam lithography. A major concern in design is the need to couple an input coil inductively to the SQUID with rather high efficiency. This problem was elegantly solved by Ketchen and Jaycox [16,17], who introduced the idea of depositing a spiral input coil on a SQUID in a square washer configuration. The coil is separated from the SQUID with an insulating layer. The version [18] of this design made at the University of California at Berkeley is shown in Fig. 3. These devices are made in batches of 36 on 50 mm-diameter oxidized silicon wafers in the following way. First, a 30 nm-thick Au (25 wt% Cu) film is deposited and patterned to form the resistive shunts. Next, we sputter a 100 nm-thick Nb film and etch it to form the SQUID loop and a strip that eventually contacts the inner end of the spiral coil. The third film is a 200 nm SiO layer with 2 μm diameter windows for the junctions, a larger window to give access to the CuAu shunt, and a window at each end of the Nb strip to provide connections to the spiral coil. The next step is to deposit and lift-off the 300 nm thick Nb spiral coil, which has 4, 20, or 50 turns. At this point, we usually dice the wafer into chips, each with a single SQUID, which is completed individually. The device is

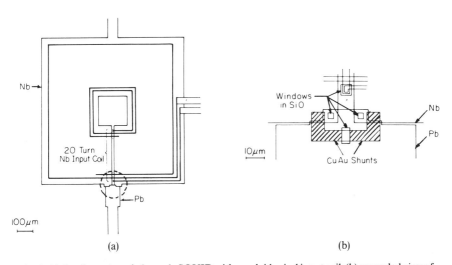

Fig. 3. (a) Configuration of planar dc SQUID with overlaid spiral input coil; (b) expanded view of junctions and shunts.

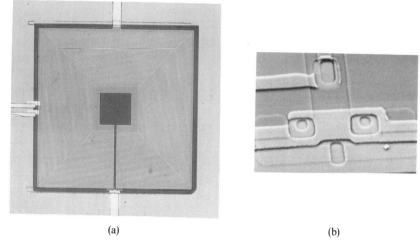

Fig. 4. (a) Photograph of planar dc SQUID made at UC Berkeley, with 50-turn input coil; the square washer is about 1 mm across. (b) Electron micrograph of junctions prior to deposition of counterelectrode; each junction is about 3 μm across.

ion-milled to clean the exposed areas of Nb and CuAu. We have two procedures for forming the oxide barrier. In one, we oxidize the Nb in a rf discharge in Ar containing 5 vol% O_2, and deposit the 300 nm Pb (5 wt% In) counterelectrode that completes the junctions and makes contact with the shunts. In the other process, we deposit approximately 6 nm of Al, form Al_2O_3 by exposing [19] it to O_2, and use Nb for the counterelectrode. The completed SQUID and the junctions are illustrated in Fig. 4. The shunt resistance R is typically 8 Ω, and the estimated capacitance C about 0.5 pF.

Jaycox and Ketchen [17] showed that a square washer (with no slit) with inner and outer edges d and w has an inductance L (loop) = 1.25 $\mu_0 d$ in the limit $w \gg d$. They gave the following expressions for the inductance L of the SQUID, the inductance L_i of the spiral coil, the mutual inductance M_i, and the coupling coefficient α^2 between the spiral coil and the SQUID:

$$L = L \text{ (loop)} + L_j, \tag{3.12}$$

$$L_i = n^2(L - L_j) + L_s, \tag{3.13}$$

$$M_i = n(L - L_j) \tag{3.14}$$

$$\alpha^2 = \frac{(1 - L_j/L)}{1 + L_s/n^2(L - L_j)}. \tag{3.15}$$

Here, L_j is the parasitic inductance associated with the junctions, n is the number of turns on the input coil, and L_s is the stripline inductance of this coil. For the SQUID just described with a 50-turn input coil, one measures $L_i \approx 800$ nH, $M_i \approx 16$ nH, and $\alpha^2 \approx 0.75$. These results are in good agreement with the predictions if one takes the predicted value L (loop) ≈ 0.31 nH and assumes $L_j \approx 0.09$ nH to give $L \approx 0.4$ nH. The stripline inductance (~ 10 nH) is insignificant for a 50-turn coil.

Reference [20–25] are a selection of papers describing SQUIDs fabricated on the Ketchen–Jaycox design. Some of the devices involve edge junctions, in which the counterelectrode is a strip making a tunneling contact to the base electrode only at the edge. This technique enables one to make junctions with a small area and thus a small self-capacitance without resorting to electron-beam lithography. However, stray capacitances are often critically important. As has been emphasized by a number of authors, parasitic capacitance between the square washer and the input coil can produce resonances that in turn induce structure on the I–V characteristics and give rise to excess noise. One way to reduce these effects is to lower the shunt resistance to increase the damping. An alternative approach is to couple the SQUID to the signal via an intermediary superconducting transformer [24], so that the number of turns on the SQUID washer and the parasitic capacitance are reduced. Knuutila et al [25] successfully damped the resonances in the input coil by terminating the stripline with a matched resistor. An alternative coupling scheme has been adopted by Carelli and Foglietti [26], who fabricated thin-film SQUIDs with many loops in parallel. The loops are coupled to a thin-film coil surrounding them.

3.3. Flux-Locked Loop

In most, although not all, practical applications, one uses the SQUID in a feedback circuit as a null-detector of magnetic flux [27]. One applies a modulating flux to the SQUID with a peak-to-peak amplitude $\Phi_0/2$ and a frequency f_m usually between 100 and 500 kHz, as indicated in Fig. 5. If the quasistatic flux in the SQUID is exactly $n\Phi_0$, the resulting voltage is a rectified version of the input signal, that is, it contains only the frequency $2f_m$ (Fig. 5a). If this voltage is sent through a lock-in detector referenced to the fundamental frequency f_m, the output will be zero. On the other hand, if the quasistatic flux is $(n + \frac{1}{4})\Phi_0$, the voltage across the SQUID is at frequency f_m (Fig. 5b), and the output from the lock-in will be a maximum. Thus, as one increases the flux from $n\Phi_0$ to $(n + \frac{1}{4})\Phi_0$, the output from the lock-in will increase steadily; if one reduces the flux from $n\Phi_0$ to $(n - \frac{1}{4})\Phi_0$, the output will increase in the negative direction (Fig. 5c).

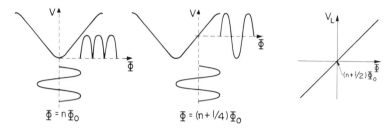

Fig. 5. Flux modulation scheme showing the voltage across the SQUID for (a) $\Phi = n\Phi_0$ and (b) $\Phi = (n + \frac{1}{4})\Phi_0$. The output V_L from the lock-in detector vs. Φ is shown in (c).

The alternating voltage across the SQUID is coupled to a low-noise preamplifier, usually at room temperature, via either a cooled transformer [28] or a cooled LC series-resonant circuit [27]. The first presents an impedance $N^2 R_d$ to the preamplifier and the second an impedance $Q^2 R_d$, where R_d is the dynamic resistance of the SQUID at the bias point, N is the turns-ratio of the transformer, and Q is the quality factor of the tank circuit. The value of N or Q is chosen to optimize the noise temperature of the preamplifier; with careful design, the noise from the amplifier can be appreciably less than that from the SQUID at 4.2 K.

Figure 6 shows a typical flux-locked loop in which the SQUID is coupled to the preamplifier via a cooled transformer. An oscillator applies a modulating flux to the SQUID. After amplification, the signal from the SQUID is lock-in detected and sent through an integrating circuit. The smoothed output is connected to the modulation and feedback coil via a large series resistor R_f.

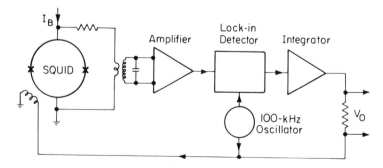

Fig. 6. Modulation and feedback circuit for the dc SQUID.

Thus, if one applies a flux $\delta\Phi$ to the SQUID, the feedback circuit will generate an opposing flux $-\delta\Phi$, and a voltage proportional to $\delta\Phi$ appears across R_f. This technique enables one to measure changes in flux ranging from much less than a single flux quantum to many flux quanta. The use of a modulating flux eliminates $1/f$ noise and drift in the bias current and preamplifier. Using a modulation frequency of 500 kHz, a double transformer between the SQUID and the preamplifier, and a two-pole integrator, Wellstood et al. [18] achieved a dynamic range of $\pm 2 \times 10^7$ Hz$^{1/2}$ for signal frequencies up to 6 kHz, a frequency response from 0 to 70 kHz (± 3 dB), and a maximum slew rate of 3×10^6 Φ_0 sec^{-1}.

3.4. Thermal Noise in the dc SQUID: Experiment

One determines the spectral density of the equivalent flux noise in the SQUID by connecting a spectrum analyzer to the output of the flux-locked loop. A representative power spectrum [29] is shown in Fig. 7: above a $1/f$ noise region, the noise is white at frequencies up to the roll-off of the feedback circuit. In this particular example, with $L = 200$ pH and $R = 8$ Ω, the measured flux noise was $S_\Phi^{1/2} = (1.9 \pm 0.1) \times 10^{-6}$ Φ_0 Hz$^{-1/2}$, in reasonable agreement with the prediction of Eqs. (3.8) and (3.9), 1.3×10^{-6} Φ_0 Hz$^{-1/2}$. The corresponding flux noise energy was 4×10^{-32} J Hz ≈ 400 $\hbar$. Many groups have

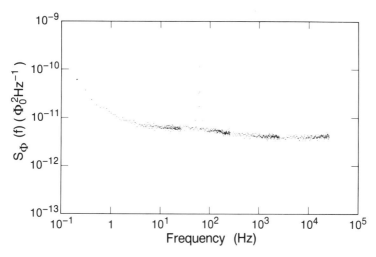

Fig. 7. Spectral density of equivalent flux noise for dc SQUID with a Pb body: $L = 0.2$ nH, $R = 8$ Ω, and $T = 4.2$ K. (Courtesy F.C. Wellstood.)

achieved noise energies that are comparable or, with lower values of L or C, somewhat better.

Rather recently, Wellstood *et al.* [30] have operated SQUIDs in a dilution refrigerator at temperatures T below 1 K, using a second dc SQUID as a preamplifier. They found that the noise energy scaled rather accurately with T at temperatures down to about 150 mK, below which the noise energy became nearly constant. This saturation was traced to heating in the resistive shunts that prevented them from cooling much below 150 mK. This heating is actually a hot electron effect [31,32]: the bottleneck in the cooling process is the rate at which the electrons can transfer energy to the phonons, which, in turn, transfer energy to the substrate. The temperature of the shunts was lowered by connecting each of them to a CuAu "cooling fin" of large volume. The hot electrons diffuse into the fins where they rapidly transfer energy to other electrons. Since the "reaction volume" is now greatly increased, the numbers of electrons and phonons interacting are also increased, and the electron gas is cooled more effectively. In this way, the effective electron temperature was reduced to about 50 mK when the SQUID was at a bath temperature of 20 mK, with a concomitant reduction in ε to about $5\hbar$. Very recently Ketchen *et al.* [33] have achieved a noise energy of about $3\hbar$ at 0.3 K in a SQUID with $L = 100$ pH and $C = 0.14$ pF.

3.5. 1/f Noise in dc SQUIDs

The white noise in dc SQUIDs is well understood. However, some applications of SQUIDs, for example neuromagnetism, require good resolution at frequencies down to 0.1 Hz or less and the level of the $1/f$ or "flicker" noise becomes very important.

There are at least two separate sources of $1/f$ noise in the dc SQUID [34]. The first arises from $1/f$ noise in the critical current of the Josephson junctions, and the mechanism for this process is reasonably well understood [35]. In the process of tunneling through the barrier, an electron becomes trapped for a while and is subsequently released. While the trap is occupied, there is a local change in the height of the tunnel barrier and hence in the critical current density of that region. As a result, the presence of a signal trap causes the critical current of the junction to switch randomly back and forth between two values, producing a random telegraph signal. If the mean time between pulses is τ, the spectral density of this process is a Lorentzian, scaling as $\tau/[1 + (2\pi f \tau)^2]$. In general, there may be several traps in the junction, each with its own characteristic time τ_i. Provided the traps are statistically independent, the superposition of these Lorentzians yields a $1/f$ power spectrum [36,37].

The second source of $1/f$ noise in SQUIDs appears to arise from the motion of flux lines trapped in the body of the SQUID [34] and is less well understood than the critical current noise. This mechanism manifests itself as flux noise; for all practical purposes the noise source behaves as if an external flux noise were applied to the SQUID. Thus, the spectral density of the $1/f$ flux noise scales as V_Φ^2, and, in particular, vanishes at $\Phi = (n \pm \frac{1}{2})\Phi_0$ where $V_\Phi = 0$. By contrast, critical current noise is still present when $V_\Phi = 0$, although its magnitude does depend on the applied flux.

The level of $1/f$ flux noise appears to depend strongly on the microstructure of the thin films. For example, SQUIDs fabricated at Berkeley with Nb loops sputtered under a particular set of conditions show $1/f$ flux noise levels of typically [34] 10^{-10} Φ_0^2 Hz^{-1} at 1 Hz. On the other hand, SQUIDs with Pb loops in exactly the same geometry exhibit a $1/f$ noise level of about 2×10^{-12} Φ_0^2 Hz^{-1} at 1 Hz, arising from critical current fluctuations. Tesche et al. [38] reported a $1/f$ noise level in Nb-based SQUIDs of about 3×10^{-13} Φ_0^2 Hz^{-1}. Foglietti et al. [39] found a critical current $1/f$ noise corresponding to 2×10^{-12} Φ_0^2 Hz^{-1}, also in Nb-based devices. Thus, we conclude that the quality of the Nb films plays a significant role in the level of $1/f$ flux noise.

There is an important practical difference between the two sources of $1/f$ noise: critical current noise can be reduced by a suitable modulation scheme whereas flux noise cannot. Several schemes have been devised [34,39,40] that involve switching both the bias current and the flux bias, and reductions in the spectral density of the $1/f$ noise due to critical current fluctuations of at least one order of magnitude have been achieved.

3.6. Alternative Read-Out Schemes

Although the flux modulation method described in Section 3.3 has been used successfully for many years, alternative schemes have recently been developed. These efforts have been motivated, at least in part, by the need to simplify the electronics required for the multichannel systems used in neuromagnetism (see Section 5.1). Fujimaki and co-workers [41] and Drung and co-workers [42] have devised schemes in which the output from the SQUID is sensed digitally and fed back as an analog signal to the SQUID to flux-lock the loop. Fujimaki et al. [41] used Josephson digital circuitry to integrate their feedback system on the same chip as the SQUID so that the flux-locked signal was available directly from the cryostat. The system of Drung and co-workers, however, is presently the more sensitive with a flux resolution of about $10^{-6}\,\Phi_0\,\text{Hz}^{-1/2}$ in a 50 pH SQUID. These workers were also able to reduce the $1/f$ noise in the system using a modified version of the modulation scheme of Foglietti et al. [39]. Although they need further development, cryogenic digital feedback schemes offer several advantages: they are compact, produce a digitized output for transmission to room temperature, offer wide flux-locked bandwidths, and need not add any noise to the intrinsic noise of the SQUID.

In yet another system, Mück and Heiden [43] have operated a dc SQUID with hysteretic junctions in a relaxation oscillator. The oscillation frequency depends on the flux in the SQUID, reaching a maximum at $(n + \frac{1}{2})\Phi_0$ and a minimum at $n\Phi_0$. A typical frequency modulation is 100 kHz at an operating frequency of 10 MHz. This technique produces large voltages across the SQUID so that no matching network to the room-temperature electronics is required. The room-temperature electronics are simple and compact, and the resolution is about $10^{-5}\,\Phi_0\,\text{Hz}^{-1/2}$ for a SQUID at 4.2 K with an inductance estimated to be about 80 pH.

4. The rf SQUID

4.1. Principles of Operation

Although the rf SQUID is still the more widely used device because of its long-standing commercial availability, it has seen very little development in recent years. For this reason, we will give only a rather brief account of its principles and noise limitations, following rather closely descriptions in earlier reviews [44,45].

The rf SQUID [4,5] shown in Fig. 8 consists of a superconducting loop of inductance L interrupted by a single Josephson junction with critical current I_0 and a nonhysteretic current–voltage characteristic. Flux quantization [1] imposes the constraint

$$\delta + \frac{2\pi\Phi_T}{\Phi_0} = 2\pi n, \qquad (4.1)$$

on the total flux Φ_T threading the loop. The phase difference δ across the junction determines the supercurrent

$$I_S = -I_0 \sin\frac{2\pi\Phi_T}{\Phi_0}, \qquad (4.2)$$

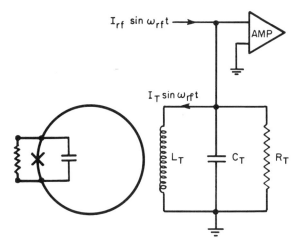

Fig. 8. The rf SQUID inductively coupled to a resonant tank circuit.

flowing in the ring. A quasistatic external flux Φ thus gives rise to a total flux

$$\Phi_T = \Phi - LI_0 \sin\frac{2\pi\Phi}{\Phi_0}. \qquad (4.3)$$

The variation of Φ_T with Φ is sketched in Fig. 9a for the typical value $LI_0 = 1.25\,\Phi_0$. The regions with positive slope are stable, whereas those with negative slope are not. A "linearized" version of Fig. 9a showing the path traced out by Φ and Φ_T is shown in Fig. 9b. Suppose we slowly increase Φ from zero. The total flux Φ_T increases less rapidly than Φ because the response flux $-LI_S$ opposes Φ. When I_S reaches I_0, at an applied flux Φ_C and a total flux Φ_{TC}, the junction switches momentarily into a nonzero voltage state and the SQUID jumps from the $k = 0$ to the $k = 1$ quantum state. If we subsequently reduce Φ from a value just above Φ_C, the SQUID remains in the $k = 1$ state until $\Phi = \Phi_0 - \Phi_C$, at which point I_S again exceeds the critical current and the SQUID returns to the $k = 0$ state. In the same way, if we lower Φ to below $-\Phi_C$ and then increase it, a second hysteresis loop will be traced out. We note that this hysteresis occurs provides $LI_0 > \Phi_0/2\pi$; most practical SQUIDs are operated in this regime. For $LI_0 \approx \Phi_0$, the energy ΔE dissipated when one takes the flux around a single hysteresis loop is its area divided by L:

$$\Delta E \approx I_0 \Phi_0. \qquad (4.4)$$

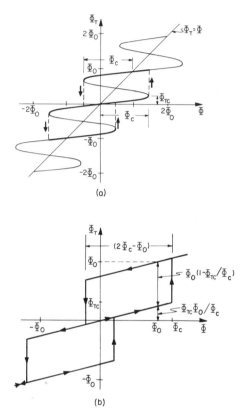

Fig. 9. The rf SQUID: (a) total flux Φ_T vs. Φ for $LI_0 = 1.25\,\Phi_0$; (b) values of Φ_T as Φ is quasistatically increased and then decreased.

We now consider the radio frequency (rf) operation of the device. The SQUID is inductively coupled to the coil of an LC-resonant circuit with a quality factor $Q = R_T/\omega_{rf}L_T$ via a mutual inductance $M = K(LL_T)^{1/2}$. Here, L_T, C_T, and R_T are the inductance, capacitance, and (effective) parallel resistance of the tank circuit, and $\omega_{rf}/2\pi$ is its resonant frequency, typically 20 or 30 MHz. The tank circuit is excited at its resonant by a current $I_{rf}\sin\omega_{rf}t$, which generates a current of amplitude $I_T = QI_{rf}$ in the inductor. The voltage V_T across the tank circuit is amplified with a preamplifier with a high input impedance. First, consider the case $\Phi = 0$. As we increase I_{rf} from zero, the peak rf flux applied to the loop is $MI_T = QMI_{rf}$ and V_T increases linearly with I_{rf}. The peak flux becomes equal to Φ_C when $I_T = \Phi_C/M$ or $I_{rf} = \Phi_C/MQ$, at A in Fig. 10. The corresponding peak rf voltage across the tank circuit is

$$V_T^{(n)} = \frac{\omega_{rf} L_T \Phi_C}{M}, \qquad (4.5)$$

where the superscript (n) indicates $\Phi = n\Phi_0$, in this case with $n = 0$. At this point, the SQUID makes a transition to *either* the $k = +1$ state or the $k = -1$ state. As the SQUID traverses the hysteresis loop, energy ΔE is extracted from the tank circuit. Because of this loss, the peak flux on the next half cycle is less than Φ_C, and no transition occurs. The tank circuit takes many cycles to recover sufficient energy to induce a further transition, which may be into either the $k = +1$ or -1 states. If we now increase I_{rf}, transitions are induced at the same values of I_T and V_T, but, because energy is supplied at a higher rate, the stored energy builds up more rapidly after each energy loss ΔE, and transitions occur more frequently. In the absence of thermal fluctuations, the "step" AB in Fig. 10 is at constant voltage. At B, a transition is induced on each positive and negative rf peak, and a further increase in I_{rf} produces the "riser" BC. At C, transitions from the $k = \pm 1$ to $k = \pm 2$ states occur and a second step begins. As we continue to increase I_{rf}, we observe a series of steps and risers.

If we now apply an external flux $\Phi = \Phi_0/2$, the hysteresis loops in Fig. 9b are shifted by $\Phi_0/2$. Thus, a transition occurs on the positive peak of the rf cycle at a flux $\Phi_C - \Phi_0/2$, whereas on the negative peak the required flux is $-(\Phi_C + \Phi_0/2)$. As a result, as we increase I_{rf} from zero, we observe the first step at D in Fig. 10 at

$$V_T^{(n+1/2)} = \frac{\omega_{rf} L_T (\Phi_C - \Phi_0/2)}{M}. \qquad (4.6)$$

As we increase I_{rf} from D to F, the SQUID traverses only one hysteresis loop, corresponding to the $k = 0$ to $k = +1$ transition at $\Phi_C - \Phi_0/2$. A further increase in I_{rf} produces the riser FG, and at G transitions at a peak rf flux $-(\Phi_C + \Phi_0/2)$ begin. In this way, we observe a series of steps and risers for $\Phi = \Phi_0/2$, interlocking those for $\Phi = 0$ (Fig. 10). As we increase Φ from zero, the voltage at which the first step appears will drop to a minimum (D) at $\Phi_0/2$ and rise to its maximum value (A) at $\Phi = \Phi_0$. The change in V_T as we increase Φ from 0 to $\Phi_0/2$, found by subtracting Eq. (4.6) from Eq. (4.5), is $\omega_{rf} L_T \Phi_0/2M$. Thus, for a small change in flux near $\Phi = \Phi_0/4$, we find the transfer function

$$V_\Phi = \frac{\omega_{rf} L_T}{M}. \qquad (4.7)$$

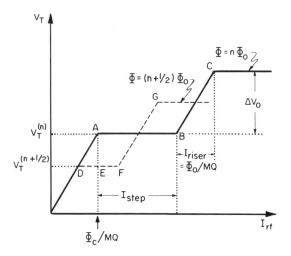

Fig. 10. V_T vs. I_{rf} in the absence of thermal noise for $\Phi = n\Phi_0, (n + \tfrac{1}{2})\Phi_0$.

At first sight, Eq. (4.7) suggests that we can make V_Φ arbitrarily large by reducing K sufficiently. However, we obviously cannot make K so small that the SQUID has no influence on the tank circuit, and we need to establish a lower bound on K. Now to operate the SQUID, we must be able to choose a value of I_{rf} that intercepts the first step for all values of Φ: this requirement is satisfied if the point F in Fig. 10 lies to the right of E, that is if DF exceeds DE. We can calculate DF by noting that the power dissipation in the SQUID is zero at D and $\Delta E(\omega_{rf}/2\pi) \approx I_0 \Phi_0 \omega_{rf}/2\pi$ at F. Thus, $\tfrac{1}{2}(I_{rf}^{(F)} - I_{rf}^{(D)})V_T^{(n+1/2)} = I_0 \Phi_0 \omega_{rf}/2\pi$ (I_{rf} and V_T are peak, rather than rms values). Furthermore, we can easily see that $I_{rf}^{(E)} - I_{rf}^{(D)} = \Phi_0/2MQ$. Assuming $LI_0 \approx \Phi_0$ and using Eq. (4.5), we find that the requirement $I_{rf}^{(E)} > I_{rf}^{(D)}$ can be written in the form

$$K^2 Q \gtrsim \pi/4. \tag{4.8}$$

If we set $K \approx Q^{-1/2}$, Eq. (4.7) becomes

$$V_\Phi \approx \omega_{rf}\left(\frac{QL_T}{L}\right)^{1/2}. \tag{4.9}$$

To operate the SQUID, one adjusts I_{rf} so that the SQUID remains biased on the first step (see Fig. 10) for all values of Φ. The rf voltage across the tank circuit is amplified and demodulated to produce a signal that is periodic in Φ. A modulating flux, typically at 100 kHz and with a peak-to-peak amplitude of $\Phi_0/2$, is also applied to the SQUID, just as in the case of the dc SQUID. The voltage produced by this modulation is lock-in detected, integrated, and fed-back as a current into the modulation coil to flux-lock the SQUID.

4.2. Theory of Noise in the rf SQUID

A detailed theory has been developed for noise in the rf SQUID [46–54]; in contrast to the case for the dc SQUID, noise contributions from the tank circuit and preamplifier are also important. We begin by discussing the intrinsic noise in the SQUID. In the previous section, we assumed that transitions from the $k = 0$ to the $k = 1$ state occurred precisely at $\Phi = \Phi_C$. In fact, thermal activation causes the transition to occur stochastically, at lower values of flux. Kurkijärvi [46] calculated the distribution of values of Φ at which the transitions occur; experimental results [55] are in good agreement with his predictions. When the SQUID is driven with an rf flux, the fluctuations in the value of flux at which transitions occur have two consequences. First, noise is introduced on the peak voltage V_T, giving an equivalent intrinsic flux noise spectral density [47,51]

$$S_\Phi^{(i)} \approx \frac{(LI_0)^2}{\omega_{rf}} \left(\frac{2\pi k_B T}{I_0 \Phi_0}\right)^{4/3}. \tag{4.10}$$

Second, the noise causes the steps to tilt (Fig. 11), as we can easily see by considering the case $\Phi = 0$. In the presence of thermal fluctuations, the

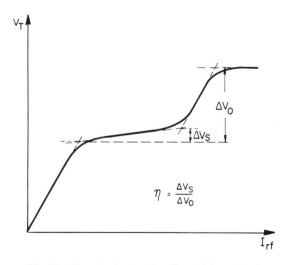

Fig. 11. V_T vs. I_{rf} showing the effects of thermal noise.

transition from the $k = 0$ to the $k = 1$ state (for example) has a certain probability of occurring at any given value of the total flux $\Phi + \Phi_{rf}$. Just to the right of A in Fig. 10, this transition occurs at the peak of the rf flux once in many rf cycles. Thus, the probability of the transition occurring in any one cycle is small. On the other hand, at B a transition must occur at each positive and negative peak of the rf flux, with unity probability. To increase the transition probability, the peak value of the rf flux and hence V_T must increase as I_{rf} is increased from A to B. Jackel and Buhrman [48] introduced the slope parameter η defined in Fig. 11 and showed that it was related to $S_\Phi^{(i)}$ by the relation

$$\eta^2 \approx \frac{S_\Phi^{(i)} \omega_{rf}}{\pi \Phi_0^2} \tag{4.11}$$

provided η was not too large. This relation is well-verified experimentally.

The noise temperature T_a of typical rf amplifiers operated at room temperature is substantially higher than that of amplifiers operated at a few hundred kilohertz and is therefore not negligible for rf SQUIDs operated at liquid ^{4}He temperatures. Furthermore, part of the coaxial line connecting the tank circuit to the preamplifier is at room temperature. Since the capacitances of the line and the amplifier are a substantial fraction of the capacitance of the tank circuit, part of the resistance damping the tank circuit is well above the bath temperature. As a result, there is an additional contribution to the noise, which we combine with the preamplifier noise to produce an effective noise temperature T_a^{eff}. The noise energy contributed by these extrinsic sources can be shown to be [48,52] $2\pi\eta k_B T_a^{eff}/\omega_{rf}$. Combining this contribution with the intrinsic noise one finds

$$\varepsilon \approx \frac{1}{\omega_{rf}} \left(\frac{\pi \eta^2 \Phi_0^2}{2L} + 2\pi\eta k_B T_a^{eff} \right). \tag{4.12}$$

Equation (4.12) shows that ε scales as $1/\omega_{rf}$, but one should bear in mind that T_a tends to increase with ω_{rf}. Nonetheless, improvements in performance have been achieved by operating the SQUID at much higher frequencies [56,57] than the usual 20 or 30 MHz. One can also reduce T_a^{eff} by cooling the preamplifier [56,58], thereby reducing T_a and reducing the temperature of the tank circuit to that of the bath. However, the best noise energies achieved for the rf SQUID are substantially higher [59] than those routinely obtained with thin-film dc SQUIDs, and for this reason workers requiring the highest possible resolution almost invariably use the latter device.

4.3. Practical rf SQUIDs

Although less sensitive than the dc SQUID, the rf SQUID is entirely adequate for a wide range of applications. It is therefore more widely used than the dc SQUID, for the simple reason that reliable, easy-to-operate devices have been commercially available since the early 1970s, notably from BTi (formerly SHE). We shall therefore confine ourselves to a brief description of the device available from this company.

Figure 12 shows a cut-away drawing of the BTi rf SQUID [40], which has a toroidal configuration machined from Nb. One way to understand this geometry is to imagine rotating the SQUID in Fig. 8 through 360° about a line running through the junction from the top to the bottom of the page. This procedure produces a toroidal cavity connected at its center by the junction. If one places a toroidal coil in this cavity, a current in the coil produces a flux that is tightly coupled to the SQUID. In Fig. 12, there are actually two such cavities, one containing the tank circuit-modulation-feedback coil and the other the input coil. This separation eliminates cross-talk between the two coils. Leads to the two coils are brought out via screw-terminals. The junction is made from thin films of Nb. This device is self-shielding against external magnetic field fluctuations and has proven to be reliable and convenient to use. In particular, the Nb input terminals enable one to connect different input circuits in a straightforward way. A typical device has a white noise energy of 5×10^{-29} J Hz^{-1}, with a $1/f$ noise energy of perhaps 10^{-28} J Hz^{-1} at 0.1 Hz.

5. SQUID-Based Instruments

Both dc and rf SQUIDs are used as sensors in a far-ranging assortment of instruments. Here we briefly discuss some of them: the selection is far from exhaustive but does include the more commonly used instruments.

Each instrument involves a circuit attached to the input coil of the SQUID. We should recognize from the outset that, in general, the presence of the input circuit influences both the signal and noise properties of the SQUID, while the SQUID, in turn, reflects a complex impedance into the input. Because the SQUID is a nonlinear device, a full description of the interactions is complicated, and we shall not go into the details here. However, one aspect of this interaction, first pointed out by Zimmerman [60], is easy to understand. Suppose we connect a superconducting pick-up loop of inductance L_p to the input coil of inductance L_i to form a magnetometer, as shown in Fig. 13a. It is

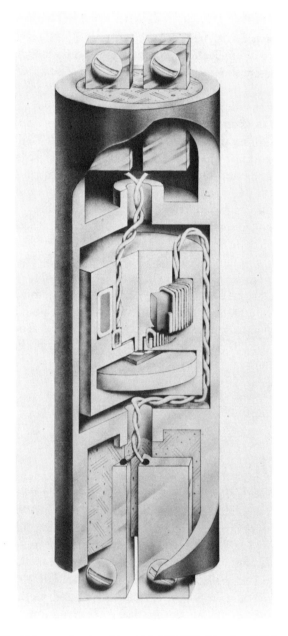

Fig. 12. Cut-away drawing of toroidal rf SQUID. (Courtesy BTi, Inc.)

easy to show that the SQUID inductance L is reduced to the value

$$L' = L\left(\frac{1 - \alpha^2 L_i}{L_i + L_p}\right), \tag{5.1}$$

where α^2 is the coupling coefficient between L and L_i. We have neglected any stray inductance in the leads connecting L_i and L_p, and any stray capacitance. The reduction in L tends to increase the transfer coefficient of both the dc SQUID (Eq. (3.9)) and the rf SQUID (Eq. (4.9)). In most cases, the reduction of L and the change in the noise properties will be non-negligible but will not have a major impact on the results presented here.

5.1. Magnetometers and Gradiometers

One of the simplest instruments is the magnetometer (Fig. 13a). A pick-up loop is connected across the input coil to make a superconducting flux transformer. The SQUID and input coil are generally enclosed in a superconducting shield. If one applies a magnetic flux $\delta\Phi^{(p)}$ to the pick-up loop, flux quantization requires that

$$\delta\Phi^{(p)} + (L_i + L_p)J_S = 0, \tag{5.2}$$

where J_S is the supercurrent induced in the transformer. We have neglected the effects of the SQUID on the input circuit. The flux coupled into the SQUID, which we assume to be in a flux-locked loop, is

$$\delta\Phi = M_i|J_S| = \frac{M_i \delta\Phi^{(p)}}{L_i + L_p}.$$

We find the minimum detectable value of $\delta\Phi^{(p)}$ by equating $\delta\Phi$ with the equivalent flux noise of the SQUID. Defining $S_\Phi^{(p)}$ as the spectral density of the equivalent flux noise referred to the pick-up loop, we find

$$S_\Phi^{(p)} = \frac{(L_p + L_i)^2}{M_i^2} S_\Phi. \tag{5.3}$$

Introducing the equivalent noise energy referred to the pick-up loop, we obtain

$$\frac{S_\Phi^{(p)}}{2L_p} = \frac{(L_p + L_i)^2}{L_i L_p} \frac{S_\Phi}{2\alpha^2 L}. \tag{5.4}$$

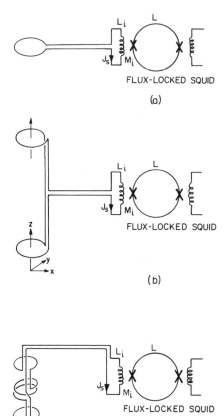

Fig. 13. Superconducting flux transformers: (a) magnetometer, (b) first-derivative gradiometer, (c) second-derivative gradiometer.

We observe that Eq. (5.4) has the minimum value

$$\frac{S_\Phi^{(p)}}{2L_p} = \frac{4\varepsilon(f)}{\alpha^2} \qquad (5.5)$$

when $L_i = L_p$. Thus, a fraction $\alpha^2/4$ of the energy in the pick-up loop is transferred to the SQUID. In this derivation, we have neglected noise currents in the input circuit arising from noise in the SQUID, the fact the the input circuit reduces the SQUID inductance, and any possible coupling between the feedback coil of the SQUID and the input circuit. Having obtained the flux resolution for $L_i = L_p$, we can immediately write down the corresponding magnetic field resolution $B_N^{(p)} = (S_\Phi^{(p)})^{1/2}/\pi r_p^2$, where r_p is the radius of the pick-up loop:

$$B_N^{(p)} = \frac{2\sqrt{2}\, L_p^{1/2} \varepsilon^{1/2}}{\pi r_p^2 \alpha}. \tag{5.6}$$

Now [61] $L_p = \mu_0 r_p [\ln(8r_p/r_0) - 2]$, where $\mu_0 = 4\pi \times 10^{-7}$ henries/meter and r_0 is the radius of the wire; for a reasonable range of values of r_p/r_0, we can set $L_p \approx 5\mu_0 r_p$. Thus, we obtain $B_N^{(p)} \approx 2(\mu_0 \varepsilon)^{1/2}/\alpha r_p^{3/2}$. This indicates that one can, in principle, improve the magnetic field resolution indefinitely by increasing r_p, keeping $L_i = L_p$. Of course, in practice, the size of the cryostat will impose an upper limit on r_p. If we take $\varepsilon = 10^{-28}$ J Hz^{-1} (a somewhat conservative value for an rf SQUID), $\alpha = 1$ and $r_p = 25$ mm, we find $B_N^{(p)} \approx 5 \times 10^{-15}$ tesla Hz$^{-1/2} = 5 \times 10^{-11}$ gauss Hz$^{-1/2}$. This is a much higher sensitivity than that achieved by any nonsuperconducting magnetometer.

Magnetometers have usually involved flux transformers made of Nb wire. For example, one can make the rf SQUID in Fig. 12 into a magnetometer merely by connecting a loop of Nb wire to its input terminals. In the case of the thin-film dc SQUID, one can make an integrated magnetometer by fabricating a Nb loop across the spiral input coil. In this way, Wellstood et al. [18] achieved a magnetic field white noise of 5×10^{-15} tesla Hz$^{-1/2}$ with a pick-up loop a few millimeters across.

Magnetometers with typical sensitivities of 0.01 pT Hz$^{-1/2}$ have been used in geophysics in a variety of applications [62], for example, magnetotellurics, active electromagnetic sounding, piezomagnetism, tectonomagnetism, and the location of hydrofractures. Although SQUID-based magnetometers are substantially more sensitive than any other type, the need to replenish the liquid helium in the field has restricted the extent of their applications. For this reason, the advent of high-temperature superconductors may have considerable impact on this field (see Section 6).

An important variation of the flux transformer is the gradiometer. Figure 13b shows an axial gradiometer that measures $\partial B_z/\partial z$. The two pick-up loops are wound in opposition and balanced so that a uniform field B_z links zero net flux to the transformer. A gradient $\partial B_z/\partial z$, on the other hand, does induce a net flux and thus generates an output from the flux-locked SQUID. Figure 13c shows a second-order gradiometer that measures $\partial^2 B_z/\partial z^2$; Fig. 14a is a photograph of a practical version.

Thin-film gradiometers based on dc SQUIDs were made as long ago [28] as

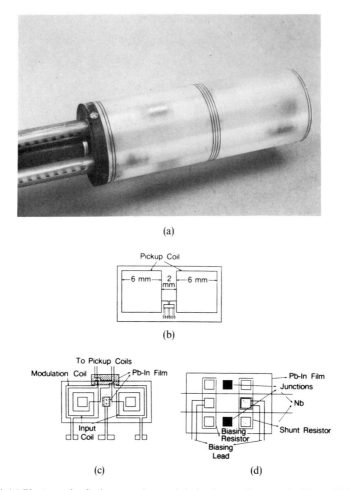

Fig. 14. (a) Photograph of wire-wound second-derivative gradiometer for biomedical applications. (Courtesy BTi, Inc.) Thin-film first-derivative gradiometer: (b) pick-up loops, (c) two-hole SQUID with spiral input coils, and (d) expanded view of the dotted circle in (c) showing junctions and resistive shunts (from [66]).

1978, and a variety of devices [25,63–67] have been reported since then. To our knowledge, all of the gradiometers made to date have been planar, and therefore measure an off-diagonal gradient, for example, $\partial B_z/\partial x$ or $\partial^2 B_z/\partial x \partial y$. A representative device is shown in Fig. 14b–d.

The most important application of the gradiometer is in neuromagnetism

[68], notably to detect weak magnetic signals emanating from the human brain. The gradiometer discriminates strongly against distant noise sources, which have a small gradient, in favor of locally generated signals. One can thus use a second-order gradiometer in an unshielded environment, although the present trend is towards using first-order gradiometers in a shielded room of aluminum and mu-metal that greatly attenuates the ambient magnetic noise. In this application, axial gradiometers of the type shown in Fig. 13a actually sense the magnetic field, rather than the gradient, because the distance from the signal source to the pick-up loop is less than the baseline of the gradiometer. The magnetic field sensitivity referred to one pick-up loop is typically 10 fT Hz$^{-1/2}$. Although great progress in this field has been made in recent years, it is generally agreed that one needs an array of 50 to 100 gradiometers to make a clinically viable system. This requirement has greatly spurred the development of integrated, thin-film devices. For example, Knuutila [69] has reported that a 24-channel first-derivative array is under construction.

There are two basic kinds of measurements made on the human brain. In the first, one detects spontaneous activity: a classic example is the generation of magnetic pulses by subjects suffering from focal epilepsy [70]. The second kind involves evoked response: for example, Romani *et al.* [71] detected the magnetic signal from the auditory cortex generated by tones of different frequencies.

There are several other applications of gradiometers. One kind of magnetic monopole detector [72] consists of a gradiometer: the passage of a monopole would link flux h/e in the pick-up loop and produce a step-function response from the SQUID. Gradiometers have recently been of interest in studies of corrosion and in the location of fractures in pipelines and other structures.

5.2. Susceptometers

In principle, one can easily use the first-derivative gradiometer of Fig. 13b to measure magnetic susceptibility χ. One establishes a static field along the z-axis and lowers the sample into one of the pick-up loops. Provided χ is nonzero, the sample introduces an additional flux into the pick-up loop and generates an output from the flux-locked SQUID. A very sophisticated susceptometer is available commercially [73]. A room temperature access enables one to cycle samples rapidly, and one can measure χ as a function of temperature between 1.8 K and 400 K in fields up to 5.5 tesla. The system is capable of resolving a change in magnetic moment as small as 10^{-8} emu.

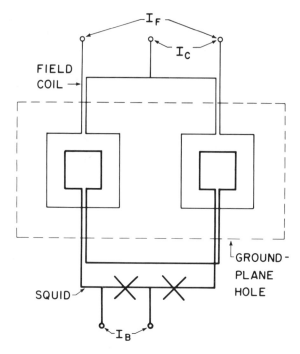

Fig. 15. Thin-film miniature susceptometer (from [74]).

Novel miniature susceptometers have been developed by Ketchen and co-workers [33,74,75]. One version is shown schematically in Fig. 15. The SQUID loop incorporates two pick-up loops wound in opposite senses and connected in series. The two square pick-up loops, 17.5 μm on a side and with an inductance of about 30 pH, are deposited over a hole in the ground plane that minimizes the inductance of the rest of the device. The SQUID is flux biased at the maximum of V_Φ by means of a control current I_C in one of the pick-up loops. One can apply a magnetic field to the two loops by means of the current I_F; by passing a fraction of this current into the center tap I_C, one can achieve a high degree of electronic balance between the two loops. The sample to be studied is placed over one of the loops, and the output from the SQUID when the field is applied is directly proportional to the magnetization. At 4.2 K, the susceptometer is capable of detecting the magnetization due to as few as 3000 electron spins. Awschalon and co-workers [33,75] have used a miniature susceptometer to perform magnetic spectroscopy of semiconductors with picosecond time-resolution.

5.3. Voltmeters

Probably the first practical application of a SQUID was to measure tiny, quasistatic voltages [76]. One simply connects the signal source—for example, a low resistance through which a current can be passed—in series with a known resistance and the input coil of the SQUID. The output from the flux-locked loop is connected across the known resistance to obtain a null-balancing measurement of the voltage. The resolution is generally limited by Nyquist noise in the input circuit, which at 4.2 K varies from about 10^{-15} V Hz$^{-1/2}$ for a resistance of 10^{-8} Ω to about 10^{-10} V Hz$^{-1/2}$ for a resistance of 100 Ω.

Applications of these voltmeters range from the measurement of thermoelectric voltages and of quasiparticle charge imbalance in nonequilibrium superconductors to noise thermometry and the comparison of the Josephson voltage–frequency relation in different superconductors to high precision.

5.4. The dc SQUID as a Radiofrequency Amplifier

Over recent years, the dc SQUID has been developed into a low noise amplifier for frequencies up to 100 MHz or more [77]. To understand the theory for the performance of this amplifier, we need to extend the theory of Section 3 by taking into account the noise in the current $J(t)$ in the SQUID loop. For a bare SQUID with $\beta = 1$, $\Gamma = 0.05$, and $\Phi = (2n + 1)\Phi_0/4$, one finds the spectral density of the current to be [78]

$$S_J(f) \approx \frac{11 k_B T}{R}. \tag{5.7}$$

Furthermore, the current noise is partially correlated with the voltage noise across the SQUID, the cross-spectral density being [78]

$$S_{VJ}(f) \approx 12 k_B T. \tag{5.8}$$

The correlation arises, roughly speaking, because the current noise generates a flux noise which in turn contributes to the total voltage noise across the junction provided $V_\Phi \neq 0$.

One can make a tuned amplifier, for example, by connecting an input circuit to the SQUID as shown in Fig. 16. In general, the presence of this circuit modifies all of the SQUID parameters and the magnitude of the noise spectral densities [79]. Furthermore, the SQUID reflects an impedance $\omega^2 M_i^2/Z$ into the input circuit [80], where Z is the dynamic input impedance of the SQUID.

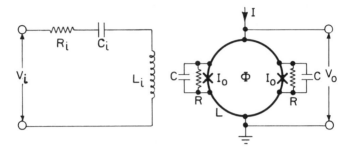

Fig. 16. Tuned radiofrequency amplifier based on dc SQUID (from [77]).

Fortunately, however, one can neglect the mutual influence of the SQUID and input circuit provided the coupling coefficient α^2 is sufficiently small, as it is under certain circumstances. For a signal at frequency f generated by a source with resistance R_i, one can optimize the noise temperature of the amplifier using standard procedures to obtain

$$T_N^{(\text{opt})} = \frac{\pi f}{k_B V_\Phi}(S_V S_J - S_{VJ}^2)^{1/2}. \tag{5.9}$$

This minimum value actually occurs off-resonance; if one wishes to operate the amplifier at the resonant frequency of the input circuit, the noise temperature is increased to

$$T_N^{(\text{res})} = \frac{\pi f}{k_B V_\Phi}(S_V S_J)^{1/2}. \tag{5.10}$$

The corresponding power gain is

$$G \approx \frac{V_\Phi}{\omega}. \tag{5.11}$$

Equations (5.10) and (5.11) can be shown to imply $\alpha^2 Q \approx 1$, that is, the coupling is weak for a high-Q input circuit. We emphasize that the noise temperatures quoted in Eqs. (5.9) and (5.10) do not include the Nyquist noise generated in R_i, which may well exceed the noise generated by the amplifier. Thus, Eq. (5.9) or (5.10) does not necessarily represent the lowest *system* noise temperature, which may well occur under quite different conditions. Nonetheless, these expressions are very useful in that they specify the performance at a particular frequency in terms of the SQUID parameters only.

Hilbert and Clarke [77] made several radiofrequency amplifiers with both tuned and untuned inputs, flux biasing the SQUID near $\Phi = (2n + 1)\Phi_0/4$. There was no flux-locked loop. The measured parameters were in good agreement with predictions. For example, for an amplifier with $R \approx 8\ \Omega$, $L \approx 0.4$ nH, $L_i = 5.6$ nH, $M_i \approx 1$ nH, and $V_\Phi \approx 3 \times 10^{10}$ sec^{-1} at 4.2 K, they found $G = 18.6 \pm 0.5$ dB and $T_N = 1.7 \pm 0.5$ K at 93 MHz. The predicted values were 17 dB and 1.1 K, respectively.

To conclude this discussion, we comment briefly on the quantum limit of the dc SQUID amplifier. At $T = 0$, Nyquist noise in the shunt resistors in Eqs. (3.7) and (3.8) should be replaced with zero point fluctuations. Koch et al. [81] performed a simulation in this limit and concluded that, within the limits of error, the noise temperature of a tuned amplifier in the quantum limit should be given by

$$T_N \approx \frac{hf}{k_B \ln 2}. \tag{5.12}$$

This is the result for any quantum-limited amplifier. The corresponding value for ε was approximately $\hbar$, but it should be emphasized that quantum mechanics does not impose any precise lower limit on ε [82]. A number of SQUIDs have obtained noise energies of $3\hbar$ or less, but there is no evidence as yet that a SQUID has attained quantum-limited performance as an amplifier.

Clarke, Hahn, and co-workers [83–86] have used tuned SQUID amplifiers in a series of experiments to observe nuclear magnetic resonance (NMR) and nuclear quadruple resonance (NQR) at about 30 MHz. As an example of the high sensitivity offered by these techniques, they were able to detect "spin noise" in ^{35}Cl nuclei in NaClO$_3$. In zero magnetic field, this nucleus has two doubly degenerate nuclear levels with a splitting of 30.6856 MHz and exhibits NQR. An rf signal at the NQR frequency equalized the populations of the two nuclear spin levels, and was then turned off to leave a zero-spin state. A SQUID amplifier was able to detect the photons emitted spontaneously as the upper state decayed, even although the lifetime per nucleus against this process was $\sim 10^6$ centuries. The detected power was about 5×10^{-21} W in a bandwidth of about 1.3 kHz.

5.5. Gravity Wave Antennas

A quite different application of SQUIDs is to detect minute displacements, notably those of Weber bar gravity wave antennas [87,88]. Roughly a dozen groups worldwide are using these antennas to search for the pulse of gravitational radiation that is expected to be emitted when a star collapses.

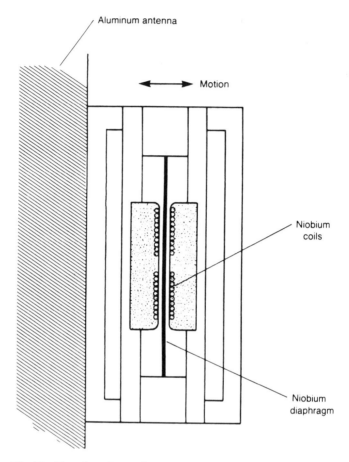

Fig. 17. Transducer for gravity wave antenna. (Courtesy P.F. Michelson.)

The radiation induces longitudinal oscillations in the large, freely suspended bar, but because the amplitude is very tiny, one requires the sensitivity of a dc SQUID to detect it. As an example, we briefly describe the antenna at Stanford University, which consists of an aluminum bar 3 meters long and weighing 4800 kg suspended in a vacuum chamber at 4.2 K. The fundamental longitudinal mode is at $\omega_a/2\pi \approx 842$ Hz, and the Q is 5×10^6. The transducer is shown schematically in Fig. 17. A circular niobium diaphragm is clamped at its perimeter to one end of the bar, with a flat spiral coil made of niobium wire mounted on each side. The two coils are connected in parallel with each other and with the input coil of a SQUID; this entire circuit is superconducting. A persistent supercurrent circulates in the closed loop formed by the two spiral coils. The associated magnetic fields exert a restoring force on the diaphragm so that by adjusting the current, one can set the resonant frequency of the diaphragm equal to that of the bar. A longitudinal oscillation of the bar induces an oscillation in the position of the diaphragm relative to the two coils, thereby modulating their inductances. As a result of flux quantization, a fraction of the stored supercurrent is diverted into the input coil of the SQUID, which detects it in the usual way.

The present Stanford antenna has a root-mean-square strain sensitivity $\langle(\delta l)^2\rangle^{1/2}/l$ of 10^{-18}, where l is the length of the bar and δl its longitudinal displacement. This very impressive sensitivity, which is limited by thermal noise in the bar, is nonetheless adequate only to detect events in our own galaxy. Because such events are rare, there is a very strong motivation to make major improvements in the sensitivity.

If the bar could be cooled sufficiently, the strain resolution would be limited only by the bar's zero-point motion and would have a value of about 3×10^{-21}. At first sight one might expect that the bar would have to be cooled to an absurdly low temperature to achieve this quantum limit, because a frequency of 842 Hz corresponds to a temperature $\hbar\omega_a/k_B$ of about 40 nK. However, it turns out that one can make the effective noise temperature T_{eff} of the antenna much lower than the temperature T of the bar. If a gravitational signal in the form of a pulse of length τ_S interacts with an antenna that has a decay time Q/ω_a, then the effective noise temperature is given approximately by the product of the bar temperature and the pulse length divided by the decay time: $T_{\text{eff}} \approx \tau_S \omega_a T/Q$. Thus, one can make the effective noise temperature much less than the temperature of the bar by increasing the bar's resonant quality factor sufficiently. To achieve the quantum limit, in which the bar energy $\hbar\omega_a$ is greater than the effective thermal energy $k_B T_{\text{eff}}$, one would have to lower the temperature T below $Q\hbar/k_B\tau_S$, which is about 40 mK for a quality factor Q of 5×10^6 and a pulse length τ_S of 1 msec. One can cool the antenna to this temperature with the aid of a large dilution refrigerator.

Needless to say, to detect the motion of a quantum-limited antenna one needs a quantum-limited transducer, a requirement that has been the major driving force in the development of ultra-low-noise dc SQUIDs. As we have seen, however, existing dc SQUIDs at low temperatures are now within striking distance of the quantum limit, and there is every reason to believe that one will be able to operate an antenna quite close to the quantum limit within a few years.

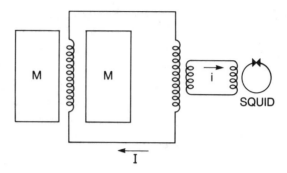

Fig. 18. Gravity gradiometer showing two proof masses M on either side of a planar spiral coil (from [90]).

5.6. *Gravity Gradiometers*

The gravity gradiometer, which also makes use of a transducer to detect minute displacements, has been pioneered by Paik [89] and Mapoles [90]. The gradiometer consists of two niobium proof masses each constrained by springs to move along a common axis (Fig. 18). A single-layer spiral coil of niobium wire is attached to the surface of one of the masses so that the surface of the wire is very close to the opposing surface of the other mass. Thus, the inductance of the coil depends on the separation of the two proof masses, which in turn depends on the gravity gradient. The coil is connected to a second superconducting coil, which is coupled to a SQUID via a superconducting transformer. A persistent supercurrent I maintains a constant flux in the detector circuit. Thus, a change in the inductance of the pick-up coil produces a change in I and hence a flux in the SQUID that is related to the gravity gradient. More sophisticated versions of this design enable one to balance the restoring forces of the two springs electronically [90] thereby eliminating the response to an acceleration (as opposed to an acceleration gradient). Sensitivities of a few Eötvös $Hz^{-1/2}$ have been achieved at frequencies above 2 Hz (1 Eötvös = 10^{-9} sec^{-2}).

Instruments of this kind could be used to map the earth's gravity gradient, and have potential in tests of the inverse square law and in inertial navigation.

6. The Impact of High-Temperature Superconductivity

The advent of the high transition temperature (T_c) superconductors [91] has stimulated great interest in the prospects for superconducting devices operating at liquid nitrogen (LN) temperatures (77 K). Indeed, a number of groups have already successfully operated SQUIDs. In this section, we shall give a brief overview of this work.

6.1. Predictions for White Noise

In designing a SQUID for operation at LN temperatures, one must bear in mind the constraints imposed by thermal noise on the critical current and inductance, $I_0 \gtrsim 10\pi k_B T/\Phi_0$ and $L \lesssim \Phi_0^2/5k_B T$. For $T = 77$ K, we find $I_0 \gtrsim 16$ μA and $L \lesssim 0.8$ nH. If we take as arbitrary but reasonable values $L = 0.2$ nH and $I_0 = 20$ μA, we obtain $2LI_0/\Phi_0 = 4$ for the dc SQUID and $LI_0/\Phi_0 = 2$ for the rf SQUID. These values are not too far removed from optimum, and to a first approximation, we can use the equations for the noise energy given in Sections 3 and 4.

For the case of the dc SQUID, the noise energy is predicted by either Eq. (3.10) or Eq. (3.11). However, since nobody has yet made a Josephson tunnel junction with high-T_c materials, it is somewhat unrealistic to use Eq. (3.11), which involves the junction capacitance, and instead we use Eq. (3.10). The value of R is an open question, and we rather arbitrarily adopt 5 Ω, which is not too different from values achieved experimentally for high-T_c grain boundary junctions [92]. With $L = 0.2$ nH, $T = 77$ K, and $R = 5$ Ω, Eq. (3.10) predicts $\varepsilon \approx 4 \times 10^{-31}$ J Hz^{-1}. This value is only about one order of magnitude higher than that found at 4.2 K for typical Nb-based, thin-film dc SQUIDs, and actually somewhat better than that found in commercially available toroidal SQUIDs. These various values are summarized in Fig. 19. If one could actually achieve the predicted resolution in a SQUID at 77 K at frequencies down to 1 Hz or less, it would be adequate for most of the applications discussed in Section 5.

For the rf SQUID, Eq. (4.10) predicts an intrinsic noise energy of about 6×10^{-29} J Hz^{-1} for $I_0 = 20$ μA, $LI_0 = 2\Phi_0$, $\omega_{rf}/2\pi = 20$ MHz, and $T = 77$ K. This value is comparable with the overall value obtained experimentally with 4.2 K devices where the effective noise temperature T_a^{eff} of the preamplifier and tank circuit is much higher than the bath temperature when the preamplifier is at room temperature (see Section 4.2). However, when one operates a SQUID at 77 K, there is no reason for T_a^{eff} to increase, and the system noise energy should be comparable with that at 4.2 K.

With regard to $1/f$ noise, in general one might expect both critical current noise and flux noise to contribute. However, it seems impractical to make any *a priori* predictions of the magnitude of these contributions.

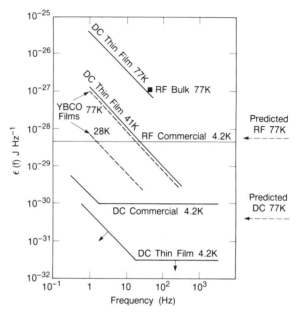

Fig. 19. Noise energy $\varepsilon(f)$ vs. frequency for several SQUIDs and for a YBCO film.

6.2. Practical Devices

Although a number of dc and rf SQUIDs have been made from YBCO, we shall describe just one of each type. It appears that the first dc SQUID was made by Koch et al. [92]. In their first devices, they patterned the films by covering the regions of YBCO to remain superconducting with a gold film, and ion implanted the unprotected regions so that they became insulators at low temperatures. The configuration is shown in Fig. 20; the estimated inductance is 80 pH. The two microbridges exhibited Josephson-like behavior, which actually arose from junctions formed by grain boundaries between randomly oriented grains of YBCO. As the quality of the films has improved, conventional patterning techniques such as lift-off and ion etching have become possible. The $I-V$ characteristics of these devices are modulated by an applied flux, although the $V-\Phi$ curves are often hysteretic and nonperiodic, probably because of flux trapped in the YBCO films. The noise energy scaled approximately as $1/f$ over the frequency range investigated, usually 1 to 10^3 Hz. The lowest noise energies achieved to date at 1 Hz are 4×10^{-27} J Hz^{-1} at 41 K and, in a different device, 2×10^{-26} J Hz^{-1} at 77 K. These values are plotted in Fig. 19.

2. SQUIDs: PRINCIPLES, NOISE, AND APPLICATIONS

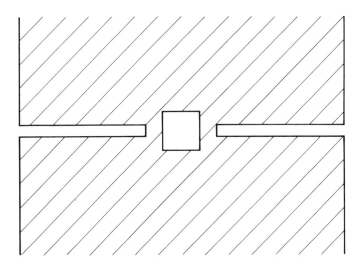

Fig. 20. Planar thin-film dc SQUID fabricated from YBCO (shaded region) deposited on substrate (redrawn from [92]).

The best characterized rf SQUID reported so far is that of Zimmerman et al [93]. They drilled a hole along the axis of a cylindrical pellet of YBCO, and cut a slot part way along a radius (Fig. 21). The pellet was glued into an aluminum holder, also with a slot, and the assembly immersed in LN. A taper pin forced into the slot in the mount caused the YBCO to break in the region of the cut; when the pin was withdrawn slightly, the YBCO surfaces on the two sides of the crack were brought together, forming a "break junction". The rf SQUID so formed was coupled to a resonant circuit and operated in the usual way. The best flux resolution was $4.5 \times 10^{-4}\, \Phi_0\, Hz^{-1/2}$ at 50 Hz, corresponding to a noise energy of $1.6 \times 10^{-27}\, J\, Hz^{-1}$ for $L = 0.25$ nH (see Fig. 19).

6.3. Flux Noise in YBCO Films

It is evident that the $1/f$ noise level in YBCO SQUIDs is very high compared with that in Nb or Pb devices at 4.2 K. Ferrari et al. [94] have investigated the source of this noise by measuring the flux noise in YBCO films. Each film, deposited on a $SrTiO_3$ chip, was patterned into a loop and mounted parallel and very close to a Nb-based SQUID (with no input coil) so that any flux noise in the YBCO loop could be detected by the SQUID. The assembly was enclosed in the vacuum can immersed in liquid helium. The SQUID was maintained at 4.2 K, while the temperature of the YBCO film could be increased by means of a resistive heater. Below T_c, the spectral density of the

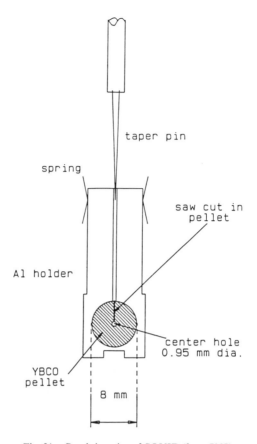

Fig. 21. Break-junction rf SQUID (from [93]).

flux noise scaled as $1/f$ over the frequency range 1 to 10^3 Hz, and increased markedly with temperature. Three films were studied, with microstructure improving progressively with respect to the fraction of grains oriented with the c-axis perpendicular to the substrate. The critical current density correspondingly increased, to a value of 2×10^6 A cm^{-2} at 4.2 K in the best film. The spectral density of the noise measured at 1 Hz is shown vs. temperature in Fig. 22. We see that in each case the noise increases rapidly as the temperature approaches T_c, and that, at a given temperature, the noise decreases dramatically as the quality of the films is improved. The noise energy estimated at 28 K and 77 K with an assumed inductance of 400 pH is shown in Fig. 19.

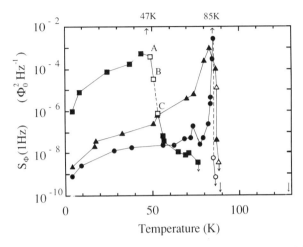

Fig. 22. Spectral density of flux noise at 1 Hz vs. temperature for three YBCO films: polycrystalline (squares), mixed a- and c-axis (triangles), and $>90\%$ c-axis (circles). Solid symbols indicate the noise is $1/f$ at 1 Hz, open that it is white or nearly white (from [94]).

The results demonstrate that YBCO films are intrinsically noisy. The noise presumably arises from the motion of flux quanta trapped in the films, possibly at grain boundaries. This mechanism is almost certainly the origin of the $1/f$ noise observed in YBCO SQUIDs, and, in general terms, it is similar to the origin of $1/f$ noise in Nb SQUIDs. It is encouraging that the noise is reduced as the microstructure of the films is improved, and it should be emphasized that there is no reason to believe the lowest noise measured so far represents a lower bound. The implications are that SQUIDs and flux transformers coupled to them should be made of very high quality films.

6.4. Future Prospects for High-T_c SQUIDs

One potential application of a high-T_c SQUID is as a geophysical magnetometer (see Section 5.1). At the moment, however, it is not entirely straightforward to predict the performance, since the devices are still evolving and it is evident from the noise measurements on YBCO films that thin-film flux transformers may introduce considerable levels of low-frequency noise. To make some kind of estimate, we assume that we can optimally couple the 77 K dc SQUID with the noise energy shown in Fig. 19 to a *noiseless* flux transformer with a thin-film pick-up loop with a diameter of 50 mm. The estimated loop inductance is about 150 nH. Using Eq. (5.6), we find a magnetic field resolution of roughly 0.1 pT Hz$^{-1/2}$ at 1 Hz, improving to 0.01 pT Hz$^{-1/2}$ at 100 Hz. Although this performance is quite good, one should realize that commercially available coils operated at room temperature offer a resolution of about 0.03 pT Hz$^{-1/2}$ over this frequency range. Furthermore, our assumption of a noise-free flux transformer is rather optimistic. Nonetheless, given the short time over which the high-T_c materials have been available, one should be rather encouraged: a relatively modest reduction in the $1/f$ noise that might be gained from improving the quality of YBCO films or even from using alternative materials might well yield a rather useful geophysical device.

One might note here that the real advantage in using liquid nitrogen as opposed to liquid helium for field applications is not really the reduction in cost, a savings that is negligible compared with the cost of mounting a field operation, but rather the very much slower boil-off rate of liquid nitrogen. The latent heat of vaporization of liquid N_2 is about 60 times that of liquid ^{4}He, so that one should be able to design cryostats of modest size with hold-times of up to a year. It is also noteworthy that liquid Ne, which boils at 28 K, has a latent heat roughly 40 times that of liquid He, and its use would also greatly extend the running time over that of liquid He, for roughly the same cost per day. We see from Fig. 22 that the $1/f$ noise in YBCO films can be considerably lower at 28 K than at 77 K, so that the lower temperature operation could be a considerable advantage.

The likely impact of high-T_c SQUIDs on the more demanding applications such as neuromagnetism is much smaller, however, at least for the near future. Here, one needs very high sensitivity at frequencies down to 0.1 Hz or less, but is not particularly concerned with the cost of liquid ^{4}He or the need to replenish it every day or two. Furthermore, low-noise, closed-circle refrigerators are just becoming available that obviate the need to supply liquid cryogens in environments where electrical power is available. Thus, it is difficult to imagine that high-T_c SQUIDs will have a significant impact in this area unless there is a major reduction in the $1/f$ noise.

In concluding this section, we note that two key problems must be solved before high-T_c SQUIDs are likely to become technologically important. The first is the development of a reproducible and reliable Josephson junction. Although great progress has been made with grain boundary junctions, it is not clear that one can base a technology on this technique. Hopefully, it will be possible to produce all YBCO junctions exhibiting Josephson tunneling in the near future. An alternative might be a superconductor–normal metal

metal–superconductor junction [95]. The second problem concerns the reduction of hysteresis and noise in thin films of high-T_c material. The motion of magnetic flux in the films is responsible for both effects, and one has to learn to produce films with lower densities of flux lines or higher pinning energies. Given the worldwide effort on the new superconductors, there is every reason to be optimistic about the long-term future of SQUIDs based on these materials.

7. Concluding Remarks

In this chapter, we have tried to give an overview of the current status of dc and rf SQUIDs. No claim is made that this account is comprehensive. There are many SQUID designs and applications that were not mentioned, but it is hoped that this chapter has given some flavor of the amazing versatility of these devices. For example, it is remarkable that a dc SQUID is the basis of both the most sensitive magnetometer available at 10^{-4} Hz and the quietest radiofrequency amplifier at 10^8 Hz.

The rf SQUID remains more widely used than the dc SQUID, simply because it is produced commercially and is thus available to people who are interested in using SQUIDs rather than in making them. One should also realize that although the kind of rf SQUID routinely used is much less sensitive than state-of-the-art thin-film dc SQUIDs, it is nonetheless entirely adequate for many applications. The fact that thin-film, integrated dc SQUIDs are not available commercially is presumably because the cost of establishing a facility to produce them is high, while the perceived market is small. However, it may well be that this situation is about to change. After all, one needs only a single major application to make reasonably priced SQUIDs available for any number of applications, and there now seem to be two such major products on the horizon. The first is in neuromagnetism: if this application is to become a clinical reality, one will need systems with as many as 100 channels, and the need for 100 channels will inevitably lead to the production of thin-film SQUIDs on a reasonable scale. The second is the advent of high-T_c thin-film SQUIDs. If these devices attain sufficient sensitivity and reliability for geophysical applications, not to mention laboratory-based applications such as voltmeters, they will be in sufficient demand to justify production on a commercial basis.

Acknowledgments

I am indebted to D. Crum for supplying Fig. 14a, P.F. Michelson Fig. 17, D. Paulsen Fig. 12, F.C. Wellstood Fig. 7, and J.E. Zimmerman Fig. 21. I wish to thank R.H. Koch and F.C. Wellstood for very helpful conversations. This work was supported by the Director, Office of Energy Research, Office of Basic Energy Sciences, Materials Sciences Division of the U.S. Department of Energy under contract number W-7405-ENG-48.

References

1. London, F. "Superfluids." Wiley, New York, 1950.
2. Josephson, B.D. Possible new effects in superconductive tunneling. *Phys. Lett.* **1**, 251–253 (1962); Supercurrents through barriers. *Adv. Phys.* **14**, 419–451 (1965).
3. Jaklevic, R.C., Lambe, J., Silver, A.H., and Mercereau, J.E. Quantum interference effects in Josephson tunneling. *Phys. Rev. Lett.* **12**, 159–160 (1964).
4. Zimmerman, J.E., Thiene, P., Harding, J.T. Design and operation of stable rf-biased superconducting point-contact quantum devices, and a note on the properties of perfectly clean metal contacts. *J. Appl. Phys.* **41**, 1572–1580 (1970).
5. Mercereau, J.E. Superconducting magnetometers. *Rev. Phys. Appl.* **5**, 13–20 (1970). Nisenoff, M. Superconducting magnetometers with sensitivities approaching 10^{-10} gauss. *Rev. Phys. Appl.* **5**, 21–24 (1970).
6. Stewart, W.C. Current–voltage characteristics of Josephson junctions. *Appl. Phys. Lett.* **12**, 277–280 (1968).
7. McCumber, D.E. Effect of ac impedance on dc voltage–current characteristics of Josephson junctions. *J. Appl. Phys.* **39**, 3113–3118 (1968).
8. Ambegaokar, V., and Halperin, B.I. Voltage due to thermal noise in the dc Josephson effect. *Phys. Rev. Lett.* **22**, 1364–1366 (1969).
9. Clarke, J., and Koch, R.H. The impact of high-temperature superconductivity on SQUIDs. *Science* **242**, 217–223 (1988).
10. Likharev, K.K., and Semenov, V.K. Fluctuation spectrum in superconducting point junctions. *Pis'ma Zh. Eksp. Teor. Fiz.* **15**, 625–629 (1972) [*JETP Lett.* **15**, 442–445 (1972)].
11. Vystavkin, A.N., Gubankov, V.N., Kuzmin, L.S., Likharev, K.K., Migulin, V.V., and Semenov, V.K. S–C–S junctions as nonlinear elements of microwave receiving devices. *Phys. Rev. Appl.* **9**, 79–109 (1974).
12. Koch, R.H., Van Harlingen, D.J., and Clarke, J. Quantum noise theory for the resistively shunted Josephson junction. *Phys. Rev. Lett.* **45**, 2132–2135 (1980).
13. Tesche, C.D., and Clarke, J. dc SQUID: Noise and Optimization. *J. Low. Temp. Phys.* **27**, 301–331 (1977).
14. Bruines, J.J.P., de Waal, V.J., and Mooij, J.E. Comment on "dc SQUID noise and optimization" by Tesche and Clarke. *J. Low Temp. Phys.* **46**, 383–386 (1982).
15. De Waal, V.J., Schrijner, P., and Llurba. R. Simulation and optimization of a dc SQUID with finite capacitance. *J. Low. Temp. Phys.* **54**, 215–232 (1984).

16. Ketchen, M.B., and Jaycox, J.M. Ultra-low noise tunnel junction dc SQUID with a tightly coupled planar input coil. *Appl. Phys. Lett.* **40**, 736–738 (1982).
17. Jaycox, J.M., and Ketchen, M.B. Planar coupling scheme for ultra low noise dc SQUIDs. *IEEE Trans. Magn.* **17**, 400–403 (1981).
18. Wellstood, F.C., Heiden, C., and Clarke, J. Integrated dc SQUID magnetometer with high slew rate. *Rev. Sci. Inst.* **55**, 952–957 (1984).
19. Gurvitch, M., Washington, M.A., and Huggins, H.A. High quality refactory Josephson tunnel junction utilizing thin aluminum layers. *Appl. Phys. Lett.* **42**, 472–474 (1983).
20. De Waal, V.J., Klapwijk, T.M., and Van den Hamer, P. High performance dc SQUIDs with submicrometer niobium Josephson junctions. *J. Low. Temp. Phys.* **53**, 287–312 (1983).
21. Tesche, C.D., Brown, K.H., Callegari, A.C., Chen, M.M., Greiner, J.H., Jones, H.C., Ketchen, M.B., Kim, K.K., Kleinsasser, A.W., Notarys, H.A., Proto, G., Wang, R.H., and Yogi, T. Practical dc SQUIDs with extremely low $1/f$ noise. *IEEE Trans. Magn.* **21**, 1032–1035 (1985).
22. Pegrum, C.M., Hutson, D., Donaldson, G.B., and Tugwell, A. DC SQUIDs with planar input coils. *IEEE Trans. Magn.* **21**, 1036–1039 (1985).
23. Noguchi, T., Ohkawa, N., and Hamanaka. K. Tunnel junction dc SQUID with a planar input coil. *In* "SQUID 85 Superconducting Quantum Interference Devices and their Applications" (H.D. Hahlbohm and H. Lubbig, eds.), pp. 761–766. Walter de Gruyter, Berlin, 1985.
24. Muhlfelder, B., Beall, J.A., Cromar, M.W., and Ono, R.H. Very low noise tightly coupled dc SQUID amplifiers. *Appl. Phys. Lett.* **49**. 1118–1120 (1986).
25. Knuutila, J., Kajola, N., Seppä, H., Mutikainen, R., and Salmi, J. Design, optimization and construction of a dc SQUID with complete flux transformer circuits. *J. Low. Temp. Phys.* **71**, 369–392 (1988).
26. Carelli, P., and Foglietti, V. Behavior of a multiloop dc superconducting quantum interference device. *J. Appl. Phys.* **53**, 7592–7598 (1982).
27. Clarke, J., Goubau, W.M., and Ketchen, M.B. *J. Low. Temp. Phys.* **25**, 99–144 (1976).
28. Ketchen, M.B., Goubau, W.M., Clarke, J., and Donaldson, G.B. Superconducting thin-film gradiometer. *J. Appl. Phys.* **44**, 4111–4116 (1978).
29. Wellstood, F.C., and Clarke, J. unpublished.
30. Wellstood, F.C., Urbina, C., and Clarke, J. Low-frequency noise in dc superconducting quantum interference devices below 1 K. *Appl. Phys. Lett.* **50**, 772–774 (1987).
31. Roukes, M.L., Freeman, M.R., Germain, R.S., Richardson, R.C., and Ketchen, M.B. Hot electrons and energy transport in metals at millikelvin temperatures. *Phys. Rev. Lett.* **55**, 422–425 (1985).
32. Wellstood, F.C., Urbina, C., and Clarke, J. Hot electron effect in the dc SQUID. *IEEE Trans. Magn.* **25**, 1001–1004 (1984).
33. Ketchen, M.B., Awschalom, D.D., Gallagher, W.J., Kleinsasser, A.W., Sandstrom, R.L., Rozen, J.R., and Bumble, B. Design, fabrication and performance of

integrated miniature SQUID susceptometers. *IEEE Trans. Magn.* **25**, 1212–1215 (1989).
34. Koch, R.H., Clarke, J., Goubau, W.M., Martinis, J.M., Pegrum, C.M., and Van Harlingen, D.J. Flicker ($1/f$) noise in tunnel junction dc SQUIDS. *J. Low. Temp. Phys.* **51**, 207–224 (1983).
35. Rogers, C.T., and Buhrman, R.A. Composition of $1/f$ noise in metal–insulator–metal tunnel junctions. *Phys. Rev. Lett.* **53**, 1272–1275 (1984).
36. Dutta, P., and Horn, P.M. Low-frequency fluctuations in solids: $1/f$ noise. *Rev. Mod. Phys.* **53**, 497–516 (1981).
37. Van der Ziel, A. On the noise spectra of semi-conductor noise and of flicker effect. *Physica* **16**, 359–372 (1950).
38. Tesche, C.D., Brown, R.H., Callegari, A.C., Chen, M.M., Greiner, J.H., Jones, H.C., Ketchen, M.B., Kim, K.K., Kleinsasser, A.W., Notarys, H.A., Proto, G., Wang, R.H., and Yogi, T. Well-coupled dc SQUID with extremely low $1/f$ noise. *In* "Proc. 17th International Conference on Low Temperature Physics LT-17," pp. 263–264. North Holland, Amsterdam, 1984.
39. Foglietti, V, Gallagher, W.J., Ketchen, M.B., Kleinsasser, A.W., Koch, R.H., Raider, S.I., and Sandstrom, R.L. Low-frequency noise in low $1/f$ noise dc SQUIDs. *Appl. Phys. Lett.* **49**, 1393–1395 (1986).
40. Biomagnetic Technologies Inc., 4174 Sorrento Valley Blvd., San Diego, CA 92121.
41. Fujimaki, N., Tamura, H., Imamura, T., and Hasuo, S. A single-chip SQUID magnetometer. *In* "Digest of Tech. Papers of 1988 International Solid-State Conference," pp. 40–41. ISSCC San Francisco. A longer version with the same title is to be published.
42. Drung, D. Digital Feedback loops for dc SQUIDs. *Cryogenics* **26**, 623–627 (1986). Drung, D., Crocoll, E., Herwig, R., Neuhaus, M., and Jutzi, W. Measured performance parameters of gradiometers with digital output. *IEEE Trans. Magn.* **25**, 1034–1037 (1989).
43. Mück, M., and Heiden, C. Simple dc SQUID system based on a frequency modulated relaxation oscillator. *IEEE Trans. Magn.* **25**, 1151–1153 (1989).
44. Clarke, J. Superconducting QUantum Interference Devices for Low Frequency Measurements. *In* "Superconductor Applications: SQUIDs and Machines" (B.B. Schwartz and S. Foner, eds.), pp. 67–124. Plenum, New York, 1977.
45. Giffard, R.P., Webb, R.A., and Wheatley, J.C. Principles and methods of low-frequency electric and magnetic measurements using rf-biased point-contact superconducting device. *J. Low. Temp. Phys.* **6**, 533–610 (1972).
46. Kurkijärvi, J. Intrinsic fluctuations in a superconducting ring closed with a Josephson junction. *Phys. Rev. B* **6**, 832–835 (1972).
47. Kurkijärvi, J., and Webb, W.W. Thermal noise in a superconducting flux detector. *In* "Proc. Applied Superconductivity Conf.," pp. 581–587. Annapolis, Maryland, 1972.
48. Jackel, L.D., and Buhrman, R.A. Noise in the rf SQUID. *J. Low. Temp. Phys.* **19**, 201–246 (1975).
49. Ehnholm, G.J. Complete linear equivalent circuit for the SQUID. *In* "SQUID Superconducting Quantum Interference Devices and their Applications" (H.D.

Hahlbohm and H. Lubbig, eds.), pp. 485–499. Walter de Gruyter, Berlin, 1977); Theory of the signal transfer and noise properties of the rf SQUID. *J. Low. Temp. Phys.* **29**, 1–27 (1977).

50. Hollenhorst, H.N., and Giffard, R.P. Input noise in the hysteretic rf SQUID: Theory and experiment. *J. Appl. Phys.* **51**, 1719–1725 (1980).
51. Kurkijärvi, J. Noise in the superconducting flux detector. *J. Appl. Phys.* **44**, 3729–3733 (1973).
52. Giffard, R.P., Gallop, J.C, and Petley, B.N. Applications of the Josephson effects. *Prog. Quant. Electron.* **4**, 301–402 (1976).
53. Ehnholm, G.J., Islander, S.T., Ostman, P., and Rantala, B. Measurements of SQUID equivalent circuit parameters. *J. de Physique* **39**, colloque C6, 1206–1207 (1978).
54. Giffard, R.P., and Hollenhorst, J.N. Measurement of forward and reverse signal transfer coefficients for an rf-biased SQUID. *Appl. Phys. Lett.* **32**, 767–769 (1978).
55. Jackel, L.D., Webb, W.W., Lukens, J.E., and Pei, S.S. Measurement of the probability distribution of thermally excited fluxoid quantum transitions in a superconducting ring closed by a Josephson junction. *Phys. Rev. B* **9**, 115–118 (1974).
56. Long, A., Clark, T.D., Prance, R.J., and Richards, M.G. High performance UHF SQUID magnetometer. *Rev. Sci. Instrum.* **50**, 1376–1381 (1979).
57. Hollenhorst, J.N., and Giffard, R.P. High sensitivity microwave SQUID. *IEEE Trans. Magn.* **15**, 474–477 (1979).
58. Ahola, H., Ehnholm, G.H., Rantala, B., and Ostman. P. Cryogenic GaAs-FET amplifiers for SQUIDs. *J. de Physique* **39**, colloque C6, 1184–1185 (1978); *J. Low Temp. Phys.* **35**, 313–328 (1979).
59. For a review, see Clarke, J. Advances in SQUID Magnetometers. *IEEE Trans. Electron. Devices* **27**, 1896–1908 (1980).
60. Zimmerman, J.E. Sensitivity enhancement of Superconducting Quantum Interference Devices through the use of fractional-turn loops. *J. Appl. Phys.* **42**, 4483–4487 (1971).
61. Shoenberg, D. "Superconductivity," p. 30. Cambridge University Press, 1962.
62. For a review, see Clarke, J. Geophysical Applications of SQUIDs. *IEEE Trans. Magn.* **19**, 288–294 (1983).
63. De Waal, V.J., and Klapwijk, T.M. Compact Integrated dc SQUID gradiometer. *Appl. Phys. Lett.* **41**, 669–671 (1982).
64. Van Nieuwenhuyzen, G.J., and de Waal, V.J. Second order gradiometer and dc SQUID integrated on a planar substrate. *Appl. Phy. Lett.* **46**, 439–441 (1985).
65. Carelli, P., and Foglietti, V. A second derivative gradiometer integrated with a dc superconducting interferometer. *J. Appl. Phys.* **54**, 6065–6067 (1983).
66. Koyanagi, M., Kasai, N., Chinone, K., Nakanishi, M., Kosaka, S., Higuchi, M., and Kado, H. An integrated dc SQUID gradiometer for biomagnetic application. *IEEE Trans. Magn.* **25**, 1166–1169 (1989).
67. Knuutila, J., Kajola, M., Mutikainen, R., and Salmi, J. Integrated planar dc SQUID magnetometers for multichannel neuromagnetic measurements. *In* "Proc. ISEC '87," p. 261.

68. For reviews, see Romani, G.L., Williamson, S.J., and Kaufman, L. Biomagnetic instrumentation. *Rev. Sci. Instrum.* **53**, 1815–1845 (1982). Buchanan, D.S., Paulson, D., and Williamson, S.J. Instrumentation for clinical applications of neuromagnetism (R.W. Fast, ed.) *Adv. Cryo. Eng.*, **33**, 97–106. Plenum, New York, 1988.
69. Knuutila. J. European Physical Society Workshop "SQUID: State of Art, Perspectives and Applications." Rome, Italy, June 22–24, 1988, unpublished.
70. Barth, D.S., Sutherling, W., Engel, J., Jr., and Beatty, J. Neuromagnetic evidence of spatially distributed sources underlying epileptiform spikes in the human brain. *Science* **223**, 293–296 (1984).
71. Romani, G.L., Williamson, S.J., and Kaufman, L. Tonotopic organization of the human auditory cortex. *Science* **216**, 1339–1340 (1982).
72. Cabrera, B. First results from a superconductive detector for moving magnetic monopoles. *Phys. Rev. Lett.* **48**, 1378–1381 (1982).
73. Quantum Design, 11568 Sorrento Valley Road, San Diego, CA 92121.
74. Ketchen, M.B., Kopley, T., and Ling, H. Miniature SQUID susceptometer. *Appl. Phys. Lett.* **44**, 1008–1010 (1984).
75. Awschalom, D.D., and Warnock, J. Picosecond magnetic spectroscopy with integrated dc SQUIDs. *IEEE Trans. Magn.* **25**, 1186–1192 (1989).
76. Clarke, J. A superconducting galvanometer employing Josephson tunneling. *Phil. Mag.* **13**, 115–127 (1966).
77. Hilbert, C., and Clarke, J. DC SQUIDs as radiofrequency amplifiers. *J. Low Temp. Phys.* **61**, 263–280 (1985).
78. Tesche, C.D., and Clarke, J. DC SQUID: Current noise. *J. Low Temp. Phys.* **37**, 397–403 (1979).
79. Hilbert, C., and Clarke, J. Measurements of the dynamic input impedance of a dc SQUID. *J. Low Temp. Phys.* **61**, 237–262 (1985).
80. Martinis, J.M., and Clarke, J. Signal and noise theory for the dc SQUID. *J. Low Temp. Phys.* **61**, 227–236 (1985); and references therein.
81. Koch, R.H., Van Harlingen, D.J., and Clarke, J. Quantum noise theory for the dc SQUID. *Appl. Phys. Lett.* **38**, 380–382 (1981).
82. Danilov, V.V., Likharev, K.K., and Zorin, A.B. Quantum noise in SQUIDs. *IEEE Trans. Magn.* **19**, 572–575 (1983).
83. Hilbert, C., Clarke, J., Sleator, T., and Hahn, E.L. Nuclear quadruple resonance detected at 30 MHz with a dc superconducting quantum interference device. *Appl. Phys. Lett.* **47**, 637–639 (1985). (See references therein for earlier work on NMR with SQUIDS.)
84. Fan, N.Q., Heaney, M.B., Clarke, J., Newitt, D., Wald, L.L., Hahn, E.L., Bielecke, A., and Pines, A. Nuclear magnetic resonance with dc SQUID preamplifiers. *IEEE Trans. Magn.* **25**, 1193–1199 (1989).
85. Sleator, T., Hahn, E.L., Heaney, M.B., Hilbert, C., and Clarke, J. Nuclear electric quadrupole induction of atomic polarization. *Phys. Rev. Lett.* **57**, 2756–2759 (1986).
86. Sleator, T., Hahn, E.L., Hilbert, C., and Clarke, J. Nuclear-spin noise and spontaneous emission. *Phys. Rev. B* **36**, 1969–1980 (1987).

87. For an elementary review on gravity waves, see Shapiro, S.L., Stark, R.F. and Teukolsky, S.J. The search for gravitational waves. *Am. Sci.* **73**, 248–257 (1985).
88. For a review on gravity-wave antennae, see Michelson, P.F., Price, J.C., and Taber, R.C. Resonant-mass detectors of gravitational radiation. *Science* **237**, 150–157 (1987).
89. Paik, H.J. Superconducting tensor gravity gradiometer with SQUID readout. *In* "SQUID Applications to Geophysics" (H. Weinstock and W.C. Overton, Jr., eds.), pp. 3–12. Soc. of Exploration Geophysicists, Tulsa, Oklahoma, 1981.
90. Mapoles, E. A superconducting gravity gradiometer. *In* "SQUID Applications to Geophysics" (H. Weinstock and W.C. Overton, Jr., eds.), pp. 153–157. Soc. of Exploration Geophysicists, Tulsa, Oklahoma, 1981.
91. Bednorz, J.G., and Muller, K.A. Possible high T_c superconductivity in the Ba-La-Cu-O system. *Z. Phys. B* **64**, 189–193 (1986).
92. Koch, R.H., Umbach, C.P., Clark, G.J., Chaudhari, P., and Laibowitz, R.B. Quantum interference devices made from superconducting oxide thin films. *Appl. Phys. Lett.* **51**, 200–202 (1987).
93. Zimmerman, J.E., Beall, J.A., Cromar, M.W., and Ono, R.H. Operation of a Y-Ba-Cu-O rf SQUID at 81 K. *Appl. Phys. Lett.* **51**, 617–618 (1987).
94. Ferrari, M.J., Johnson, M., Wellstood, F.C., Clarke, J., Rosenthal, P.A., Hammond, R.H., and Beasley, M.R. Magnetic flux noise in thin film rings of $YBa_2Cu_3O_{7-\delta}$. *Appl. Phys. Lett.* **53**, 695–697 (1988).
95. Mankiewich P.M., Schwartz, D.B., Howard, R.E., Jackel, L.D., Straughn, B.L., Burkhardt, E.G., and Dayem, A.H. Fabrication and characterization of an $YBa_2Cu_3O_7/Au/YBa_2Cu_3O_7$ S–N–S microbridge. *In* "Fifth International Workshop on Future Electron Devices—High Temperature Superconducting Devices, June 2–4, 1988," pp. 157–160. Miyaki-Zao, Japan.

CHAPTER 3

Computing

HISAO HAYAKAWA

Department of Electronics Engineering
Nagoya University
Furo-cho, Chikusa-ku, Nagoya, Japan

1. Introduction 101
2. Advantages of Josephson Switching Devices 103
3. Josephson Switching Gates 105
4. Logic Gates 107
    4.1. Magnetically Coupled Gates 107
    4.2. Direct Coupled Gates 111
    4.3. Hybrid Gates 114
    4.4. Switching Delay of Logic Gates 117
5. Memory Cells 118
    5.1. Non-Destructive Read-Out (NDRO) Memory Cells 118
    5.2. Destructive Read-Out (DRO) Memory Cells 119
    5.3. Other Memory Cells 121
6. Circuits 123
    6.1. OR–AND Cells 123
    6.2. AC-Power and Sequential Logic 125
    6.3. Memory Circuits 128
    6.4. Small-Scale Computer Circuits—Microprocessors 130
    References 132

1. Introduction

Since Josephson predicted the possibility of tunneling of superconducting pair electrons in a superconductor–insulator–superconductor (SIS) system [1], the technology of so-called superconducting electronics has made much progress. Superconducting electronics based on the Josephson effect cover a large number of applications in both analog and digital electronics. The analog applications include SQUIDs, mm-wave mixers, and voltage standards are now almost at the point of real use. Digital applications, the goal being to construct high-speed computer systems, are still under development because the scale of the technology is much larger than that of analog applications. However, superconducting digital technology has been a leading technology for the other applications such as SQUIDs, mixers, and voltage standards. The fabrication technology for high-quality junctions and its integration technology have been developed mainly for digital applications, but these technologies are now being used for developing high-performance analog devices, i.e., integrated dc SQUIDs.

The idea of applying superconductors to switching device was suggested in the mid-1950s. A device called a "cryotron" was the first one in which the transition between the superconductive and normal state in a superconducting thin-film wire was used for switching a current path [2]. However, the switching speed of the cryotron was rather slow, more than 1 ns, which was easily exceeded by transistors. The reason for this slowness is that the super–normal transition involves a thermal process rather than an electronic one.

Not very long after Josephson's prediction, the first Josephson switching device was demonstrated by J. Matisoo of IBM in 1966 [3]. The device exhibited a sub-nanosecond switching speed, which was a very attractive result at the time. Following this, IBM made a systematic effort to develop a Josephson computer until they announced that they were stopping the project in 1983. In the course of IBM's Josephson project, essential components for the Josephson computer, including logic and memory circuits and systems have been built, and technologies for fabricating and packaging the devices were also developed and advanced.

Stimulated by IBM's results, other parts of the world, mainly Europe and Japan, became interested in Josephson digital applications and research and development efforts were started in the mid-1970s. Especially in Japan, major computer companies were very much interested in the Josephson technology as a post-semiconductor technology and started research and development at this time. The Ministry of International Trade and Industry (MITI) of the Japanese government arranged a large-scale project for the development of a high-speed scientific computer system in 1981, in which the Josephson device—in addition to GaAs and HEMT (High Electron Mobility Transistor) devices—was selected to be a candidate for the high-speed device to be used in the future computer system. Electrotechnical Laboratory, Fujitsu, NEC, and Hitachi joined the project to develop Josephson devices. At the same time, Nippon Telephone and Telegram Corporation (NTT) also started R&D of the Josephson digital technology. MITI's project has continued even after IBM stopped its Josephson Project. These Japanese efforts have led to much progress in the Josephson technology of junction materials, devices, and systems.

3. COMPUTING

The most important innovation in Josephson technologies is that now circuits can be fabricated with all Nb or NbN junctions, instead of Pb-alloy junctions. The introduction of all refractory junctions has completely changed the device fabrication process, allowing the construction of circuits with LSI complexities. (Currently, these technologies are used not only in digital applications but also in analog applications, contributing to the development of high-performance devices in this area as well.)

In this chapter, the operating principles of Josephson logic and memory devices and the recent achievements of Josephson digital circuits will be reviewed.

2. Advantages of Josephson Switching Devices

The most attractive feature of Josephson junctions, which is the cause of the drive to develop Josephson digital devices, is that Josephson junctions can switch with a fast speed and a low power dissipation. In Fig. 1, delay-power characteristics of Josephson devices are compared with various semiconductor devices. As shown in Fig. 1, the switching delays of Josephson devices

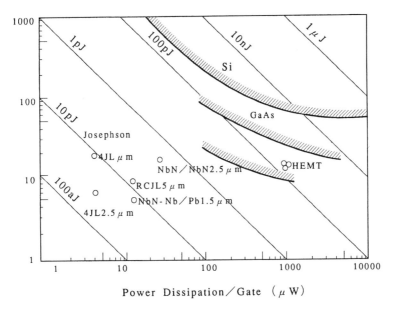

Fig. 1. Rough comparison of device performances based on the delay–power relation for various devices. Josephson devices are superior in both propagation delay and power dissipation to any semiconductor devices.

are scattered around values less than 10 ps, which is about one order of magnitude faster than those of any other semiconductor devices. Furthermore, the power dissipations are several μW, two or three orders of magnitude less than those of semiconductor devices.

The low power dissipation of Josephson devices is exceptionally important to make the system speed high. To decrease the system delay, it is necessary to have extremely densely packed devices. The very low power dissipation of Josephson devices makes it possible to achieve the high density packaging of devices, which decreases the total interconnect, or wiring length, resulting in high system speed.

Another advantage of Josephson computing is that superconducting matched striplines can be used for wiring chips and packagings. The matched superconducting stripline can transmit signals without loss or dissipation up to the frequency corresponding to the gap energy. This means that superconducting striplines provide an ideal means of transmitting the fast pulses with sub-10 ps rise times generated in Josephson switches.

Of course, superconducting striplines could be used for wiring semiconductor devices. However, it seems to be difficult for semiconductor devices to use superconducting wiring at liquid He temperature, 4.2 K, because the power dissipation of semiconductor devices is too high to cool the devices with liquid He. The recent discovery of a high-T_c oxide superconductor whose T_c is higher than 77 K, liquid N_2 temperature, may make it possible to use superconducting striplines for wiring semiconductor devices operated at 77 K.

The low temperature operation of Josephson devices has both positive and negative meanings. The use of the low temperature atmosphere for operating Josephson systems gives rise to the difficulty of smooth connections with room-temperature systems, and also the higher refrigeration and maintenance costs. However, the low temperature operation provides some advantages for a large computing system. For example, the low power dissipation of Josephson devices are caused by the low temperature operation. If they are operated at higher temperatures, the logic swing has to be increased, hence, higher power consumptions are needed, in order to overcome thermal noises. It should be noted that the very low power consumption of Josephson devices is assured by the very low temperature operation. At the low temperature, i.e., liquid He temperature, electrochemical reactions such as electromigration are suppressed, which makes the reliability of systems high; this is another advantage of low temperature operation.

By having these advantages, Josephson devices with fast switching speeds of less than 10 ps and extremely low power dissipation of several μW have a potential to realize a high speed computing system whose performance cannot be matched by semiconductor systems.

3. Josephson Switching Gates

Usually, Josephson gates make use of switching in a tunnel junction between a superconducting state and a voltage state corresponding to the logical states "0" and "1", respectively. Figure 2 schematically illustrates the operating principle of Josephson gates. At first, a single Josephson junction is biased with a bias current I_g below the critical current I_0 of the junction. If an input current I_c is added to I_g so as to exceed I_0, the junction switches into the voltage state along a load line. Since the load resistance R_L is chosen to be much smaller than the sub-gap resistance of the junction, almost all of the current through the junction is transferred to the load after the junction switches. This is the basic principle of the Josephson switch.

The switching of a junction is illustrated in Fig. 2a. The switching delay mainly consists of two components: one is the so-called turn-on delay and the other is the rise time required for charging the junction capacitance.

The turn-on delay is a time delay required for raising the rate of change of the junction phase with the initial overdrive current. The turn-on delay τ_t is defined by the time for the phase to increase to $t_0 = \pi/2 + 0.5$ (rad). To estimate the turn-on delay, we assume a situation in which the junction is biased $I_g = I_0$; hence, the junction phase is $\theta = \pi/2$ at the initial stage, and then the bias current is increased by ΔI. The time needed for the phase to reach $\pi/2 + 0.5$ is written as

$$\tau_t \approx \sqrt{\frac{C_J \Phi_0}{I_0 \Delta I}}, \tag{3.1}$$

where C_J and Φ_0 are the junction capacitance and the flux quantum, respectively [4]. Equation (3.1) indicates that the turn-on delay is proportional to $\sqrt{C_J}$ and inversely proportional to $\sqrt{\Delta I}$. This means that smaller junction capacitance and higher overdrive of the current decreases the turn-on delay.

Another delay component is the rise time: the time required to charge up the junction capacitance. The rise time is usually defined by

$$\tau_r = RC_J, \tag{3.2}$$

where R is the effective resistance of the parallel combination of the junction resistance R_J and the load resistance R_L. In order to transfer a large current to

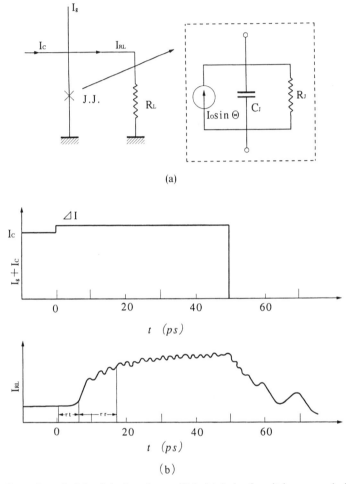

Fig. 2. Operating principle of the Josephson switch. (a) A simple switch composed of a single Josephson junction and the equivalent circuit of the Josephson junction (insertion). I_g is a bias current. By injecting the control current I_c, the junction switches into the voltage state. (b) The output current waveform when the junction switches. The switching delay is composed of two components; one is the turn-one delay τ_t and the other is the rise time τ_r.

the load, the load resistance is usually chosen to be much smaller than the junction sub-gap resistance, so that the load line crosses the $I-V$ curve below the gap voltage. Taking $R_L \ll R_J$, the rise time can be written as

$$\tau_r = R_L C_J. \tag{3.3}$$

Figure 2b shows a typical switching waveform obtained with a computer simulation by assuming a junction whose parameters are $I_0 = 100\,\mu A$, $R_{SG} = 150\,\Omega$, $R_N = 15\,\Omega$, $C_J = 1$ pF, and $R_L = 10\,\Omega$. In this calculation, the bias current is initially set so as to be $I_g = I_0$, and then step current $\Delta I = 0.1\,I_0$ is applied to switch the junction. The total delay $\tau = \tau_r + \tau_t = 10$ ps $+ 6$ ps $= 16$ ps is obtained.

As described above, the switching delay—both the turn-on and the rise time—directly depends on the junction capacitance C_J. To obtain a smaller switching delay, and hence a faster switching speed, it is essential to reduce the junction capacitance C_J, i.e., to make the junction size small.

4. Logic Gates

The switching device illustrated in Fig. 2a cannot be used in logic circuits because there is no input–output current isolation. If gates without current isolation are used in circuits, the current switched from one gate can flow in the backward direction as well as in the forward direction so that it becomes very difficult to get directionality of the logic signal in the circuit.

In addition to the current isolation function, there are several other items required in the design of logic gates. The important criteria for logic gates are the following:

(1) larger gain (for larger fan-outs and large operating margin),
(2) smaller size (for large scale integration),
(3) fast switching,
(4) smaller power consumption (for dense packaging).

A number of logic gates have been proposed by various workers. These logic gates are classified into two types: one is the magnetically coupled gate and the other is the direct coupled gate.

4.1. Magnetically Coupled Gates

The magnetically coupled gate is essentially based on SQUIDs that are controlled by the magnetic field produced by overlaying control lines.

A typical gate of this type is the so-called three-junction interferometer logic (JIL) gate [5]. Figure 3 shows the equivalent circuit of the JIL gate and its threshold characteristics. When a sufficient input current is applied to the control line, the gate switches to the voltage state. This corresponds to an

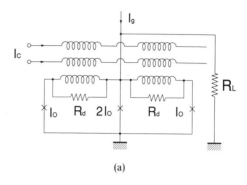

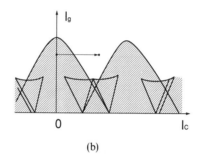

Fig. 3. (a) The equivalent circuit configuration of a three-junction interferometer logic gate and (b) its threshold characteristics. R_d's are damping resistors. Hatched region in the threshold curve indicates the superconducting state [5].

action whereby the operating point moves along the arrow in Fig. 3b. Two-junction SQUIDs can also be used for logic gates. However, three-junction SQUIDs are more widely used than two-junction SQUIDs because of their wider operating margin. By choosing the critical current of three junctions as $I_1 = I, I_2 = 2I$, and $I_3 = I$, side lobes between the main lobes in the threshold characteristics can be well suppressed so that the wider operating window is obtained when the gate goes to the voltage state. The loop inductance L is typically chosen as $LI_0 = 0.2-0.3\,\Phi_0$.

The SQUID gate with the tunnel junction whose equivalent circuit consists of serial combinations of the loop inductance L and the junction capacitance C_J easily induces an LC resonance oscillation with a frequency of $\omega_r = 1/\sqrt{LC_J}$ when the gate goes to the voltage state. This resonance oscillation prevents the gate from generating a proper output voltage when it switches. In

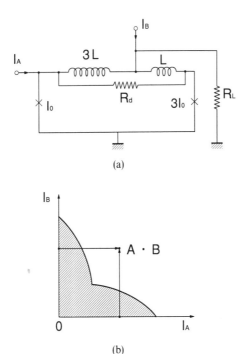

Fig. 4. (a) The equivalent circuit configuration of the Current Injection Device (CID) and (b) its threshold characteristics. The threshold characteristic is symmetric for two inputs I_A, I_B, indicating the CID is useful as an AND gate [7].

order to suppress the resonance oscillation, damping resistors are inserted to shunt the loop inductances as shown in Fig. 3a [6].

Another important gate based on the SQUID is the so-called Current Injection Device (CID) [7]. The equivalent circuit and the threshold curve of the CID are shown in Fig. 4. The CID is operated by direct current injections to a SQUID loop. The threshold characteristic becomes symmetric for both input currents if the circuit parameters in Fig. 4 are chosen as

$$L_1 I_{01} = L_2 I_{02} \tag{4.1}$$

$$(L_1 + L_2)I_{02} = \Phi_0 \tag{4.2}$$

where I_{01} and I_{02} are the critical currents of junctions J_1 and J_2, respectively. In the CID, the gate switches into the voltage state with the large operating window if both inputs are simultaneously applied to the gate as shown in

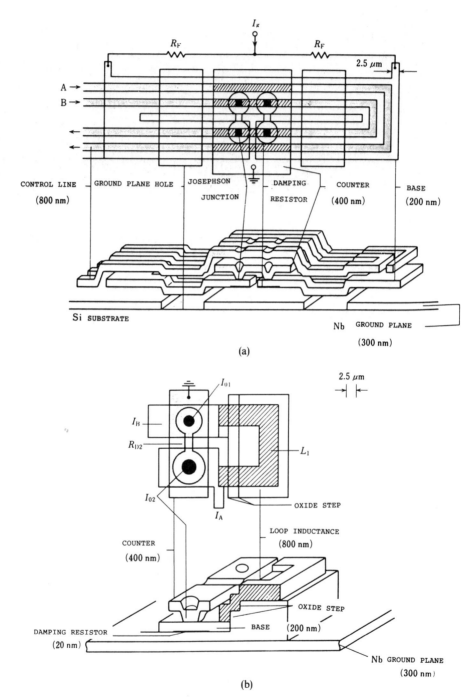

Fig. 5. (a) Thin-film configurations of the three-junction interferometer gate and (b) the current injection device. In order to make the device size small, the effective loop inductances are increased by making holes in the ground plane and oxide steps on the ground plane.

3. COMPUTING

Fig. 4b. This means that the CID is useful for operation as a two-input AND gate. However, the CID has no current isolation function so that the CID has to be used in combination with a current isolation gate; for example, JILs are used in front of the CID as current isolation gates. The thin-film structures of these SQUID-based gates are shown in Fig. 5 [7].

In SQUID gates such as JIL and CID, a relatively large inductance is needed to keep gains or sensitivity high. This large inductance makes the area of the gate large, which limits the number of gates integrated on a chip. A ground plane hole and a step of insulation, illustrated in Fig. 5, are formed to increase loop inductances on those portions, resulting in a decreased loop length and hence a smaller gate size.

4.2. Direct Coupled Gates

Direct coupled gates are controlled by the direct injection of input currents. The CID described above is also controlled by the injection current. However, the CID is essentially based on a SQUID so that a magnetic field produced by the current flowing in the loop plays an important role for switching the gate. For this reason, the CID is classified as a magnetically coupled gate in this chapter.

An important feature of direct coupled gates is that inductances can be eliminated from the gates, resulting in an advantage for making the gate size small. In direct coupled gates, it is important to utilize the current isolation function. The first device with a current isolation function was made of two junctions and a resistor, named JAWS (Josephson Atto Weber Switch) [8]. Figure 6 shows the equivalent circuit and the threshold characteristic of the

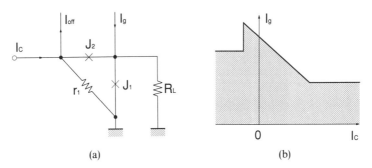

Fig. 6. (a) The equivalent circuit of the JAWS gate and (b) its threshold curve. Input–output current isolation is achieved by adding a junction J_2 and a small resistor r_1 [8].

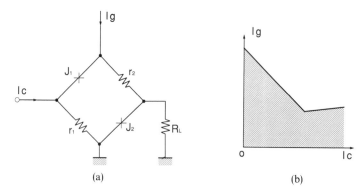

Fig. 7. (a) The equivalent circuit of the Direct Coupled Logic (DCL) gate and (b) its threshold curve [9].

JAWS gate. The operating principle of the JAWS gate is as follows. A gate current is biased to the gate somewhat below the critical current of the junction J_1. At this time, the gate current flows through J_1 and the gate remains in the superconducting state. When an input current I_c is applied to the gate through J_2, the current through J_1 is increased to $I_g + I_c$, which is enough to switch J_1 into the voltage state. After J_1 switches, the gate current I_g flows through J_2 and a small resistor r_1 to the ground, which results in switching J_2 to the voltage state. At this stage, since both J_1 and J_2 have switched to highly resistive states, I_g is transferred to a load resistor and I_c flows to the ground through r_1, completing the current isolation.

The gate illustrated in Fig. 7 called DCL (Direct Coupled Logic) is another example of the current isolation gate [9]. The operating principle of the DCL gate is quite similar to that of JAWS.

In the JAWS and DCL current isolation gates, the threshold curves separating the superconducting state from the voltage state have a slope of -1, i.e., $|\Delta I_g/\Delta I_c| = 1$. If the slope becomes steeper, i.e., $|\Delta I_g/\Delta I_c| > 1$, higher sensitivity can be obtained. The gates shown in Fig. 8 were developed to have higher sensitivities by adding another junction branch to the current isolation gates described above. These gates are called RCJL (Resistor Coupled Josephson Logic) [10] and RCL (Resistor Coupled Logic) [11].

Gates with modified threshold curves were also designed. Fig. 9a,b shows the RCJL AND and RCJL 2/3 gates [10]. As shown in Fig. 9, the threshold

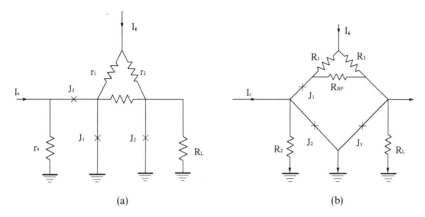

Fig. 8. (a) The circuit configurations of the Resistor Coupled Josephson Logic (RCJL) and (b) the Resistor Coupled Logic (RCL) gate. By putting in another branch of a Josephson junction (J_2 for RCJL, J_3 for RCL), the sensitivities or gains can be higher than those of the JAWS and DCL gates [10][11].

curves are modified to have symmetric characteristics for input currents, which are useful to make AND and 2/3-Majority functions.

Another direct coupled type logic family called 4JL (4 Junctions Logic) was demonstrated by Takada et al. [12]. The 4JL gate is composed of four junctions that are closely coupled in a loop as shown in Fig. 10. The essential feature of the 4JL gate is that the loop inductance can be eliminated; hence, the threshold characteristic is determined only by the phase relations of junctions. The current isolation function is achieved by putting a small resistor at an input terminal and making junction J_1 switch at the final stage of the switching sequence. In order to make the gate sensitivity high, critical currents of junctions in the right branch are taken to be larger than those in the left branch (typically $I_{03,4} = 3I_{01,2}$). An AND gate based on the 4JL concept can also be designed by making the threshold characteristic symmetric for two input currents [13]. Fig. 11 shows the 4JL AND gate and its threshold characteristics.

The essential advantage of the direct coupled gate is that the gate size can be smaller than those of the magnetically coupled SQUID gates because inductance loops are eliminated in the direct coupled gates. In fact, the 4JL gate fabricated with a 2.5 μm^2 minimum rule has a gate area as small as 1200 μm^2. On the other hand, the JIL gate designed with the same rule has an area of 4400 μm^2 [7].

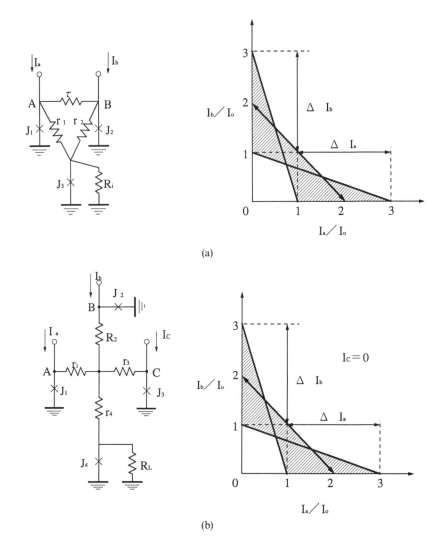

Fig. 9. The circuit configuration of the RCJL logic family: (a) an AND gate and (b) a 2/3 gate [10].

4.3. Hybrid Gates

There is a series of gates that is not classified as belonging to either of the two categories described above. This gate family was developed recently; it operates in a "hybrid" manner with both magnetic and direct coupled gates. Figure 12 shows a gate of this type called MVTL (Modified Variable

3. COMPUTING

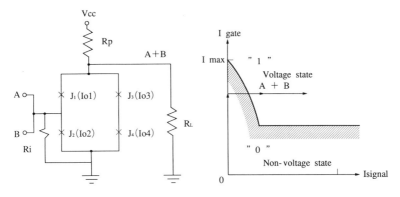

Fig. 10. (a) The equivalent circuit configuration of the 4JL gate and (b) its threshold curve [12].

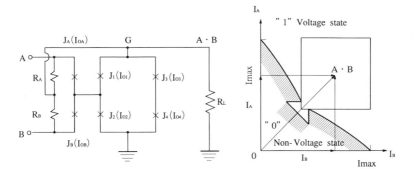

Fig. 11. (a) The equivalent circuit configuration of the 4JL AND gate and (b) its threshold curve [13].

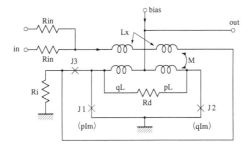

Fig. 12. The equivalent circuit configuration of the Modified Variable Threshold Logic (MVTL) gate [14].

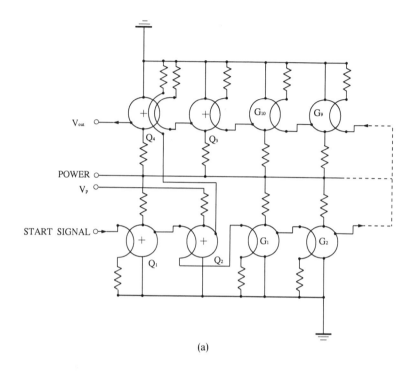

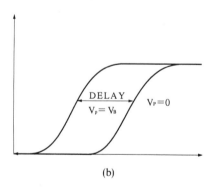

Fig. 13. (a) An example of the circuit configuration for measuring the switching delays of Josephson gates. (b) By comparing a bypassing signal ($V_p = V_B$) and a signal propagating through the chain ($V_p = 0$), the switching delay of the gate can be estimated [17].

Threshold Logic) [14]. An input current is first fed into a control line of an asymmetric two-junction SQUID and then injected into the SQUID. The input–output current isolation function is performed by connecting a junction J_3 and a small resistor R_i onto the current injection terminal of the SQUID. This gate is controlled by both the magnetic field and the injection current so that the sensitivity of the gate is high. Gates called HTCID* (High Tolerance Current Injection Device) [15] and CCL* (Counter Coupled Logic) [16] operate along the same lines as the CID.

4.4. Switching Delay of Logic Gates

Switching delays have been measured in a number of logic gates described above. The logic delay is usually measured using a chain of gates as shown in Fig. 13 [17]. When a start signal switches a triggering gate Q_1, the output of Q_1 is fed into two gates: one is fed to the first gate of the chain G_1, the other is fed to a gate Q_2 whose output current triggers an output gate Q_4. If the gate current of Q_2 is applied, the short-cut signal triggers Q_4, generating an output voltage. If the gate current is not applied to Q_2, a signal propagating along the chain triggers Q_4. By comparing the time difference between this short-cut signal and the propagating signal, the logic delay per gate can be estimated. In Table 1, the logic delay of gates measured so far are listed [7,18–23]. The fastest delay obtained is 2.5 ps/gate in the MVTL fabricated with 1.5 μm Nb/AlO$_x$/Nb junctions [23]. However, this data does not necessarily mean that the MVTL is superior to the other gates in speed. Most of these delay measurements were made at an earlier stage of development, in which only Pb-alloy technology was available for fabricating gates. Recently, fabrication

Table 1. Switching Speeds for Various Logic Gates

Gate	Rule (μm)	Switching time (ps)	Power dissipation	Junction	Ref.
CIL	2.5	13	2 μW	Pb-alloy	[7]
JAWS	5	15	$I_g = 90$ μA	Pb-alloy	[18]
RCJL	5	10.3	11.7 μW	Pb-alloy	[19]
RCL	2	4.2	$I_g = 0.72$ mA	Pb-alloy	[20]
4JL	2.5	7	4 μW	Pb-alloy	[21]
DCL	1.5	5.6	4 μW	NbN/Pb-In	[22]
MVTL	1.5	2.5	4 μW	Nb/AlO$_x$/Nb	[23]

technology based on all hard junctions has greatly improved, so that gate delays in gates fabricated by modern technology may be much faster than those made with the older technology.

The power dissipation of a gate is quite dependent on the gate current. In actual Josephson logic circuits, most of the power is dissipated at dropping resistors inserted between a power line and gates to assure a constant current operation of gates. Usually the dropping resistor should be chosen to be more than five times larger than the load resistor, i.e., $R_d \geq 5R_L$. Assuming $I_g = 200\,\mu A$, $R_L = 10\,\Omega$, and $R_d = 50\,\Omega$, the power dissipation of a gate is about $2.4\,\mu W$. The different values of power dissipation of the gates shown in Table 1 may be due to differences in gate current levels.

5. Memory Cells

In order to build complete superconducting digital systems, memories are inevitably important. In semiconductor devices, memories are made using charges stored in capacitances. In superconducting devices, on the other hand, persisting currents or magnetic fluxes in superconducting loops (inductances) are used for the storage of information. Josephson gates are used as switches to get magnetic fluxes in or out of the loops.

Various types of memory cells have been proposed; these are generally divided into two categories: non-destructive read-out (NDRO) memory and destructive read-out (DRO) memory.

5.1. *Non-Destructive Read-Out (NDRO) Memory*

The non-destructive read-out cell has the capability to read stored information without changing the cell state by reading operations. The NDRO memory is suitable for a cache memory that communicates directly with the CPU, so that speed is essentially important.

The earliest memory cell of this type was originated by Anacker [24] and then experimentally demonstrated by Zappe [25]. Much effort has been made to develop memory cells with wider operating margins. An improved NDRO memory cell is shown in Fig. 14 [26]. The cell makes use of a three-junction SQUID gate as a write gate and a two-junction SQUID gate as a sense gate. In order to write "1", a supply current I_y and a control current I_x are applied. In this sequence, the write gate switches to the voltage state, driving I_y to flow in the right branch of the loop. After I_y and I_x are turned off, the clockwise circulating current indicates the cell is in the state "1".

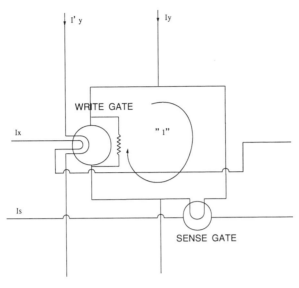

Fig. 14. The equivalent circuit configuration of the Non-Destructive Read-Out (NDRO) memory cell. The three-junction interferometer gate is used for the write gate and the asymmetric two-junction interferometer for the sense gate [26].

In order to read the cell, the gate current I_s of the sense gate and I_y are applied. If the cell is in "1", i.e., a circulating current is stored in the cell, the sense gate goes to the voltage state. If a "0" had been stored, the sense gate remains in the superconducting state. Thus, information in the cell can be read in a non-destructive manner. For the sense gate, an asymmetrically fed SQUID gate is used to widen the margin for sensing operations. This NDRO memory cell usually contains several flux quanta, typically $3\Phi_0$.

5.2. Destructive Read-Out (DRO) Memory Cells

A two-junction SQUID has a threshold characteristic with overlapping regions of vortex modes where either of the two states is stably maintained. The DRO memory cell is designed to store a single flux quantum in the two-junction SQUID by using a mode overlapping region in the threshold characteristics [27]. This memory stores a single flux quantum so that the size of the cell can be smaller, which is suitable for constructing a main memory.

The structure of the memory cell is shown in Fig. 15. The memory operation is performed using the mode overlapping region of the threshold characteristic

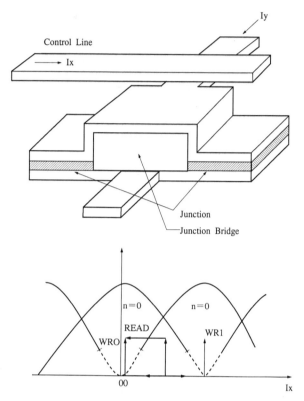

Fig. 15. The device configuration of the Destructive Read-Out (DRO) memory cell and its threshold characteristic. The DRO memory cell makes use of the mode overlap region in the threshold curve for writing and reading information [27].

as shown in Fig. 15 (lower figure). There are two transition modes on the threshold curve when the operating point moves across the curve. One is the so-called vortex transition and the other is the voltage transition. For the vortex transition, the vortex mode changes without generating voltages (only a spike voltage is generated) when an operating point crosses the regions on the threshold curve indicated by dotted curves in Fig. 15. On the other hand, for the voltage transition, the cell generates voltages when an operation point crosses regions on the threshold curve indicated by solid curves in Fig. 15. To write "1", I_y and I_x are applied to move the operating point across the vortex transition region as shown by the arrow WR1 in Fig. 15. If the cell had been in "0" state, the cell contains one flux of quantum by this operation. To write "0",

3. COMPUTING

I_y and I_x must be of opposite polarity to move the operation point across the vortex transition as shown by the arrow WR0 in Fig. 15. To read the cell, I_y and I_x are applied to move the operation point across the voltage transition region as shown by the arrow R1. In this operation, the cell generates a voltage if the cell had been in "0", but nothing happens if the cell had been in "1". By reading the cell, the information stored in the cell is usually destroyed so that a "refresh" operation is needed to rewrite the cell after each reading operation.

5.3. Other Memory Cells

Many suggestions for different types of memory cells have been made. Two of them will be briefly introduced in this section.

5.3.1. Variable Threshold Memory Cells

The variable threshold memory cell [28] basically has the same structure as a two-junction SQUID in which one flux quantum is stored. The equivalent circuit of the cell is illustrated in Fig. 16. As shown in Fig. 16, one junction in a

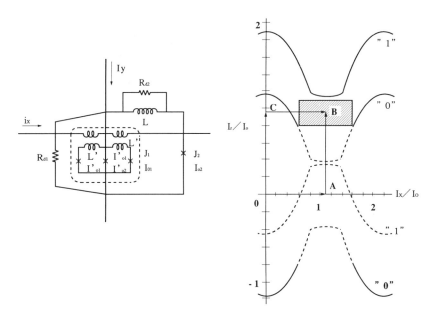

Fig. 16. The equivalent circuit of the Variable Threshold Memory Cell and its threshold curve [28].

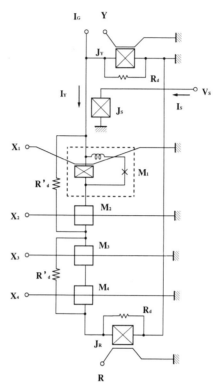

Fig. 17. The bit-line configuration of the Variable Threshold Memory Cell. By using this configuration, the memory cell can be made non-destructive read-out [29].

two-junction SQUID is replaced by a three-junction SQUID gate, which is coupled to a control current I_x. In this configuration, the vortex mode changes along the I_y-axis rather than the I_x-axis as shown in Fig. 16. To write "1", I_x and then I_y are applied to move the operation point across the mode boundary $(0 \rightarrow A \rightarrow B)$ as shown in Fig. 16. The dotted curves on the threshold curve represent the vortex transition. To read the cell, I_y and then I_x are applied to move the operation point across the boundary of the voltage transition $(0 \rightarrow C \rightarrow B)$. If the cell had been in "0", the cell generates a voltage. This reading sequence is basically destructive. However, the cell can be made into an NDRO cell by adding an external circuit in cell arrangements [29]. Figure 17 shows a bit-line arrangement of the cells for non-destructive read-out. The bit-line consists of a large superconducting loop containing the cells and set and reset gates. In the reading sequence, a cell that had been in "0" generates a voltage, transferring I_y to the set gate. By this operation, I_y is automatically

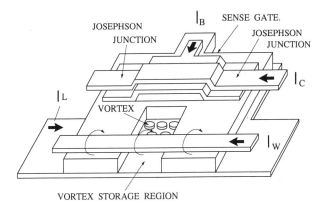

Fig. 18. The device configuration of the Abrikosov memory cell. Vortices are introduced in a superconducting thin film by the magnetic field procedure by the current I_w [30].

removed from the bit-line with the application of I_x, resulting in rewriting "0" in the cell.

5.3.2. Abrikosov Vortex Memory Cells

It is well known that vorticies (Abrikosov vorticies) are introduced and trapped in type II superconductors by the application of a magnetic field. A memory that makes use of these Abrikosov vorticies as information bits has been suggested and experimentally demonstrated [30]. Figure 18 shows the schematic structure of this memory. A control line is used to generate the magnetic field to get vorticies into a superconducting thin film. In order to sense these vorticies, a two-junction SQUID is placed just above the thin film in which the vorticies are stored. To write "0", vorticies with the opposite polarity are introduced in the film. In this type of memory, it is important to select superconducting thin films with lower surface potentials for vorticies to reduce the writing current and the writing time. A Pb-In-Au film with fine grains and amorphous Be film have been examined for this purpose [31].

6. Circuits

6.1. OR–AND Cells

Logic circuits are made using the logic gates described in Section 4. Josephson logic gates are usually operated in the "latching" manner, which makes it difficult to use inverter gates in arbitrary places in circuits, so that the dual rail

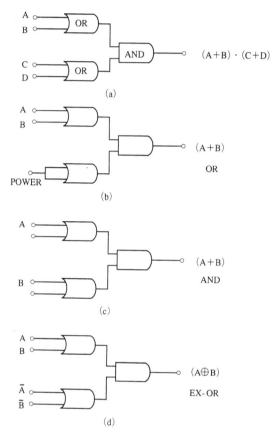

Fig. 19. The logic functions of the OR–AND unit cell. By using the dual rail system, the OR–AND cell can be operated as (b) OR, (c) AND, or (d) Exclusive-OR.

system is employed. In dual rail logic, complemental signals always accompany the true signals in logic operations.

Using the dual rail system, one can construct a unit cell composed of two OR gates and one AND gate as shown in Fig. 19. In this configuration, the OR gates also act as the input–output isolation gates since the AND gate usually has no isolation functions. Using this unit cell, OR, AND, and Exclusive-OR functions are easily obtained (Fig. 19b–d), by which all of the logic functions can be performed. Figure 20 shows a circuit configuration of the OR–AND unit cell based on the 4JL logic family [32].

Integrating these OR–AND unit cells, a number of logic circuits such as adder and multiplier circuits have been made. The logic circuits demonstrated so far are summarized in Table 2 [32–37].

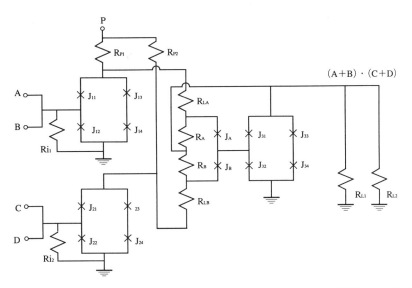

Fig. 20. The equivalent circuit of the OR-AND unit cell based on the 4JL logic family [32].

Table 2. Various Logic Circuits

Circuits	Gate family	Junction	Number of gates	Performance	Ref.
8 bit adder	4JL	Pb-alloy	300	add time 300 ps	[32]
8 bit adder	4JL	NbN/oxide/NbN	364	add time 700 ps	[33]
4 bit adder	RCJL	Pb-alloy	56	add time 172 ps	[34]
4 bit multiplier	4JL	NbN/oxide/NbN	652	mult. time 1 ns	[33]
4 bit multiplier	RCJL	Pb-alloy	249	mult. time 280 ps	[35]
4 bit multiplier	JTHL	Nb/Al oxide/Nb	104	mult. time 210 ps	[36]
16 bit multiplier	MVTL	Nb/Al oxide/Nb	828	mult. time 1.1 ns (8.7 ps/gate)	[37]

These circuits are useful to evaluate the performances of logic gates. Recent results with a 16 bit × 16 bit multiplier fabricated by modern integration technology based on Nb junctions showed that the Josephson gate can be operated with a sub-10 ps gate delay (8.7 ps) in an LSI circuit [37].

6.2. AC-Power and Sequential Logic

Usually once Josephson logic gates switch to the voltage state, they do not reset to the superconducting state when the input signal is turned off (latching

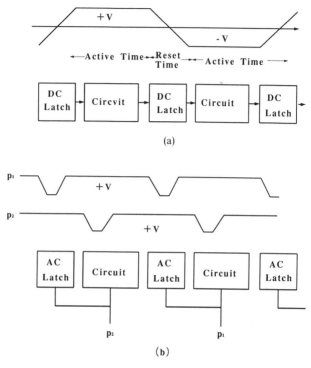

Fig. 21. Power systems for Josephson logic circuits: (a) single-phase bipolar power system and (b) two-phase unipolar power system.

operation). In order to reset the gate, it is necessary to turn off the power of the gate. This means that an ac-power supply is needed to operate Josephson logic circuits; this situation is thus quite different from semiconductor logic devices.

Two kinds of power supplies have been suggested: one is the single-phase bipolar power system [38] and the other is the multiphase (2- or 3-phase) unipolar power system. These power supply methods are schematically illustrated in Fig. 21.

In the single power system (Fig. 21a), logic operations are performed during both the positive and negative portions of the power current, and the gates are reset during the transition time from positive to negative polarities. When bipolar power is used, it is important to take into account the so-called punch-through phenomenon [39]. In latching gates, there is an increased probability of failure to reset the gates when the transition time becomes fast. This effect is known as "punch-through". In order to make the punch-through probability

3. COMPUTING

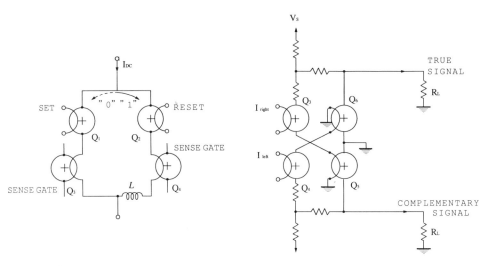

Fig. 22. DC-powered latches: the master latch (left) and the slave latch (right) [40].

low, the transition time has to be long enough. This puts a limitation on the clock frequency.

During the resetting time, all of gates in the circuit reset to the superconducting state, so that a dc latch is needed to store the calculated data in the preceding circuit. The dc latch is operated by another power supply providing dc power. Figure 22 shows a typical dc latch circuit [40]. The latch circuit consists of a master latch and a slave latch. The calculated data in the circuit are stored in a superconducting loop of the master latch. The information stored in the master latch is sensed by the slave latch at the rise time of the next cycle of power.

The multiphase power system is another choice for powering Josephson logic devices. Figure 21b shows a two-phase-unipolar-powered logic circuit [41]. As shown in the figure, two-phase power each cycle overlapping the other are fed into the logic circuits alternatively. In this case, the dc-powered latch is not necessary to store data calculated in the preceding circuit since the data can be read out to the next circuit operated by the second phase power cycle before the first phase power cycle is turned off. A latch circuit for this power system can be composed of latching gates. Figure 23 shows an example of a latch circuit based on the 4JL logic family with a two-phase power system [41]. One of the advantages of the two-phase power system is that a timed inverter can easily be constructed, as shown in Fig. 23. The two-phase power

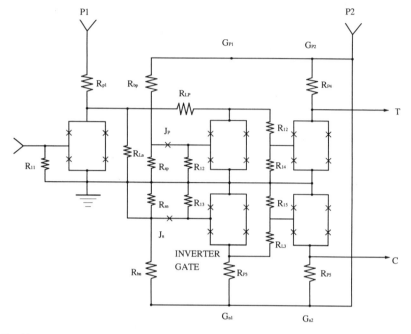

Fig. 23. The direct coupled latch circuit based on the 4JL logic family. The latch is operated by a two-phase power system, so that a timed inverter can easily be made [41].

operation makes it possible to take enough time for resetting the circuits, resulting in a lower punch-through probability without degrading the effective calculation speed. This is another advantage of the two-phase power system.

Taking advantage of the two-phase power system and using latch circuits, a number of sequential logic circuits such as shift resisters [42], counters [43], and multiplexers have been fabricated and successfully operated. These circuits are important components for constructing computer CPUs.

6.3. Memory Circuits

In order to make a fully decoded memory chip, a well-defined integration technology that has the capability to fabricate circuits with a LSI complexity is needed. The memory circuit consists of a memory cell array and peripheral logic circuits, including address latches, X, Y-decoders, and a timing control circuit. A typical memory circuit arrangement is illustrated in Fig. 24 [44].

3. COMPUTING

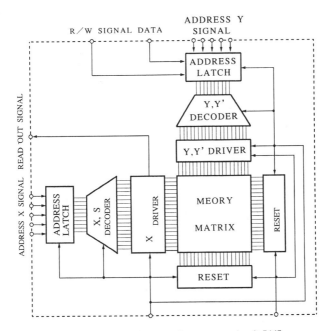

Fig. 24. A block diagram of a memory circuit [44].

The address latches generate true and complemental signals of address signals. A cell to write and read information is accessed by two crossing lines selected by the X and the Y decoders. The decoder is composed of several stages of AND gates by which one of lines is selected.

Many efforts have been made to achieve fully decoded memory circuits. Using Pb-alloy integration technologies, Yamamoto et al. [45] have fabricated a 1 Kbit RAM (Random Access Memory) with an NDRO cell array built using a 5 μm Pb-alloy technology [45]. The access time of the memory was measured to be 3.3 ns. A 2.5 μm Pb-alloy technology has been adopted to design a 4 Kbit NDRO RAM chip [46]. A 1 Kbit array cross-section in the 1 K × 4 bit design was operated. From the measurements of individual circuit components, the access time of a 4 Kbit RAM chip was estimated to be about 800 ps. A 1 Kbit arrangement based on the variable threshold memory cell (see Section 5.3) has also been integrated using a 5 μm Pb-alloy technology [29]. This memory circuit is operated by words rather than bits, resulting in a simplified peripheral circuit structure. The access time was 1.1 ns.

Although many efforts have been made aimed to obtaining full memory operation, only several bits were accessed in all of these memories based on the

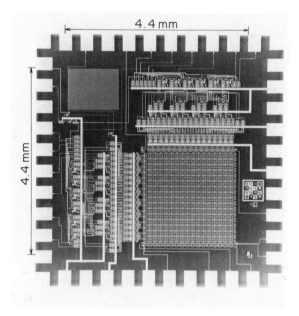

Fig. 25. A photograph of the 1 Kbit NDRO memory circuit fabricated with a 3.3 μm Nb/Al oxide/Nb junction technology [Courtesy NEC Co.].

Pb-alloy technology. This is due to the nonuniformity of circuit parameters that is inherent in the Pb-alloy technology. Recently developed Nb-based integration technology with increased uniformity and stability is expected to improve the circuit performance of memory LSI chips. Recently, a 1 Kbit NDRO cell arrangement has been built using a 3.3 μm Nb-based technology [47]. The access time was measured to be 570 ps at the power dissipation of 13 mW. 40% of the cells in the 1 Kbit RAM were successfully operated, indicating the Nb technology is much superior to the Pb-alloy technology. Figure 25 shows a photograph of the 1 Kbit RAM chip based on the Nb technology.

6.4. Small-Scale Computer Circuits—Microprocessors

In order to evaluate the feasibilities of Josephson computer systems, it is important to operate processors in which circuit components for computer systems are functionally arranged.

The first attempt to operate this kind of circuit was made by Mukherjee of IBM [48]. He made a simple signal processing circuit including 12 latches and

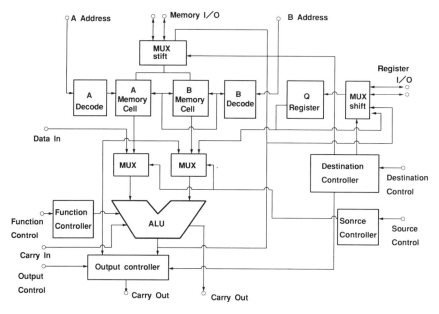

Fig. 26. The block diagram of the 4 bit Josephson microprocessor [50].

16 AND gates based on the CIL logic family. This circuit processes data by 10 bit key codes and has been operated with a cycle time of 665 ps.

As integration technology based on Nb junctions has progressed, efforts have been put into research on building more complicated sequential circuits.

Hatano et al. [49] have operated a small-scale data processor model in which a decoded 64 bit ROM (Read Only Memory), ALU (Exclusive-OR operation only), and registers (Flip–Flop) are integrated on the same chip using 2.5 μm Nb technology [49]. A critical path operation was performed: data were taken from a resister, passed through ROM and ALU, and finally stored in a resister. The delay of the critical path was 650 ps.

More recently, a 4 bit microprocessor, which is representative of a realistic computer model, was operated by Kotani et al. [50] of Fujitsu. A circuit consisting of a 64 bit NDRO RAM, an ALU, and control circuits was built on a chip integrating 1841 gates using 2.5 μm Nb technology. The circuit block diagram is shown in Fig. 26. A critical path operation, a carry signal processed in the ALU transferred to the RAM, was performed with a maximum clock frequency of 770 MHz, which is 10 times faster than the same version of a GaAs processor. Figure 27 shows a photograph of this microcomputer chip.

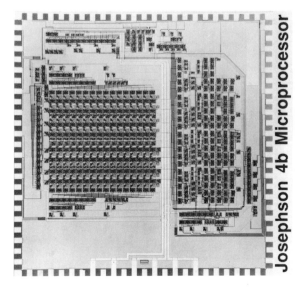

Fig. 27. A photograph of the 4 bit Josephson microprocessor chip [Courtesy Fujitsu Ltd.].

References

1. Josephson, B.D. *Phys. Lett.* **1**, 251 (1962).
2. Buck, D.A. *Proc. IRE* **44**, 482 (1956).
3. Matisoo, J. *Appl. Phys. Lett.* **9**, 167 (1966).
4. Harris, E.P. *IEEE Trans. Magn.* **15**, 562 (1979).
5. Klein, M., and Herrel, D.J. *IEEE J. Solid-State Circuits* **13**, 593 (1978).
6. Zappe, H.H., and Landman, B.S. *J. Appl. Phys.* **49**, 344 (1978).
7. Gheewala, T.R. *IBM J. Res. Dev.* **24**, 130 (1980).
8. Fuiton, T.A., Pei, S.S. and Dunbleberger, L.N. *Appl. Phys. Lett.* **34**, 1876 (1979).
9. Gheewala, T.R., and Mukherjee, A. *In* "Tech. Digest International Electron Device Meeting (IEDM)," p. 482, 1979.
10. Sone, J., Yoshida, T., and Abe, H. *Appl. Phys. Lett.* **40**, 886 (1982).
11. Hohkewa, K., Okada, M., and Ishida, A. *Appl. Phys. Lett.* **39**, 653 (1981).
12. Takada, S., Kosada, S., and Hayakawa, H. *In* "Proc. 11th Conf. International Solid State Devices, Tokyo, 1979." *Japan. J. Appl. Phys.* Suppl. **19-1**, 607 (1981).
13. Hayakawa, H. *Japan. J. Appl. Phys.* Suppl. **22**, 447 (1983).
14. Fujimaki, N., Kotani, S., Hasuo, S., and Yamaoka, T. *Japan. J. Appl. Phys.* **24**, L1 (1985).
15. Beha, H., and Jackel, H. *IEEE Trans. Magn.* **17**, 3423 (1981).
16. Hasuo, S., Suzuki, H., and Yamaoka, T. *IEEE Trans. Magn.* **17**, 583 (1981).
17. Gheewala, T.R. *IEEE J. Solid-State Circuits* **14**, 787 (1979).
18. Pei, S.S. *Appl. Phys. Lett.* **40**, 739 (1982).

19. Sone, J., Yoshida, T., Tahara, S., and Abe, H. *Appl. Phys. Lett.* **41**, 886 (1982).
20. Nakano, J., Mimura, Y., Nagata, K., Hasumi, Y., and Waho, T. *In* "Extended Abtracts of 16th Conf. Solid-State Devices and Materials," p. 636. Kobe, 1984.
21. Nakagawa, H., Odake, T., Sogawa, E., Takada, S., and Hayakawa, H. *Japan. J. Appl. Phys.* **22**, L297 (1983).
22. Hatano, Y., Nishino, T., Tarutani, Y., and Kawabe, U. *Appl. Phys. Lett.* **44**, 1095 (1984).
23. Kotani, S., Imamura, T., and Hasuo, H. *In* "IEEE IEDM Technical Digest," p. 865, 1987.
24. Anacker, W. *IEEE Trans. Magn.* **5**, 968 (1969).
25. Zappe, H.H. *IEEE J. Solid-State Circuits* **10**, 12 (1975).
26. Faris, S.M., Henkels, W.H., Valsamakis, E.A., and Zappe, H.H. *IBM. J. Res. Dev.* **24**, 146 (1980).
27. Zappe, H.H. *Appl. Phys. Lett.* **25**, 424 (1974).
28. Kurosawa, I., Yagi, A., Nakagawa, H., and Hayakawa, H. *Appl. Phys. Lett.* **43**, 1067 (1983).
29. Kurosawa, I., Nakagawa, H., Yagi, A., Takada, S., and Hayakawa, H. *In* "Extended Abstracts of 16th Conf. Solid-State Devices and Materials," p. 619. Kobe, 1984.
30. Miyahara, K., Mukaida, M., and Hohkawa, K. *Appl. Phys. Lett.* **47**, 754 (1985).
31. Miyahara, K., Mukaida, M., Tokumitsu, M., Kubo, S., and Hohkawa, K. *IEEE Trans. Magn.* **23** (1987).
32. Nakagawa, H., Ohigashi, H., Kurosawa, I., Sogawa, E., Takada, S., and Hayakawa, H. *In* "Extended Abstracts of 15th Conf. on Solid-State Devices and Materials" p. 137. Tokyo, 1983.
33. Kosaka, S., Shoji, A., Aoyagi, M., Shinoki, F., Tahara, S., Ohigashi, H., Nakagawa, H., Takada, S., and Hayakawa, H. *IEEE Trans. Magn.* **21**, 102 (1985).
34. Sone, J., Yoshida, T., Tahara, S., and Abe, H. *In* "Technical Digest Int. Electron Device Meeting (IEDM)," p. 765, 1982.
35. Sone, J., Tsai, J.S., and Abe, H. *IEEE Solid-State Circuits* **20**, 1056 (1985).
36. Hatano, H., Harada, Y., Yamashita, K., and Kawabe, U. *In* "ISSCC Digest of Tech. Papers," p. 196, 1986.
37. Kotani, S., Fujimaki, N., Morohashi, S., Ohara, S., and Hasuo, S. *IEEE J. Solid-State Circuits* **22**, 98 (1987).
38. Jones, H.C., and Gheewala, T.R. *IEEE J. Solid-State Circuits* **17**, 1201 (1982).
39. Jewett, R.E., and Van Duzer, T. *IEEE Trans. Magn.* **17**, 599 (1981).
40. Davidson, A. *IEEE J. Solid-State Circuits* **13**, 583 (1978).
41. Nakagawa, H., Kurosawa, I., Takada, S., and Hayakawa, H. *In* "Extended Abstracts of 17th Conf. Solid State Devices and Materials," p. 123. Tokyo, 1985.
42. Fujimaki, N., Kotani, S., Imamura, T., and Hasuo, S. *IEEE J. Solid-State Circuits* **22**, 886 (1987).
43. Nakagawa, H., Kurosawa, I., Takada, S., and Hayakawa, H. *IEEE Trans. Magn.* **23**, 739 (1987).
44. Wada, Y., Hidaka, M., Nakagawa, S., and Ishida, I. *IEEE J. Solid-State Circuits* **22**, 892 (1987).

45. Yamamoto, M., Yamauchi, Y., Miyahara, K., Kuroda, K., Yanagawa, F., and Ishida, A. *IEEE Electron. Device Lett.* **4**, 150 (1983).
46. Henkels, W.H., Brown, K.H., Rajeerkumar, T.V., Geppert, L., Allan, J.W., Lee, Y.H., and Yeh, J.T. *In* "Proc. Inf. Conf. Computer Design," p. 580. New York, 1983.
47. Wada, Y., Nagasawa, S., Ishida, I., Hidaka, M., Tsuge, H., and Tahara, S. *In* "ISSCC Digest of Technical Papers," p. 84, 1988.
48. Mukherjee, A. *IEEE Electron Device Lett.* **3**, 29 (1982).
49. Hatano, Y., Yano, S., Hirano, M., Tarutani, Y., and Kawabe, U. *In* "Extended Abstracts of 1987 Int. Superconductivity Electronics Conf.," p. 239. Tokyo, 1987.
50. Kotani, S., Fujimaki, N., Imamura, T., and Hasuo, S. *In* "IEEE ISSCC Digest of Technical Papers," p. 150, 1988.

CHAPTER 4

Josephson Arrays as High Frequency Sources

JAMES LUKENS
Department of Physics
State University of New York
Stony Brook, New York

1. Introduction 135
2. Single-Junction Sources 137
   2.1. Small Junctions 137
   2.2. Fluxon Oscillators (Wide Junctions) 140
   2.3. Radiation Linewidth 141
3. Arrays 142
4. Phase-Locking 146
5. Phase-Locking in Arrays 149
   5.1. Locking Strength within the RSJ Model 149
   5.2. Radiation Linewidth of Arrays 153
   5.3. Effects of Capacitance on Phase-Locking in Arrays 156
6. Distributed Arrays 159
7. Prospects for the Future 163
   Acknowledgments 165
   References 165

1. Introduction

There is a rapidly developing need for compact submillimeter sources for use in such applications as satellite communications and receivers for astronomical observation. Fundamental solid-state oscillators such as Gunn or IMPATT diodes are presently limited to the millimeter wave range, leaving such bulky and power hungry sources as carcinotrons and CO_2 lasers, which operate in some parts of the submillimeter range.

Josephson junctions are natural voltage-controlled oscillators, and it has been recognized since the discovery of the Josephson effect that these junctions have the potential for filling this source gap at least up to frequencies of several terahertz. As the Josephson equations imply, the average frequency of supercurrent oscillation v_0, which we call the Josephson frequency, in a junction is related to the dc component $\bar{V}$ of the voltage across the junction only through fundamental constants,

$$\bar{V} = \frac{h}{2e} v_0, \qquad (1.1)$$

where e is the charge on the electron and h is Planck's constant, so $2e/h = 483$ GHz/mV. Indeed this remarkable result has been shown [1,2] to be independent of the materials or structure of the junction to better than 1 part in 10^{16}. One would thus hope that Josephson junctions would be useful tunable sources operating up to the superconducting gap frequency—a few THz for conventional superconductors and perhaps tens of THz for the new high-T_c materials.

Unless the junction voltage is constant in time, i.e., $V(t) = \bar{V}$, Eq. (1.1) does not necessarily imply a pure sinusoidal oscillation with frequency proportional to the dc voltage. In practice, at high frequencies, it is usually easier to control the junction's bias current than its voltage. Indeed, there are many examples in the literature, such as chaotic behavior or periodic mode-locking to resonant structures, where, even though Eq. (1.1) is satisfied, almost none of the power generated by the junction is at the frequency v_0. Although a great deal of work in many laboratories has gone into the development of practical sources based on the Josephson effect, the serious problems related to the very low power and source impedance of individual junctions in addition to that of obtaining spectrally pure oscillation at the Josephson frequency have been difficult to surmount. One technique for overcoming these problems, which will be the focus of this chapter, is to use arrays of junctions in place of single junction sources. The discussion presented here has evolved from a lecture at the Nato Advanced Study Institute on Superconducting Electronics [3].

First, the properties of single junction sources will be reviewed in order to get a perspective on the problems to be solved. After a brief look at the performance expected from idealized arrays, perturbative techniques for analyzing arrays will be developed based on the resistively shunted junction (RSJ) model. While this model is clearly only approximately correct for real tunnel junctions, it can produce analytic results useful for achieving an insight into the design of coherent arrays. These results have proven to provide a very good description of many of the experiments to date. The final parts of this chapter will present a number of recent results for practical array sources where the junctions are distributed over many wavelengths as well as speculations on future directions for research.

2. Single-Junction Sources

2.1. Small Junctions

The properties of single-junction sources will be described first in order to see where arrays may be useful. The RSJ model (Fig. 1a) with $C = 0$ will be used to describe the junction's behavior since, for this case, analytic solutions exist which provide a useful insight. See, for example, [4], Chapter 4. When the bias current I_b is increased above I_c, the junction's phase ϕ begins to increase with time, producing an oscillating supercurrent (since $I_s = I_c \sin \phi$) and consequently an oscillating voltage. For I_b near I_c, the voltage waveform is nearly a spike, having a large harmonic content. As I_c is increased the higher harmonic content of the waveform decreases, giving a nearly sinusoidal wave for $\bar{V} \geq V_c \equiv I_c R_J$. The amplitudes of the harmonics are given by

$$\tilde{V}_n \equiv V_c \tilde{v}_n = V_c \frac{2\bar{v}}{[(1 + \bar{v}^2)^{1/2} + \bar{v}]^n}, \qquad (2.1)$$

where $\bar{v} \equiv \bar{V}/V_c$ and $i \equiv I_b/I_c$ and for the RSJ model $\bar{V} = (i^2 - 1)^{1/2}$. Also, $\omega_c \equiv 2\pi V_c/\Phi_0$. This waveform approaches a nearly pure sinusoid for $\bar{v} > 1$. Since for most applications one would like a reasonably sinusoidal source, we will impose the constraint that $\bar{v} \geq 1$ at the desired operating frequency.

Since the junction impedance at the Josephson frequency is $Z \approx R_J$ for $\bar{v} \geq 1$, at high frequencies the junction can be viewed as an oscillator of

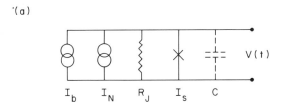

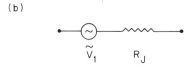

Fig. 1. (a) Equivalent circuit model for resistively shunted junctions (RSJ) with bias current I_b, noise current I_N, shunt resistance R_J, capacitance C, and supercurrent $I_s = I_c \sin \phi$. (b) Equivalent circuit for frequencies near v_0.

amplitude $\tilde{V}_1$ at the Josephson frequency in series with R_J as shown in Fig. 1b. The power available to a matched load from a single junction is then

$$P_1(v) = \frac{1}{8}\frac{\tilde{V}_1^2}{R_J} \to \frac{1}{8}I_c^2 R_J, \quad \bar{v} \geq 1. \tag{2.2}$$

There are limits on both I_c and R_J that limit the maximum power obtainable at a given frequency. The discussion so far has assumed that the junction is one dimensional; that is, the phase difference between the electrodes is independent of the position on the electrode transverse to the direction of tunneling current flow. This, in general, means that the dimensions of the junction must be limited to assure a constant current density. Since the natural scale over which current density varies in the junction is the Josephson penetration depth $\lambda_J^2 = \hbar/(2e\mu_0 dJ_c)$ (where d is the magnetic thickness of the barrier), keeping the junction's dimensions of order λ_J will assure the required uniformity. For an in-line junction with critical current density J_c (shown in Fig. 2) over a ground plane, the current is confined (on the x-axis) within about $2\lambda_J$ of the end of the junction giving a maximum effective critical current in zero field of $I_c \approx 2\lambda_J w J_c$, where w is the width of the junction [5]. As long as the current feeding the junction is uniform across the width of the junction (y-axis), a solution exists in which the current density within the junction is also constant across its width for arbitrarily wide junctions.

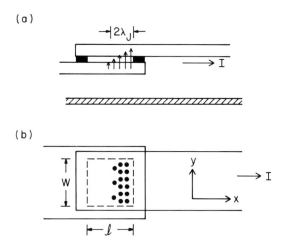

Fig. 2. Diagram of an in-line junction. (a) Edge view showing $J_c(x)$ concentrated within $2\lambda_J$ of the end of the junction. (b) Top view. Tunnel barrier is enclosed by dashed lines. Regions of non-zero tunneling current are shown by ●.

4. JOSEPHSON ARRAYS AS HIGH FREQUENCY SOURCES

At some point instabilities will develop in this uniform solution and limit the junction's width in practice. A very conservative estimate for this maximum stable width would be $2\lambda_J$. Then, since $\lambda_J^2 \propto 1/J_c$, the maximum critical current for a given material would be independent of J_c and equal to about 4 mA for niobium. At the other extreme, very wide in-line junctions are known to have flux flow across the junction. One might expect this sort of instability to develop when the electromagnetic wavelength λ_{em} in the junction was equal to twice the junction width, that is, at the first zero field step, since for this condition there will be where strong coupling between the Josephson and fluxon oscillations. Indeed, it has been observed [6] that the phase-locking of junctions in voltage standard arrays to external radiation decreases abruptly for wider junctions. If the first instability is the first zero field step, then the maximum junction width will be given by

$$w_{em} = \frac{1}{2v(\mu_0 dC_s)^{1/2}}, \qquad (2.3)$$

where C_s is the specific capacitance. This width is independent of J_c but varies inversely with frequency. Thus, the maximum critical current for this condition would depend on the critical current density but would decrease with frequency. For example, a niobium junction with a critical current density of 10^5 A/cm^2 would have a maximum stable critical current of $I_c \approx 7.6$ mA/v [THz].

Having fixed I_c, R_J should be adjusted using a shunt resistor depending on the desired operating frequency and limited, of course, by the material-dependent intrinsic $I_c R_J$ product of the junction. The condition $\bar{v} \geq 1$ for a sinusoidal waveform limits the maximum resistance for a given I_c and frequency v. On the other hand, if R_J is reduced below the value for which $\bar{v} = 1$, Eq. (2.2) indicates that the power will decrease. Thus, one should select the shunt resistance such that $R_J \approx v\Phi_0/I_c$ giving

$$P_1(v) \approx \frac{1}{8} v \Phi_0 I_c. \qquad (2.4)$$

Figure 3 illustrates the source resistance, critical current, and available power as a function of frequency for single-junction sources subject to the width constraints discussed above. For junctions with $w = 2\lambda_J$, one sees that the maximum single junction power is proportional to frequency with

$$P_1(v) \approx 1(v \text{ [THz]}) \, \mu W \qquad (2.5)$$

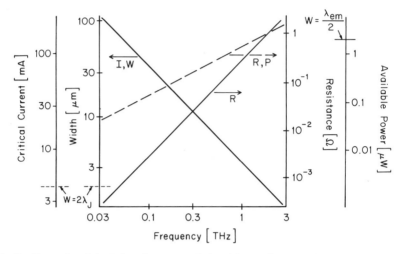

Fig. 3. Properties of single-junction sources designed for maximum available power at a given frequency subject to the width constraints $w = 2\lambda_J$ (--- narrow) and $w = 0.5\ \lambda_{em}$ (—— wide); for both cases, $J_c = 10^5$ A/cm^2 and length is $2\lambda_J = 2\ \mu$m. For narrow junctions, the frequency-independent values of I_c and w are indicated on the axes ($\cdots$). The available power of the wide junction is also frequency independent as shown on the power axis.

Thus, for example, one might expect a maximum power of about 1 μW at 1 THz. While this is sufficient for some applications, it is about the limit with proportionately less power available at lower frequencies. Also the source impedance for this 1 μW source would be less than an ohm. This would require a substantial transformer ratio for typical loads and could be a problem if wide tuning were desired. Junctions with $w = \lambda_{em}/2$ have the potential for delivering above a microwatt at lower frequencies but at the expense of even lower source impedances.

2.2. Fluxon Oscillators (Wide Junctions)

There are several groups that have reported significant power levels at frequencies up to several hundred gigahertz from fluxon propagation in very wide junctions. While the focus of this chapter will be on series arrays of junctions small enough to have a spacially uniform current density, we will briefly review these results for fluxon oscillators for the perspective that they provide on the possible advantages and disadvantages of this type of source.

Significant power has been observed from two different types of fluxon oscillators. The first, which uses unidirectional flow, has been studied mainly in Japan [7,8]. In this type of oscillator, a magnetic field is applied in the plane

4. JOSEPHSON ARRAYS AS HIGH FREQUENCY SOURCES 141

of the junction, along the x-axis (Fig. 2) to form flux vortices or fluxons in the junction. The bias current then forces the vortices to flow across the junction (along the y-axis) along the transmission line formed by the base and counter electrodes. If the bias current is increased until the vortices are moving near the speed of light in the junction, voltage and thus the frequency become insensitive to small changes in the bias, i.e., the junction has very low differential resistance. The frequency can be tuned over a wide range by varying the vortex density in the junction by changing the applied field. It is estimated [8] that of the order of a microwatt of power should be available from such oscillators at frequencies ranging from about 100 GHz to 1 THz, and powers of this magnitude coupled to an SIS detector at the end of the junction transmission line have been reported up to 400 GHz. A problem associated with this type of oscillator is that the characteristic impedance, which is that of the transmission line formed by the junction, is in general quite low. Transformers have been developed [7] to couple power more efficiently to higher impedance loads and have succeeded in coupling about 0.1 μW to a 1 Ω load at 200 GHz. The penalty for using these resonant transformers, however, is that the tuning range is severely restricted.

The second type of fluxon oscillator, which has been studied primarily by groups in Europe, has a similar geometry but uses a high Q resonant junction. These resonant fluxon oscillators are operated on their zero field steps, the operating frequency being determined by the dimensions of the junction rather than an applied field. Two groups have recently reported detecting power levels of the order of 0.1 μW from such junctions. One group [9] reported radiation at 75 GHz using on-chip detection with small junctions. The oscillator junctions had a transmission line impedance of 1 Ω and an estimated available power of about 0.1 μW. Simulations show a steplike structure in junction's phase vs. time indicating a large harmonic content to the radiation as one might expect from the picture of fluxons shuttling back and forth in the junction. A second group [10,11] reported radiation near 10 GHz from an array of resonant fluxon oscillators in which a significant degree of phase-locking among the junctions was observed. In this case, the radiation was coupled to a detector outside of the cryostat. Power levels of over 0.1 μW in a 50 Ω load were reported. In smaller arrays, where more complete locking could be achieved, linewidths of several kilohertz were observed.

2.3. Radiation Linewidth

A final consideration related to single junction sources is their linewidth. The linewidth of the Josephson radiation is determined by frequency modulation due to low frequency voltage noise across the junction [12] with frequencies up to about the linewidth Δv being important. In terms of the current noise in the junction,

$$\Delta v = \frac{1}{2}\left(\frac{2\pi}{\Phi_0}\right)^2 S_I(0) R_d^2. \tag{2.6}$$

Here $S_I(0)$ is the low frequency current spectral density, and R_d is the differential resistance at the operating voltage. If $S_I(0)$ is just the Johnson noise current of the junction resistance R_J and $R_d \approx R_J$, then $\Delta v \approx 160$ MHz per ohm of junction resistance at 4 K. This is only a rough guide, since $S_I(0)$ will in general be increased due to such things as $1/f$ noise prevalent in high J_c junctions, as well as to down-converted quantum noise from near the Josephson frequency. On the positive side, since only low frequency noise is important in determining Δv, one can in principle make the linewidth arbitrarily small by shunting the junction at low frequencies without reducing its high frequency impedance. This technique has been successfully used [13], although it can have drawbacks such as the introduction of instabilities or chaotic behavior.

Another technique for linewidth reduction is to reduce the differential resistance of the junction at the operating point through coupling to a resonant structure. An example of this is the fluxon oscillations in resonant wide junctions discussed above. These junctions can have very low differential resistance when biased on their zero field steps. The situation with unidirectional flux-flow is more complex. While the differential resistance is also small at the operating point, the junctions are tunable by varying the flux, so the linewidth will to some extent be affected by fluctuations of the flux linking the junctions, perhaps more than by spacially uniform fluctuations in the bias current.

One can summarize the properties of single junction sources by saying that, in general, such sources have either too little power, too low an impedance, too broad a linewidth, or all of the above, although the power and impedance begin to become useful for some applications as terahertz frequencies are approached. Next we will take a brief, rather elementary look at small junction arrays to see to what extent the replacement of single junctions by arrays of junctions might solve these problems.

3. Arrays

Interest in Josephson arrays was sparked in the late 1960s by experiments of Clark [14] and a paper by Tilley [15] who predicted superradiance in such arrays, much as in a collection of atoms in a cavity. One signature of this

4. JOSEPHSON ARRAYS AS HIGH FREQUENCY SOURCES 143

superradiance was a prediction that the output power would scale as the square of the number of junctions. This led to a hope—rather naive in retrospect—that significant power levels could be obtained from Josephson junctions simply by connecting a large number of junctions together without worrying in detail about just how they were coupled. The initial experiments were done by Clark [14,16] on two-dimensional arrays of superconducting balls, which were Josephson-coupled through their oxide coating as shown in Fig. 4b. Indeed, evidence of interactions among the junctions was seen; however, experiments of this type have never produced significant levels of power. As will be seen later, when the details of how junctions phase-lock is discussed, for arrays of this type power is mostly dissipated in the array itself. The development of Josephson-effect arrays, involving hundreds of workers, has been covered in detail in two review papers by Jain et al. [17] and Lindelof and Hansen [18] and more recently in a book by Likharev [4]. Readers are referred to these sources for a comprehensive review.

In more recent work, including successful attempts to obtain increased power from arrays of junctions, the junctions are simply treated as classical oscillators, i.e., the junction's current and phase are taken to be classical variables as in the discussion of single-junction sources above. This is the approach taken throughout this chapter. It is worth emphasizing the distinction between the present work based on the classical picture and the initial discussion of Josephson arrays in terms of superradiance, since much confusion has been caused over the years by not fully appreciating this distinction. This confusion has been compounded by the fact that there have been observations of the power from arrays increasing as N^2, as predicted for superradiance. As far as we know, all these observations can be explained in terms of purely classical circuit analysis, as will be seen, for example, below.

Probably the simplest example of the advantages to be gained from arrays of junctions can be seen by considering the one-dimensional array shown in Fig. 4a. Here a number of junctions are connected, e.g., by a transmission line, in series with each other and with the load R_L to be driven, so that the rf current generated by a junction flows through all of the other junctions and the load. Representing each junction by its high frequency equivalent circuit (Fig. 1b), which to first order is not affected by the rf current, one sees that the array impedance can be matched to the load by taking the number of junctions to be $N = R_L/R_J$, thus solving the low impedance problem of single junction sources. For now, it is simply assumed that all of the junctions will oscillate at the same frequency and in-phase. The discussion of the conditions needed for such phase-locking will occupy much of the remaining sections of this chapter. It will also be assumed for now that the circuit dimensions are

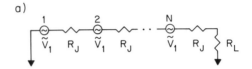

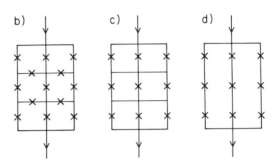

Fig. 4. (a) High frequency equivalent circuit for a one-dimensional array. (b)–(d) Possible junction connections for two-dimensional arrays.

small compared to the wavelength in the transmission line at the frequency of operation so that the lumped circuit approximation can be used. The constraints on I_c and R_J discussed in Section 2 can then be used to estimate the number of junctions needed to match a 50 Ω load to an array designed to operate near a frequency v, giving

$$N = \frac{R_L}{R_J} \approx \begin{cases} \dfrac{100}{v\,[\mathrm{THz}]}, & w = 2\lambda_J, \\ \dfrac{200}{v\,[\mathrm{THz}]^2}, & w = \dfrac{\lambda_{em}}{2}. \end{cases} \quad (3.1)$$

Here the two estimates for N correspond to the limits on the width of the small junctions as discussed in Section 2, taking $J_c = 10^5$ A/cm^2.

The available power from this one-dimensional array is just N times that available from a single junction in the array. Which for the 50 Ω load is

$$P_N = NP_1 \approx \frac{1}{8} I_c^2 R_L \approx \begin{cases} 0.1\ \mathrm{mW}, & w = 2\lambda_J, \\ \dfrac{0.4\ \mathrm{mW}}{v\,[\mathrm{THz}]^2}, & w = \dfrac{\lambda_{em}}{2}. \end{cases} \quad (3.2)$$

If it is possible to match an array of any size to a load using a transformer, then the dependence of power on array size is linear in the number of junctions. The

4. JOSEPHSON ARRAYS AS HIGH FREQUENCY SOURCES

enormous impedance mismatch between a single junction and a typical load makes the use of transformers problematical, especially if much tuning range is to be preserved. Without transformers, situations occur where, when the number of junctions is varied, the load power varies as N^2. For example, this happens when the number of junctions is increased but the total array impedance is small compared to that of the load. An N^2 dependence also occurs if an array of junctions with $I_c < I_{cmax}$ is matched to the load by reducing R_J as N increases such that NR_J and $I_c R_J$ remain constant. Thus, for many practical situations, the power from arrays is expected to increase as N^2. These situations are clearly purely classical in nature but are sometimes confused with superradiance.

Next, one could ask whether anything is gained by replacing the one-dimensional array by a two-dimensional array, as in Figs. 4b–d. We imagine making such an array by replacing each junction in Fig. 4a by a parallel (transverse to the rf current) string of junctions, M junctions wide as in Fig. 4c. If all of these junctions were identical and all oscillated in phase, this would be equivalent to replacing each junction in Fig. 4a with a junction having $I'_c = MI_c$ and $R'_J = R_J/M$. One would then need M times as many of these series junctions to match the load; thus, the power delivered to the load would be increased by a factor M^2.

As an example, at 1 THz approximately 100 junctions with $w = 2\lambda_J$ would be required in a one-dimensional array to match a 50 Ω load, producing a power of 0.1 mW. If a two-dimensional array were used with a width $M = 100$, then the matched array would have 10^6 junctions and deliver 1 W of power. It is not difficult to fabricate a million-junction array with modern lithographic techniques. The real question is whether all of the junctions could be made to oscillate in phase as assumed above, particularly since the motivation for thinking about a two-dimensional array is that the useful critical current of a single junction is limited due to phase instabilities that arise at larger values of I_c. Similar power estimates are obtained for the configuration shown in Fig. 4d.

The arrays discussed above (Figs. 4a, c, d) could all be called linear arrays since for proper operation the phase should vary only in one direction, even in the two-dimensional arrays. It is important to distinguish this situation from truly two-dimensional arrays (Fig. 4b) where the phase varies in both dimensions. There has been a great deal of very interesting work, primarily to study phase transitions, in these latter arrays. This work will be completely ignored here since it really does not address the problems related to using Josephson arrays as radiation sources.

4. Phase-Locking

It should be clear from the brief discussion above that the real key to the usefulness of arrays is how, or if, the junctions phase-lock. Even if all of the junctions are identical, one must still ask if the "uniform phase" condition (in which all junctions have the same phase relative to the locking current) is a solution, and if so, is it a stable solution. If there is such a stable solution, the next problem is to find out what happens if all of the junctions are not identical. In real arrays, there is always some degree of scatter in the junction parameters, e.g., the critical current, as well as random noise, which tend to make the junctions of the array oscillate at different frequencies.

It is possible to get much insight into both the stability and strength of phase-locking in arrays by considering the well-known phenomenon of a single-junction phase-locking to external radiation. We still start by using perturbation theory to study the effects of external radiation on an RSJ for which analytic solutions are available. Later, the effects of junction's capacitance in the low β_c ($\beta_c \equiv \omega_c R_J C \leq 1$) limit will be included. The perturbation techniques used are standard and have been applied to Josephson junctions by several authors [19,20]. Here the key ideas of the theory will be reviewed briefly and then applied to the phase-locking problem. To begin, a quantity related to the junction phase ϕ, called the "linearized phase", is defined by

$$\Theta = \hat{\omega} t, \qquad (4.1)$$

where $\hat{\omega}$ is the junction's frequency averaged over a time long compared to a period of a Josephson oscillation, yet short enough to respond to the low frequency noise and modulation. The $\hat{}$ symbol is used in general to indicate averaging over the time scale which is long compared to $1/\omega$. The success of the perturbation theory depends on the wide separation of the Josephson frequency from the low frequency currents that are important in fixing the linewidth and oscillation frequency.

The essential results of the perturbation theory are shown schematically in Fig. 5 where the junction is represented by an equivalent circuit with two parts, one for the high frequency (HF) (near ω) behavior, and the second modeling the low frequency (LF) response. The high frequency circuit consists of the Josephson oscillator (with amplitude $\tilde{V}_1$ given by Eq. (2.1) and frequency ω) and the source impedance R_J. This HF section is coupled to the LF section through ω, which is determined by the LF voltage through the Josephson equation (1.1).

4. JOSEPHSON ARRAYS AS HIGH FREQUENCY SOURCES

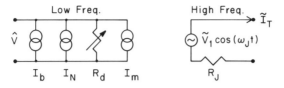

Fig. 5. Equivalent circuit for the RSJ model from perturbation theory. Low frequency (LF) section models response for $\omega \ll \omega_J$ ($\omega_J = 2\pi \hat{V}/\Phi_0$) and is coupled to the high frequency (HF) section by the mixing current I_m—see text. The HF section contains the Josephson oscillator $\tilde{V}_1 \cos \omega_J t$ and source impedance R_J. The perturbation consists of an rf current $\tilde{I}_T$ with $\omega \sim \omega_J$ flowing through the HF terminals.

The perturbations that we wish to consider are caused by an rf current $\tilde{I}_T$ with a frequency near ω flowing through the HF terminals. This in turn affects the LF voltage (and thus ω) through the presence of a "mixing current",

$$I_m = \alpha \widehat{(2\tilde{I}_T \cos \Theta)}, \tag{4.2}$$

in parallel with the bias and noise currents on the LF side. Here α is the conversion coefficient, which is given in the RSJ model as

$$\alpha = \frac{1}{2(1 + \bar{v}^2)^{1/2}}. \tag{4.3}$$

The cause of this perturbation might, for example, be either an external rf current source or a load placed across the HF terminals, or both. So

$$\hat{\omega} = \hat{\omega}_u(\hat{I} + I_m), \tag{4.4}$$

where the subscript u refers to the value of the variable ω in the absence of the HF perturbation. In other words, in the presence of a HF perturbation, the junction will oscillate at the same frequency as an unperturbed junction biased with a current equal to the sum of the bias and mixing currents in the perturbed junction.

In order to apply this technique to understand the phase-locking of a junction to external radiation, we take the perturbing rf current to be that due to an external current source with amplitude I_e and frequency ω_e near ω, so

$$\tilde{I}_T = I_e \cos \omega_e t. \tag{4.5}$$

This gives a mixing current

$$I_m = \alpha I_e \cos \delta\Theta, \tag{4.6}$$

where $\delta\Theta \equiv \Theta - \omega_e t$.

Equation (4.4) is actually a differential equation for Θ which can be rewritten by expanding $\omega(I)$ about I_b using the differential resistance of the unperturbed junction and remembering (Eq. (4.1)) that $\dot{\Theta} = \hat{\omega}$. Thus,

$$\frac{\Phi_0}{2\pi R_d} \dot{\Theta} - \alpha I_e \cos \delta\Theta = \frac{\Phi_0}{2\pi R_d} \omega_u(\hat{I}). \tag{4.7}$$

If a new variable, $\theta \equiv \Theta - \omega_e t - \pi/2$ is defined, then (4.7) becomes

$$\frac{\Phi_0}{2\pi R_d} \dot{\theta} + I_L \sin \theta = \delta I, \tag{4.8}$$

where

$$I_L = \alpha I_e \tag{4.9}$$

and

$$\delta I = I_b - I_{be}. \tag{4.10}$$

That is, δI is the difference between the actual bias current and the bias current I_{be} that would make the unperturbed junction oscillate at frequency ω_e. This equation is just the familiar equation for the phase of a RSJ with critical current I_L, resistance R_d, and bias current δI; hence, the solutions are well known.

The main result that we need is the locking strength, that is, the range of bias current over which $\dot{\theta} = 0$. Equation (4.8) clearly has a constant θ solution for $-I_L \leq \delta I \leq I_L$, with $0 \leq \Theta - \omega_e t \leq \pi$. Thus, as seen in Fig. 6a, $\bar{V}$ remains constant over a range of bias currents $2I_L$ about the bias current I_{be} for which

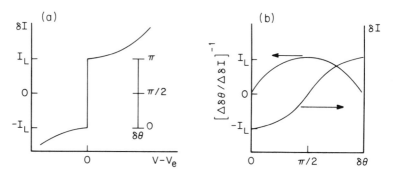

Fig. 6. (a) Junction I–V curve in neighborhood of a radiation-induced step. (b) Phase-locking stability along step, $[\partial(\delta\Theta)/\partial(\delta I)]^{-1}$. Note greatest stability is for $\delta\Theta = \pi/2$. The region $0 > \delta\Theta > -\pi$ is unstable.

the unperturbed junction would have frequency ω_e. Note that this locking strength could also be expressed in terms of the variation in I_c (since I_{be} is a function of I_c), which is possible at fixed bias without losing phase-lock. This latter view is more relevant for arrays in which we may wish to bias a string of junctions with a common current and ask how large a scatter (e.g., in I_c) can be tolerated. In this sense, a junction is most strongly locked when biased in the center of the current step where the difference between the phase of the junction's oscillation and that of the external radiation is $\pi/2$, i.e., $\delta\Theta = \pi/2$. For this bias, the greatest deviation of I_c is possible in an arbitrary direction. The condition that the phase shift be $\pi/2$ for strongest locking has important implications for the design of arrays, as we shall see below. Another measure of locking strength is the variation in $\delta\Theta$ with δI. This is shown in Fig. 6b. Again one sees that $\delta\Theta$ is most stable with respect to changes in δI for $\delta\Theta = \pi/2$ and becomes completely unstable at the edges of the step, $\delta\Theta = 0$ or π. Note that for the range of negative $\delta\Theta$, where the current leads the oscillator phase, $d(\delta\dot\Theta)/d(\delta\Theta) > 0$; hence, the phase-locked solution is unstable.

5. Phase-Locking in Arrays

5.1. Locking Strength within the RSJ Model

A detailed analysis of phase-locking in arrays has been carried out in Jain et al. [17], Chapter 6, as well as in Likharev [4], Chapter 13. The discussion presented here in terms of phase-locking to external radiation will, it is hoped, be intuitive while minimizing the mathematical complications. For large arrays, this approach gives nearly the same results as does the more exact analysis. To begin, consider a series array of identical junctions, modeled by their HF equivalent circuits and connected in a loop through a load Z_ℓ as shown in Fig. 7. If all of the oscillators have the same phase, then the rf current that flows in series through all of the junctions and the load is

$$\tilde{I}_\ell = \frac{\tilde{V}_1}{R_J z_c}, \qquad (5.1)$$

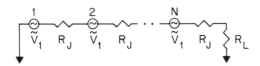

Fig. 7. Equivalent HF circuit for an array terminated in load Z_ℓ.

where the coupling impedance per junction z_c is given in terms of the load impedance Z_ℓ and the junction impedance R_J as

$$z_c = \frac{1}{NR_J}(NR_J + Z_\ell). \tag{5.2}$$

To calculate how much the critical current of one of the junctions can be varied without having it come unlocked from the array, we can just treat this current as external radiation assuming that the array is large enough that a variation of the phase of a single junction will have a negligible effect on $\tilde{I}_\ell$.

When all of the oscillators are running in phase, the relative phase of an oscillator and the locking current is fixed by the loop impedance. Since this impedance always contains a real part equal to the sum of the junctions' resistances plus the load resistance, it is clear that the ideal situation of having the locking current lag the oscillator phase by $\pi/2$ cannot be achieved for the circuit shown in Fig. 7 unless $\text{Im}(Z_\ell) \to \infty$. Thus, in optimizing $\text{Im}(Z_\ell)$ for the maximum locking strength, there is a tradeoff between the amplitude of the locking current and its phase, the phase being given by

$$p_\ell = \tan^{-1}\left[\frac{-\text{Im}(z_c)}{\text{Re}(z_c)}\right]. \tag{5.3}$$

We now wish to see how far the bias current (or critical current) of the kth junction can be varied from the mean for the array without the junction coming unlocked. The mixing current for this junction I_{mk} is

$$I_{mk} = I_L \cos(p_\ell + \delta\Theta_k), \tag{5.4}$$

where $\delta\Theta_k \equiv \bar{\Theta} - \Theta_k$, $\bar{\Theta}$ being the mean phase of the oscillators, and

$$I_L = I_c \frac{\alpha \tilde{v}_1}{|z_c|}. \tag{5.5}$$

Note that the product $\alpha \tilde{v}_1$ is a maximum near $\bar{v} = 1$, since $\tilde{v}_1 \propto \bar{v}$ for $\bar{v} \ll 1$ and $\alpha \propto 1/\bar{v}$ for $\bar{v} \gg 1$.

As for the case of a single junction locked to external radiation, phase-locking will be maintained for $-p_\ell \leq \delta\Theta_k \leq \pi - p_\ell$. Since $p_\ell \neq \pi/2$, it will be possible to shift I_{ck} farther in one direction than in the other. In a real array, one would likely have a symmetric, roughly gaussian distribution of critical currents with the operating frequency of the array determined by the mean I_c of the distribution. In that case, the maximum width that this distribution could have and still maintain complete locking would be set by the lesser of the

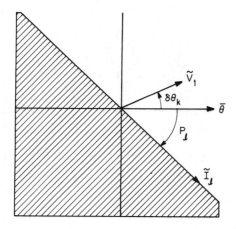

Fig. 8. Relative phase of the mean oscillator voltage (real axis), load current $\tilde{I}_\ell$, and kth junction with different bias or critical current for the array modeled in Fig. 7. Hatched region is unstable.

two deviations, i.e., by

$$\delta I_{k\max} = I_L \min\{1 + \cos p_\ell, 1 - \cos p_\ell\}. \tag{5.6}$$

These phase relations are illustrated in Fig. 8.

Just as with a junction locking to external radiation, the stable situation is when the locking current lags the Josephson oscillator, i.e., the load must be inductive. The inductance that gives the largest locking strength can easily be determined by maximizing Eq. (5.6) with respect to L. If the load impedance is $Z_\ell = R_\ell + jL\omega$, then the locking strength is a maximum for

$$L\omega = \sqrt{3}\,(NR_J + R_\ell), \tag{5.7}$$

so for strong locking, the coupling impedance must have a large reactive component with an inductive character. One can also see the importance of this reactance by calculating the variation in $\delta\Theta_k$ with changes in I_{bk} from Eq. (5.4). Near equilibrium ($\delta\Theta_k = 0$), this variation is

$$\left[\frac{\partial(\delta\Theta_k)}{\partial(\delta I_k)}\right]^{-1} \propto |\tilde{I}_\ell|\sin p_\ell \propto \frac{\sin p_\ell}{|z_c|}, \tag{5.8}$$

i.e., the phase stability is proportional to $\mathrm{Im}(y_c)$, where $y_c = 1/z_c$. Subject to this constraint the maximum power will be delivered to R_ℓ when $R_\ell = NR_J$.

Thus, $L\omega = 2\sqrt{3}\, NR_J$ gives $p_\ell = \pi/3$ and a value for $\delta I_{k\max}$ of

$$\frac{\delta I_{k\max}}{I_c} = \frac{\alpha \tilde{v}_1}{8}. \tag{5.9}$$

Using the RSJ values for α and $\tilde{v}_1$ with $\bar{v} = 1$ gives

$$\frac{\delta I_{k\max}}{I_c} \approx .04. \tag{5.10}$$

Therefore, the total spread in I_{bk} for this type of array can be about 8% before junctions will start to unlock. Since I_b/I_c is about 1.4 (in the RSJ model) for $\bar{v} = 1$, this implies a permissible variation in I_c of about 10%. This number is a maximum since the spread of I_c for the junctions near the center of the distribution will produce some scatter in their phases with a consequent small reduction in the locking current $\tilde{I}_\ell$.

If we are concerned about the unlocking of the first few junctions in the tails of a large distribution, the estimate above is rather close, as can be shown from computer simulations [21]. It is worth noting that the interaction range of the junctions in this type of array is essentially infinite, i.e., the interaction of the kth and ℓth junctions does not depend on their separation. As a consequence, the unlocking of several junctions in a large array has a negligible effect on the phase-locking among the remaining junctions. It may be undesirable to have even one junction unlocked, however, since if its frequency is close enough to that of the array, mixing will occur, which will modulate the array frequency to some degree. As the width σ of the I_c distribution is increased and additional junctions unlock, $\bar{I}_\ell$ will begin to decrease, causing yet more junctions to unlock and leading to a rapid uncoupling of the array with increasing σ. Computer simulations on a 40-junction array show that this "catastrophic" failure occurs for a value of σ about twice that at which the junctions in the tail of the distribution first unlock.

We conclude this section with some brief comments on the prospects for two-dimensional linear arrays. As discussed above, when RSJs are connected in an inductive loop, their rf voltage tends to add in-phase around the loop. For the two-dimensional arrays shown in Figs. 4b,c, the lowest impedance path seen by a junction is the inductive path through the junction in parallel with it. For these configurations, the tendency is for the rf polarities to change along a parallel chain of junctions with the result that circulating currents are set up within the chain. Hence, power is dissipated internally instead of being coupled to the load. For the two-dimensional array in Fig. 4d, on the other hand, the lowest impedance path for all of the series chains is the (presumably inductive) path through the load. Consequently, one would expect a constant phase transverse to the current flow, as desired, for this array. Capacitive coupling between the chains might further stabilize this situation. These stability arguments are developed in much greater detail in Jain *et al.* [17].

5.2. *Radiation Linewidth of Arrays*

Each junction of an array has an intrinsic noise current $S_{I_k}(0)$ that tends to vary the junction's voltage and thus the array's frequency. In the discussions above, we determined the phase-locking effect of the array on a single junction of the array by treating the remaining N-1 junctions as a fixed frequency external source. In this approximation, the frequency of the perturbed junction clearly would not change as long as it remained phase-locked to the array. Thus, no information is provided about the linewidth. The more exact treatment needed to calculate the linewidth is given in Jain *et al.* [17], Chapter 6. Again, one starts from Eq. (4.4). However, the mixing current for the kth junction I_{mk} is now calculated using the rf current obtained by explicitly summing the rf voltage due to all of the other junctions, whose frequencies can now depend on the bias current of the kth junction. Considering an array of N identical junctions as in Fig. 9, this analysis yields an almost intuitive result: Changing the effective low frequency bias current $I_{ek} \equiv I_{bk} + I_{nk} + I_{mk}$ of the kth junction while keeping that of the other N-1 junctions fixed, i.e., $R_s \gg NR_J$, changes the low frequency voltage of the kth junction (and thus the rest of the array since all junctions are phase-locked) by a factor $1/N$ of

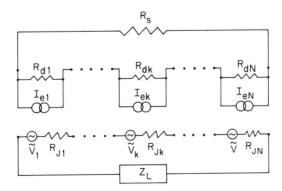

Fig. 9. LF and HF equivalent circuit of junctions coupled in linear array showing LF shunt R_s and HF coupling load Z_L.

that obtained if the junction were not coupled to the array. That is,

$$\delta V_A = \frac{1}{N} R_{dk} \delta I_{ek}. \tag{5.11}$$

From Eq. (2.6), $\Delta v \propto R_d^2$, so this reduction of the differential resistance due to phase-locking would narrow the linewidth by a factor N^2. However, there are N noise sources, so their incoherent sum results in an effective noise current with low frequency spectral density $S_I(0) \propto N$, giving for the linewidth Δv_A of the phase-locked array,

$$\Delta v_A = \frac{1}{N} \Delta v_J, \tag{5.12}$$

where Δv_J is the linewidth of a junction when it is not coupled to the array. For example, while a single 0.1 Ω junction would ideally have a linewidth of about 16 MHz, an array of, say, 1000 such junctions should have a 16 kHz linewidth. If the junctions' resistance is varied along with N so that the total array impedance NR_J remains constant, the linewidth would vary as $1/N^2$, e.g., a single 100 Ω junction should have a linewidth of 16 GHz! As we shall see next, even greater reductions in linewidth are possible if the effects of low frequency shunting are taken into account.

Consider the case where the low frequency shunt $R_s \ll NR_d$. Then one can see from Fig. 9 that the noise voltage generated by the junctions' noise currents is largely shorted by R_s giving an intrinsic linewidth Δv_{A_i} of

$$\Delta v_{A_i} \approx \frac{1}{N^3} \Delta v_J \left(\frac{R_s}{R_d}\right)^2. \tag{5.13}$$

If one takes $R_s \approx R_d$, this implies a dramatic linewidth reduction in large arrays—by a factor of N^3! This soon reaches the point where the intrinsic noise will be dominated by coherent noise sources, e.g., by the Johnson noise voltage generated by R_s. Since this voltage is divided equally across all the junctions in a phase-locked array, $S_v(0) \propto 1/N^2$, giving a factor N^2 reduction in linewidth Δv_{A_s} due to the shunt. So,

$$\Delta v_{A_s} \approx \frac{1}{N^2} \Delta v_s, \quad \Delta v_s \equiv \frac{1}{\pi} \left(\frac{2\pi}{\Phi_0}\right)^2 k_B R_s T_s, \tag{5.14}$$

where T_s is the temperature of the shunt and Δv_s is the linewidth a single junction would have if its voltage noise were that produced by R_s. This is

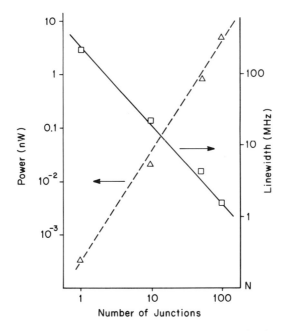

Fig. 10. Power (△) and line width (□) as a function of junction number for a low impedance (≪ 50 Ω) one-dimensional array of microbridge junctions, with $v_0 = 10$ GHz, $Z_L = 50$ Ω.

worth noting since it implies that, in large arrays, there may not be a penalty in linewidth for using noisy, e.g., high J_c, junctions.

A possibly severe penalty for the use of low impedance shunts is the large increase in bias and cooling power that they can require. Since, roughly speaking, only noise up to frequencies Δv contributes to Δv, a shunt in a large array must only be effective at very low frequencies. A rather large inductance should thus be tolerable in series with the shunt resistor, making it possible to place this resistor, for example, at 77 K. While this would increase the linewidth due to the shunt, this linewidth would still be so small as to be acceptable for most applications.

Experimental evidence for the predicted dependence of power and linewidth on array size is shown in Fig. 10 taken from the result Jain *et al.* [17]. Here arrays of 0.1 Ω microbridge junctions incorporated in a 50 Ω microstrip were measured. The power is seen to increase as N^2 as is expected for the coherent state of this array, since the array impedance is always much less than that of the load (see Section 2). Since the low frequency shunting in this array is negligible, the linewidth is expected to vary as $1/N$ as observed.

5.3. Effects of Capacitance on Phase-Locking in Arrays

So far only junctions with no capacitance have been considered. In this section, we will examine what changes in the behavior of arrays one might expect if $\beta_c \neq 0$. One must be aware that capacitance is dangerous. A clue to this is seen in the locking of a single junction to external radiation where the situation corresponding to a capacitive load, i.e., the current phase leading the oscillator, is unstable. It has been shown that for an array like that in Fig. 7, the uniform phase solution is unstable if Z_ℓ is capacitive [17,4]. Instead, the rf voltage tends to sum to zero around the loop. Further, there are many examples of chaotic behavior in capacitive junctions subject to applied radiation or external loads. In spite of this, there are several reasons to consider using capacitive junctions in arrays.

Potential advantages of capacitive junctions are the following: By far the most advanced technology for making Josephson junctions is for tunnel junctions where capacitance is unavoidable. Also, as we shall see below, the presence of a small shunt capacitance can, under certain conditions, enhance the locking strength in the array. Indeed, there have been very successful examples of locking large arrays of high capacitance tunnel junctions to external radiation [6], as well as demonstrations that even junctions with $\beta_c \gg 1$ can phase lock and generate radiation [22-26,13]. Furthermore, it has been shown that tunnel junctions generate significant power levels, at least up to the sum of the gap frequencies [27]. Fortunately, a very general technique for analyzing the stability of the uniform solution in arrays with arbitrary β_c and Z_ℓ has been developed by Hadley et al. [28,29]. This should be of great help in designing arrays of capacitive junctions to avoid the "dangerous" regions of parameter space.

In this section, the analysis above for arrays of junctions with $C = 0$ will be extended, using perturbation theory, to the case of small C, i.e., for $\beta_c \leq 1$, in order to develop some insight into the effects of capacitance. This low C region near $\bar{v} \geq 1$, where one can still hope to obtain meaningful results from perturbation theory, is also the region in which the analysis of Hadley et al. [28,29] indicates that the uniform solution should have the greatest stability.

In order to include the effects of junction capacitance, a capacitor will be connected across the HF terminals of the junction and treated as an additional perturbation. The details of this analysis, which is summarized below, are shown more completely in Lukens et al. [3]. The effects of the various perturbations to the junction are additive since the circuits are linear. The direct effect of the shunt capacitor on the junction will be to change the voltage

4. JOSEPHSON ARRAYS AS HIGH FREQUENCY SOURCES

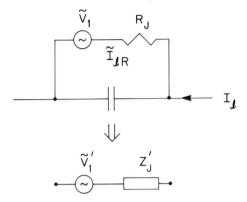

Fig. 11. (a) HF equivalent circuit of a junction with a capacitor for perturbation. (b) Equivalent circuit as seen by the rest of the array.

$\bar{V}$ obtained for a given bias current. This will be ignored since to first order it does not influence the locking behavior of the junction, but just means that a slightly different bias must be used to achieve the desired frequency. The most important effects of the capacitance are to change the effective impedance and rf voltage of the junction as seen by the rest of the circuit. This is illustrated in Fig. 11.

These junctions are now connected in a loop in series with a resistive load having a resistance, in units of NR_J, of r_ℓ. The phase shift between the junctions' oscillators and the loop current $\tilde{I}_\ell$ produced by the shunt capacitance is

$$p_\ell = \tan^{-1}\left(\frac{-\beta_c \bar{v} r_\ell}{r_\ell + 1}\right),$$

which has the same sign as that due to an inductance in series with the load. Thus, the phase relationship between the loop current and the oscillators is that required for stable locking, even with a purely resistive load. The fraction of $\tilde{I}_\ell$ that flows through R_J, $\tilde{I}_{\ell R}$, is responsible for the phase-locking. $\tilde{I}_{\ell R}$ is further shifted with respect to $\tilde{I}_\ell$ by a phase $\tan^{-1}(-\beta_c \bar{v})$ giving a total phase shift $p_{\ell R}$ between the oscillators and $\tilde{I}_{\ell R}$ that can exceed $\pi/2$. These phase relations are illustrated in Fig. 12 and show a potential advantage of the shunt capacitance over a series inductance. Recall from Section 5.1 that the locking strength and hence the acceptable scatter in I_c was substantially reduced with an inductance, since it was not possible to have a $\pi/2$ phase shift between the mean phase of oscillators and the locking current. With a shunt capacitance

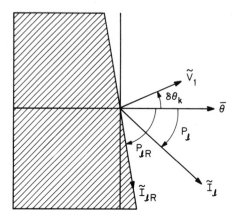

Fig. 12. Relative phases of the oscillators, loop current $\tilde{I}_\ell$, and locking current $\tilde{I}_{\ell R}$ for an array of capacitive junctions.

one can achieve this optimum phase shift by choosing

$$\beta_c \bar{v} = \left(\frac{1 + r_\ell}{r_\ell}\right)^{1/2}.$$

There is a large parameter space in $\bar{v}$, β_c and Z_ℓ that can be explored to optimize the power and locking strength for a given application. To get a feel for the performance of these arrays with $\beta_c > 0$, let us take the purely resistive load that maximizes the load power for given β_c and $\bar{v}$. This gives

$$r_\ell = [1 + (\beta_c \bar{v})^2]^{-1/2}. \tag{5.15}$$

The desired $\pi/2$ phase shift is then obtained for $\beta_c \bar{v} = \sqrt{3}$. For these values the locking strength is

$$\frac{\delta I_k}{I_c} = \frac{\alpha \tilde{v}_1}{2\sqrt{3}}. \tag{5.16}$$

This is more than twice that for the RSJ array from Eq. (5.9), indicating that complete locking should still be possible with a total spread in I_c of greater than 20%. For smaller values of $\bar{v}$, the limits of perturbation theory are being pushed, so the exact values need to be compared with computer simulations. We note that the estimate from perturbation theory is in line with the result of simulations done by Hadley et al. [28,29] on 100-junction arrays with $\beta_c \approx 0.75$ and $i \approx 2.3$, where locking was still observed with a scatter greater than 15% in R_J, C, and I_c.

6. Distributed Arrays

In all of the discussions above, it has been assumed that the dimensions of the arrays were much less than the wavelength λ. As a result, the lumped circuit approximation could be used. To see if this is realistic, note that if the entire array is to have a length less than $\lambda/8$ the junction spacings must be

$$s = \frac{1}{8} \frac{v_p}{vN} \approx 0.1 \ \mu\text{m}, \tag{6.1}$$

where $v_p \approx 10^8$ m/s is the propagation velocity in the superconducting transmission line connecting the junctions, and the value of vN from Eq. (3.1) has been used. Unfortunately, 0.1 μm spacing is about two orders of magnitude closer than is practical to place the junctions in the array when such things as heating and the limits of lithography are considered. We conclude that in order to achieve maximum power, even from one-dimensional arrays, the junctions must be distributed over a wavelength or more.

The analysis of phase-locking above has shown that the phase of the junctions' oscillations relative to the locking current flowing in the coupling circuit is crucial. In general, an oscillator in a transmission line will generate waves propagating in both directions. This makes it impossible to maintain the same phase relationship between all of the oscillators and the locking current when the junctions are placed at arbitrary positions along the transmission line. There have been several proposals [17,30,31] for placing junctions along a transmission line such that they will phase-lock. The simplest approach, described in this section, is just to place the junctions at wavelength intervals along the transmission line. Hence, all junctions see the same impedance and the same relative phase. The analysis of this circuit at the frequency v_λ, where the spacing is equal to λ, is identical to the lumped circuit analysis above. The disadvantage of this approach is that it is only valid at a discrete set of frequencies. It is therefore not clear that such an array will be continuously tunable over a large range of frequencies.

Figure 13 shows a schematic of such a distributed array. The junctions are placed at λ intervals along a serpentine microstrip transmission line. An independently biased detector junction is placed immediately after a load resistor in the line. By measuring the range of detector bias current over which the detector phase locks to the array-generated locking current flowing through the load resistor, the power to the load can be determined for each operating frequency of the array. The ends of the array are terminated with $\lambda/4$ stubs so that, to the junctions, the array appears grounded through the

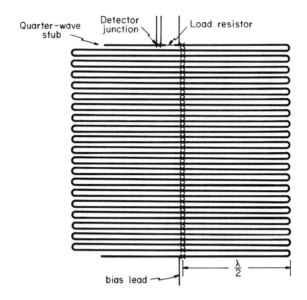

Fig. 13. Layout of distributed array. Oscillator junctions (×) are placed at wavelength intervals along the serpentine microstrip. Load resistor and detector junction to monitor the load current are shown at upper left.

load resistor. Additional length can be added to the stub in order to add a reactive component to the load to optimize the locking strength.

The current amplitude of the Josephson step in the detector junction of such an array is shown in Fig. 14 as a function of the average frequency of the array junctions as determined by measuring the dc voltage across the array [32]. This array contained 40 junctions, which were biased in series, with the bias current flowing through the microstrip. The junctions were separated by 1 mm of microstrip giving an expected value for v_λ of 100 GHz. Indeed a sharp peak in rf current I_ℓ through the detector is observed at 108 GHz indicating the presence of phase-locking near this frequency. The peak value of the rf current, however, is that expected if only seven of the forty junctions were locked in-phase. It is possible to directly measure the distribution of critical currents of the junctions in the array; the resulting distribution is in fact too broad for complete locking. From the measured maximum rf current and the distribution of I_c's, one concludes that only seven or eight of the junctions could be phase-locked. Thus, the data show that the junctions that are sufficiently uniform to lock lock with very nearly the same phase, as predicted.

4. JOSEPHSON ARRAYS AS HIGH FREQUENCY SOURCES

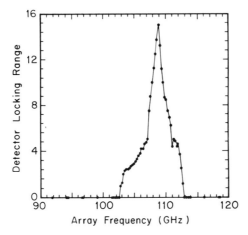

Fig. 14. RF current in transmission line (as measured by detector locking range) vs. average frequency junctions in array.

Since our fabrication process does not yield sufficiently uniform critical currents to insure that locking would be achieved if all junctions were biased with the same current, a "parallel" bias scheme has been used. The superconducting bias leads inject (remove) current at alternate bends in the microstrip, as seen in Fig. 15, with the result that each junction is part of two interlocking dc SQUIDS. All junctions then have the same average voltage, which alternates in polarity along the microstrip. This forces the bias current to divide so as to compensate to first order for the variations in the junctions' critical currents. The rf locking current is still crucial, however, since without it the phase of each junction would be essentially random due to random flux linking the SQUIDs. Further, noise currents would cause voltage (and frequency) fluctuations among the junctions. Phase-locking for this type of parallel biasing has been analyzed in detail in Jain *et al.* [17]. It is primarily as discussed above for series-biased junctions except that the effective scatter in I_c approximately equals Φ_0/L, where L is the inductance of the dc SQUID. For this circuit $\delta I_c \approx 5\mu A$. The junctions are resistively shunted Pb-alloy tunnel junctions having an area of about 1.5 μm^2. These resistively shunted junctions, which have $I_c = 2.5$ mA, have been described in detail elsewhere [31].

The junctions' separation in this array is 350 μm giving a value for v_λ of 350 GHz. The power (as determined from the rf current through the detector and the 23 Ω load resistor) vs. frequency is shown in Fig. 16. Significant power is observed starting at about v_λ; however, the array also generates over 1 μW of

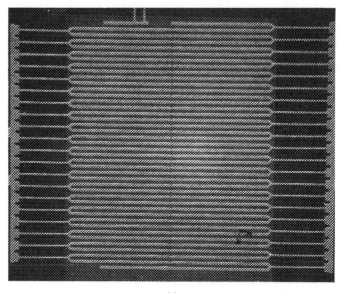

(a)

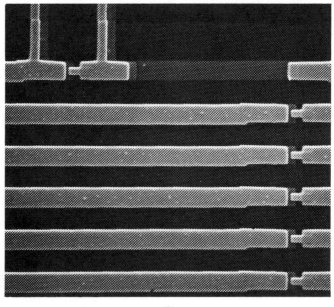

(b)

Fig. 15. (a) Micrograph of distributed array. (b) Blowup of (a), top left, showing detector junction, load resistor, and several oscillator junctions on the right. The vertical separation of the oscillators is 10 μm.

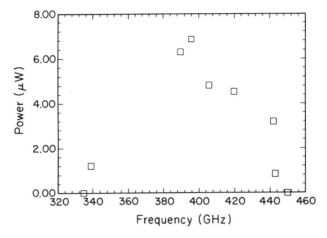

Fig. 16. Power delivered to 23 Ω load resistor vs. array frequency for 40-junction array. Junction parameters: $I_c = 2.5$ mA and $R = 0.5$ Ω.

power at a number of discrete frequencies over a band from 340 GHz to 440 GHz, where the junctions are not separated by integer wavelengths. The maximum power of 7 μW is consistent with all 40 junctions of the array locked in-phase. Other arrays with lower critical current junctions are continuously tunable through this band delivering more than 1 μW of power [33,34], again consistent with all junctions in-phase.

7. Prospects for the Future

As of this writing, the distributed arrays described above are the most powerful fundamental solid state sources at 400 GHz. Rather straightforward, though technically complex, extensions of that work should result in sources operating at over 1 THz with nearly 1 mW of power. Clearly, much work— both experimental and theoretical—needs to be done. This is especially true when considering wide ($w \approx \lambda_{em}/2$) junctions and two-dimensional arrays, where there are essentially no experimental results at present. Probably the most immediate problem is to couple the submillimeter radiation off-chip to determine its spectral purity. Fortunately, very broadband antennas have recently been demonstrated in this frequency range [35,36] for use with superconducting SIS mixers. Also, more theoretical work is needed to understand the unexpectedly wide tuning range seen in many of the distributed arrays.

Table 1. Prospective Sources at 1 Terahertz[a]

Array dimension	1	1	2
Junction (array) width	$2\lambda_J$ ($2\lambda_J$)	$\lambda_{em}/2$ ($\lambda_{em}/2$)	$\lambda_{em}/2$ ($\lambda_t/2$)
J_c [A/cm^2]	—	2×10^6	2×10^6
I_c [mA]	4	40	40
Number of junctions	100	1000	10^6
Power [mW]	0.1	10	10^4
Linewidth (no LF shunt)	5 MHz	50 kHz	50 Hz
Linewidth (1 Ω LF shunt)	20 kHz	200 Hz	<1 Hz

[a] Estimates are based on parameters for nonhysteretic NbAlNb junctions with effective length $2\lambda_J$. A 50 Ω load is assumed. λ_t is the wavelength in the transmission line.

Table 1 summarizes the range of powers and linewidths that might be expected from several different types of arrays, based on the discussions in this chapter. These range from one-dimensional, unshunted arrays of narrow junctions in the upper left of the table, to two-dimensional, linear arrays of the type shown in Fig. 4c with low frequency shunts, at the lower right. Essentially, the projections in the upper left are rather straightforward extrapolations of present design, while, moving to the lower right, one ventures into a territory with progressively more ideas that have yet to be tested. It would, for example, be surprising if a 1 Hz linewidth at one terahertz were ever achieved.

Finally, it is worth considering what impact the new copper oxide superconductors (HTS) might have on the development of these array sources. The most straightforward application of these materials could be as superconducting ground planes and transmission lines. For a given type of conventional superconductor, e.g., niobium, transmission lines become very lossy at half of the upper frequency limit for the junctions' oscillation. Thus, the upper frequency limit might be doubled using HTS microstrip, e.g., to about 3 THz using NbN junctions. While making high quality tunnel junctions of HTS material may be some time in the future, it is worth remembering that the array oscillators do not require good tunneling characteristics. Junctions, e.g., microbridges, with a normal metal barrier should work well; preliminary reports of such thin film junctions which show the Josephson effect have already been presented [37]. While the use of HTS junctions may well extend the upper frequency limit of Josephson effect sources to above 10 THz, the most important advantage initially should be to permit operation at 77 K and thus enhance compatibility with many semiconductor systems. From all of this, we see that Josephson sources have a great, perhaps unique, potential for filling a real void throughout the submillimeter wave band.

Acknowledgments

I would like to acknowledge the essential contributions of all those at Stony Brook who have worked so hard over the years on the development of array sources and especially to thank A.K. Jain, K.-L. Wan, and J.E. Sauvageau, whose work on distributed arrays pushed Josephson sources into the realm of respectability. Special thanks is also due to K.K. Likharev for his valuable and stimulating contributions to our understanding of arrays made during his summer at Stony Brook. Various aspects of the work on arrays at Stony Brook have been supported at different times by the Office of Naval Research, the Air Force Office of Scientific Research, and the Strategic Defense Initiative through the Air Force terahertz technology program.

References

1. Tsai, J.S., Jain, A.K., and Lukens, J.E. High-precision test of the universality of the Josephson voltage-frequency relation. *Phys. Rev. Lett.* **51**, 316–319 (1983).
2. Jain, A.K., Lukens, J.E., and Tsai, J.-S. Test for relativistic gravitational effects on charged particles. *Phys. Rev. Lett.* **58**, 1165–1168 (1987).
3. Lukens, J.E., Jain, A. K., and Wan, K.-L. Application of Josephson effect arrays for submillimeter sources. *In* "Proceedings of the NATO Advanced Study Institute on Superconducting Electronics" (M. Nisenoff and H. Weinstock, eds.). Springer-Verlag, Heidelberg, to be published, 1989.
4. Likharev, K.K. "Dynamics of Josephson Junctions and Circuits." Gordon and Breach, New York, 1986.
5. Basovaish, S., and Broom, R.F. Characteristics of in-line Josephson tunneling gates. *IEEE Trans. Magn.* **11**, 759–762 (1975).
6. Kautz, R.L., Hamilton, C.A., and Lloyd, F.L. Series-array Josephson voltage standards. *IEEE Trans. Magn.* **23**, 883–890 (1987).
7. Yoshida, K., Qin, J., Enpuku, K. Inductive coupling of a flux-flow type Josephson oscillator to a stripline. *IEEE Trans. Magn.* **25**, 1084–1087 (1989).
8. Nagatsuma, T., Enpuku, K., Irie, F., and Yoshida, K. Flux-flow-type Josephson oscillator for millimeter and submillimeter wave region. *J. Appl. Phys.* **54**, 3302–3309 (1983).
9. Cirillo, M., Modena, I., Carelli, P., and Foglietti, V. Millimeter wave generation by fluxon oscillations in a Josephson junction. *J. Appl. Phys.*, to be published (1989).
10. Monaco, R., Pagano, S., and Costabile, G. Superradiant emission from an array of long Josephson junctions. *Phys. Lett. A* **131**, 122–124 (1988).
11. Pagano, S., Monaco, R., and Costabile, G. Microwave oscillator using arrays of long Josephson junctions. *IEEE Trans. Magn.* **25**, 1080–1083 (1989).
12. Likharev, K.K., and Semenov, V.K. Fluctuation spectrum in superconducting point junctions. *JETP Lett.* **15**, 442–445 (1972). Reprinted from *2hETF Pis. Red.* **15**, No. 10, 625–629.
13. Smith, A.D., Sandell, R.D., Silver, A.H., and Burch, J.F. Chaos and bifurcation in

Josephson voltage-controlled oscillators. *IEEE Trans. Magn.* **23**, 1267–1270 (1987).
14. Clark, T.D. Experiments on coupled Josephson junctions. *Phys. Lett. A* **27**, 585–586 (1968).
15. Tilley, D.R. Superradiance in arrays of superconducting weak links. *Phys. Lett. A* **33**, 205–206 (1970).
16. Clark, T.D. Electromagnetic properties of point-contact Josephson junction arrays. *Phys. Rev. B* **8**, 137–162 (1973).
17. Jain, A.K., Likharev, K.K., Lukens, J.E., and Sauvageau, J.E. Mutual phase-locking in Josephson junction arrays. *Phys. Rep.* **109**, 309–426 (1984).
18. Lindelof, P.E., and Hansen, J.B. Static and dynamic interactions between Josephson junctions. *Rev. Mod. Phys.* **56**, 431–459 (1984).
19. Forder, P.W. A useful simplification of the resistively shunted junction model of a Josephson weak-link. *J. Phys. D* **10**, 1413–1436 (1977).
20. Kuzmin, L.S., Likharev, K.K., and Ovsyannikov, G.A. Mutual synchronization of Josephson contacts. *Radio Eng. & Electron. Phys.* **26**, No. 5, 102–110 (1981).
21. Sauvageau, J.E. Phase-locking in distributed arrays of Josephson junctions. Ph.D. dissertation, State University of New York at Stony Brook, 1987.
22. Finnegan, T.F., and Wahlsten, S. Observation of coherent microwave radiation emitted by coupled Josephson junctions. *Appl. Phys. Lett.* **21**, 541–544 (1972).
23. Lee, G.S., and Schwarz, S.E. (1984). Numerical and analytical studies of mutual locking of Josephson tunnel junctions. *J. Appl. Phys.* **55**, 1035–1043 (1984).
24. Lee, G.S., and Schwarz, S.E. Mutual phase locking in series arrays of Josephson tunnel junctions at millimeter-wave frequencies. *J. Appl. Phys.* **60**, 465–468 (1986).
25. Kuzmin, L.S., Likharev, K.K., and Soldatov, E.S. Experimental study of mutual phase locking in Josephson tunnel junctions. *IEEE Trans. Magn.* **23**, 1051–1053 (1987).
26. Krech, V.W., and Reidel, M. Synchronisationseffekte in Anordnungen aus zwei Josephson Verbindungen mit endlichem McCumber-Parameter. *Ann. Phys. (Leipzig)* **44**, 329–339 (1987).
27. Robertazzi, R.P., and Buhrman, R.A. NbN Josephson tunnel junctions for terahertz local oscillators. *Appl. Phys. Lett.* **24**, 2441–2443 (1988).
28. Hadley, P., Beasley, M.R., and Wiesenfeld, K. Phase-locking of Josephson junction arrays. *Appl. Phys. Lett.* **52**, 1619–1621 (1988).
29. Hadley, P., Beasley, M.R., and Wiesenfeld, K. Phase-locking of Josephson junction series arrays. *Phys. Rev. B* **38**, 8712–8719 (1988).
30. Davidson, A. New wave phenomena in series Josephson junctions. *IEEE Trans. Magn.* **17**, 103–106 (1981).
31. Sauvageau, J.E., Jain, A.K., Lukens, J.E., and Ono, R.H. Phase-locking in distributed arrays of Josephson oscillators. *IEEE Trans. Magn.* **23**, 1048–1050 (1987).
32. Sauvageau, J.E., Jain, A.K., and Lukens, J.E. Millimeter wave phase-locking in distributed Josephson arrays. *Int. J. of Infrared & Millimeter Waves* **8**, 1281–1286 (1987).

33. Wan, K.-L., Jain, A.K., and Lukens, J.E. Submillimeter wave generation using Josephson junction arrays. *IEEE Trans. Magn.* **25**, 1076–1079 (1989).
34. Wan, K.-L., Jain, A.K., and Lukens, J.E. Submillimeter wave generation using Josephson junction arrays. *Appl. Phys. Lett.*, to be published (1989).
35. Büttenbach, T.H., Miller, R.E., Wengler, M.J., Watson, D.M., and Phillips, T.G. A broad-band low-noise SIS receiver for submillimeter astronomy. *IEEE Trans. Microwave Theory Tech.* **36**, 1720–1726 (1988).
36. Li, X., Richards, P.L., and Lloyd, F.L. SIS quasiparticle mixers with bow tie antennas. *Int. J. Infrared & Millimeter Waves* **9**, 101–133 (1988).
37. Schwartz, D.B., Mankiewich, P.M., Howard, R.E., Jackel, L.D., Straughn, B.L., Burkhardt, E.G., and Dayem, A.H. The observation of the ac Josephson effect in a $YBa_2Cu_3O/Au/YBa_2Cu_3O_7$ junction. *IEEE Trans. Magn.* **25**, 1298–1300 (1989).

CHAPTER 5

Quasiparticle Mixers and Detectors

QING HU
P.L. RICHARDS

Department of Physics
University of California
and
Materials and Chemical Sciences Division
Lawrence Berkeley Laboratory
Berkeley, California

1. Introduction 169
2. Photon-Assisted Tunneling 172
3. Quasiparticle Mixing 174
4. Mixer Noise 177
5. Imbedding Admittance and Computer Modeling 179
6. LO Power and Saturation 181
7. Series Arrays of Junctions 182
8. Types of Junctions 183
9. Quasiparticle Mixer Measurements 184
   9.1. Performance of Waveguide SIS Mixers and Receivers 185
   9.2. Quasioptical SIS Receivers 188
10. Quasiparticle Direct Detector 190
   10.1. Theory 190
   10.2. Detector Performance 191
   Acknowledgments 193
   References 193

1. Introduction

The beautiful phenomena of superconducting tunneling were discovered nearly 30 years ago. Shortly thereafter, interest developed in the use of these effects for the detection and mixing of infrared and millimeter wave radiation. The first attempts made use of Josephson pair tunneling. The experience with

Josephson detectors and mixers will be mentioned briefly before turning to the more successful quasiparticle detectors and mixers.

Josephson currents respond to very high frequencies, are very nonlinear, and oscillate at a frequency $\omega_J = 2eV/\hbar$, which is proportional to the applied voltage V. They occur at liquid He temperatures, so devices based on them can be expected to have low noise. The Josephson effect devices explored included a direct detector, both an internally pumped and externally pumped heterodyne mixer, several types of internally and externally pumped parametric amplifiers and an oscillator. Of these, only the externally pumped parametric amplifier and the millimeter wave oscillator are under active development today. Neither has yet seen extensive applications outside of the laboratory.

The Josephson effect detector and mixer require tunnel junctions whose capacitance is negligible at the energy gap frequency. In practice, this means point contact junctions, which are unstable and difficult to reproduce. Even with a reliable junction technology, however, these devices would not be very satisfactory. For both devices, the response to applied power scales as ω^{-2}, so decreases rapidly with increasing frequency. As a consequence, the direct detector cannot compete with the widely used composite bolometer at submillimeter wavelengths where sensitive direct detectors are needed. The noise in the Josephson heterodyne mixer is significantly larger than might be expected from the operating temperature because harmonic mixing at all frequencies up to the energy gap downconverts noise into the output. Also, because of the ac Josephson oscillations, both the Josephson detector and mixer are sensitive to the embedding admittances over a wide range of frequencies.

A different type of superconducting mixer was explored that is based on the nonlinearity of single particle or quasiparticle tunneling through a Schottky barrier between a superconductor and a semiconductor. This super Schottky diode gave very low noise at microwave frequencies. Because of the series resistance of the semiconductor, however, extension of this good performance to millimeter wavelengths proved difficult.

Faced with these problems, workers in the field turned to the nonlinearity of the quasiparticle tunneling in thin film superconductor–insulator–superconductor (SIS) tunnel junctions. The first SIS quasiparticle mixer made had lower noise than the best Josephson mixer ever made [1]. The quantum theory of mixing [2] developed to explain the properties of the super Schottky diode gave a complete theoretical treatment of the SIS quasiparticle mixer and direct detector. Within two years of its invention, the SIS mixer was in active use on radio telescopes [3]. This application has expanded and continued to the present.

The practical success of the SIS quasiparticle mixer rests on its low noise and high conversion efficiency, but also on its ability to use tunnel junctions that have significant capacitance at the signal frequency. Such junctions can be produced by optical lithography and have been developed for Josephson effect digital applications. Quasiparticle devices are biased close to the gap voltage so that the ac Josephson frequencies are comparable to the gap frequency. These high frequency Josephson currents are shorted by the junction capacitance, so do not play an essential role in the operation of the SIS quasiparticle mixer or direct detector. If mixer operation is required at a significant fraction of the gap frequency, however, Josephson effects can be troublesome. Some work has been done on mixing in superconductor–insulator–normal metal (SIN) junctions, which have a weaker nonlinearity, but no pair tunneling. This SIN mixer is similar to the super Schottky diode, but with less series resistance than for the semiconductor.

This review will focus on the SIS mixer. The SIS direct detector will also be described. Frequent reference will be made to the extensive review by Tucker and Feldman [4] when describing early work. Readers interested in more information about Josephson devices are referred to the early review by Richards [5] and the update by Richards and Hu [6].

Frequent reference will be made throughout this review to well-known properties and figures of merit for infrared and millimeter wave devices. Direct detectors are characterized in terms of the dark current or the detector noise limit to the noise equivalent power (NEP), the responsivity S, and the dynamic range, or saturation power P_{sat}. Heterodyne receivers are described by the mixer noise temperature T_M, the mixer gain G, the IF amplifier noise temperature T_{IF}, the receiver noise temperature $T_R = T_M + T_{IF}/G$, and the saturation power P_{sat}. Receivers based on direct detectors are used for broad bandwidths and high frequencies. Receivers that use heterodyne mixers are used for narrow bandwidths and low frequencies. The crossover between these approaches is at $\sim 300\,\text{GHz}$ for a fractional bandwidth of 10^{-1} and $\sim 2\,\text{THz}$ for a fractional bandwidth of 10^{-5}. Coherent mixers, as well as direct detectors that are coupled to the electromagnetic field by an antenna, are sensitive to only one electromagnetic mode with throughput $A\Omega = \lambda^2$. Here A is the area of the focal spot and Ω the solid angle of the focus. Other detectors, such as the superconducting bolometer, have a sensitive area of arbitrary size so can accept any number of electromagnetic modes. Readers not familiar with these concepts may wish to refer to a general review of detection technology [7].

2. Photon-Assisted Tunneling

The first step in understanding the performance of quasiparticle detectors and mixers is to compute the response of the quasiparticle current to a coherent local oscillator (LO) source with constant amplitude voltage V_{LO}.

$$V = V_0 + V_{LO} \cos \omega t. \quad (2.1)$$

After complicated calculations [4], the tunneling current as a function of time can be written

$$I(t) = a_0 + \sum_{m=1}^{\infty} (2a_m \cos m\omega t + 2b_m \sin m\omega t), \quad (2.2)$$

where

$$2a_m = \sum_{n=-\infty}^{\infty} J_n(\alpha)[J_{n+m}(\alpha) + J_{n-m}(\alpha)] I_{dc}\left(V_0 + \frac{n\hbar\omega}{e}\right)$$

and $\quad (2.3)$

$$2b_m = \sum_{n=-\infty}^{\infty} J_n(\alpha)[J_{n+m}(\alpha) - J_{n-m}(\alpha)] I_{KK}\left(V_0 + \frac{n\hbar\omega}{e}\right).$$

Here α is a dimensionless parameter proportional to the local oscillator voltage $\alpha = eV_{LO}/\hbar\omega$, and I_{KK} is the Kramers–Kronig transform of the dc $I-V$ curve, which is defined as

$$I_{KK}(V) = P \int_{-\infty}^{\infty} \frac{dV'}{\pi} \frac{I_{dc}(V') - V'/R_N}{V' - V}, \quad (2.4)$$

where $P\int_{-\infty}^{\infty}$ is the Cauchy principle value. A dc $I-V$ curve for a Nb/Pb-alloy junction is shown in Fig. 1a.

The time dependent current $I(t)$ from Eqs. (2.2)–(2.4) depends only on the $I-V$ curve of the unpumped junction $I_{dc}(V)$ and the amplitude V_{LO} of the pump. The static response of the pumped junction is

$$I = a_0 = \sum_{n=-\infty}^{\infty} J_n^2(\alpha) I_{dc}\left(V_0 + \frac{n\hbar\omega}{e}\right). \quad (2.5)$$

This $I-V$ curve has the form of a sum over the dc $I-V$ curves $I_{dc}(V)$ observed without the LO pump, each displaced in voltage by an amount $n\hbar\omega_{LO}/e$, where $n = 0, \pm 1, \ldots$. The amplitude of each term in the sum is $J_n^2(\alpha)$, where $\alpha = eV_{LO}/\hbar\omega$. The sharp onset of quasiparticle tunneling thus causes a series of steps as is shown in Fig. 1b. These steps are due to photon-assisted tunneling.

5. QUASIPARTICLE MIXERS AND DETECTORS

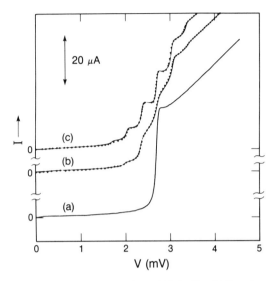

Fig. 1. Line (a) is the experimental $I-V$ curve for a Nb/Pb-alloy SIS junction at $T \ll T_c$ traced in the direction of decreasing current. The Josephson current at zero voltage seen when the $I-V$ curve is traced in the direction of increasing current is not shown. Dotted line (b) is the same $I-V$ curve pumped at 90 GHz with an LO source whose admittance is high compared with R_N^{-1} (constant voltage source). Solid line (b) is the photon-assisted tunneling calculated from the dc $I-V$ curve and Eq. (2.2). Dotted line (c) is the same junction pumped from a 90 GHz LO source whose admittance is small compared with R_N^{-1}. Solid line (c) is the photon-assisted tunneling calculated from the quantum theory of mixing with the admittance at the LO port adjusted for the best fit [8].

Whenever a multiple $n\hbar\omega$ of the photon energy makes up the difference $(eV - 2\Delta)$ there will be an enhancement $(eV < 2\Delta)$ or reduction $(eV > 2\Delta)$ of the tunneling current. The quasiparticle is said to tunnel with the absorption or emission of one or more photons. Distinct photon-assisted tunneling steps are seen only when the onset of quasiparticle tunneling is sharp on the voltage scale $\hbar\omega/e$. At low frequencies, or for rounded $I-V$ curves, Eq. (2.5) approaches the classical limit of a dc $I-V$ curve averaged over the LO voltage swing.

From Eqs. (2.2) and (2.3), we can see that the tunneling current has nondissipative out-of-phase components as well as dissipative in-phase components. These out-of-phase currents are a quantum-mechanical "sloshing" between coupled states whose energies cannot be matched through the absorption or emission of an integral number of photons. They produce a quantum susceptance that has no analogy in classical resistive mixers. This quantum susceptance is only important when the onset of $I_{dc}(V)$ is sharp in the voltage scale $\hbar\omega/e$. It can be either inductive or capacitive, depending on the junction properties and the bias conditions. The quantum susceptance complicates the design and optimization of SIS mixers.

When the admittance of the LO source is small, the voltage amplitude V_{LO} of the pump depends on the RF admittance of the junction which, in turn, is a function of the bias voltage for a constant pump power. Negative resistance can occur on photon-assisted tunneling steps when the RF admittance of the pumped junction varies in such a way that V_{LO} decreases with increasing dc bias. The quantum theory of quasiparticle detection and mixing [4] includes the formalism required to calculate pumped $I-V$ curves with arbitrary LO source admittance. The nonlinear quantum susceptance, which is the counterpart of the nonlinear quasiparticle conductance, plays an important role in this theory. The results of such a calculation are shown in Fig. 1c for a case in which negative resistance is seen [8].

3. Quasiparticle Mixing

A heterodyne mixer is generally used to downconvert signals from some high RF frequency to a lower frequency where amplification and further signal processing is convenient. The classical mixer makes use of the nonlinear resistance of a Schottky barrier diode that is strongly pumped by a local oscillator at ω_{LO}. The mixer produces a linear response at the intermediate frequency ω_{IF} when a small signal is supplied at the signal frequency $\omega_S = \omega_{LO} + \omega_{IF}$ or the image frequency $\omega_I = -\omega_{LO} + \omega_{IF}$. In general, currents flow in the mixer at all frequencies $\omega_m = m\omega_{LO} + \omega_{IF}$, $m = 0, \pm 1, \ldots$, where $\omega_S = \omega_1$, $\omega_I = \omega_{-1}$, and $\omega_{IF} = \omega_0$. The mixer performance depends on the admittances Y_m that terminate the pumped junction at all of these frequencies, or ports, as is shown in Fig. 2. In practical mixers, the junction capacitance usually provides a high enough admittance that the RF voltage is zero for all ports $|m| > 1$. In this case, a three-port theory, which includes only ω_S, ω_I, and ω_{IF} is a very useful approximation.

The classical theory of microwave mixers assumes that the instantaneous current $I(t)$ in a nonlinear resistance is determined by the instantaneous voltage $V(t)$. The single sideband (SSB) conversion efficiency or gain of such a mixer is always less than 0.5. The quantum theory of mixing [4] includes classical mixer theory as a limiting case when the $I-V$ curve is not sharp on the voltage scale $\hbar\omega/e$. When quantum effects are important, this theory predicts many unusual properties for quasiparticle mixers, including large gain and very low noise.

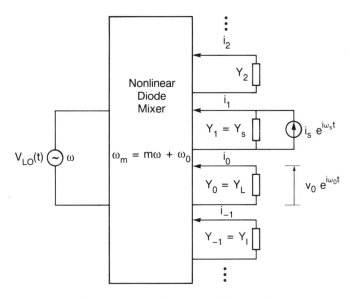

Fig. 2. Functional circuit for a mixer made using a nonlinear diode pumped at ω_{LO}. The imbedding admittances of the mixer at frequencies $\omega_m = m\omega_{LO} + \omega_0$, where $\omega_0 = \omega_{IF}$, $\omega_1 = \omega_S$, and $\omega_{-1} = \omega_I$, are given by Y_m. There is a small signal applied at $\omega_1 = \omega_S$. Noise can drive the mixer at other frequencies.

The quantum mixer theory is formulated in terms of an admittance matrix $Y_{mm'}$ that relates the current i_m in port m to the voltage $V_{m'}$ in port m'. A prescription is given for calculating $Y_{mm'}$ from V_{LO} and the dc I–V curve of the junction [4].

$$Y_{mm'} = G_{mm'} + iB_{mm'}, \qquad (3.1)$$

where

$$G_{mm'} = \frac{e}{2\hbar\omega_{m'}} \sum_{n,n'=-\infty}^{\infty} J_n(\alpha) J_{n'}(\alpha) \delta_{m-m', n'-n} \left\{ \left[I_{dc}\left(V_0 + \frac{n'\hbar\omega}{e} + \frac{\hbar\omega_{m'}}{e}\right) \right.\right.$$
$$\left.\left. - I_{dc}\left(V_0 + \frac{n'\hbar\omega}{e}\right) \right] + \left[I_{dc}\left(V_0 + \frac{n\hbar\omega}{e}\right) - I_{dc}\left(V_0 + \frac{n\hbar\omega}{e} - \frac{\hbar\omega_{m'}}{e}\right) \right] \right\}$$
$$(3.2)$$

and

$$B_{mm'} = \frac{e}{2\hbar\omega_{m'}} \sum_{n,n'=-\infty}^{\infty} J_n(\alpha) J_{n'}(\alpha) \delta_{m-m', n'-n} \left\{ \left[I_{KK}\left(V_0 + \frac{n'\hbar\omega}{e} + \frac{\hbar\omega_{m'}}{e}\right) \right.\right.$$
$$\left.\left. - I_{KK}\left(V_0 + \frac{n'\hbar\omega}{e}\right) \right] - \left[I_{KK}\left(V_0 + \frac{n\hbar\omega}{e}\right) - I_{KK}\left(V_0 + \frac{n\hbar\omega}{e} - \frac{\hbar\omega_{m'}}{e}\right) \right] \right\}.$$
$$(3.3)$$

The coupled mixer gain $G_{\pm 1}$ is defined as the power coupled to the IF load divided by the power available from an RF source at the signal port (+1) or the image port (−1). It can be expressed in terms of the Z-matrix $[Z_{mm'}] = [Y_{mm'} + Y_m \delta_{mm'}]^{-1}$,

$$G_{\pm 1}(\text{SSB}) = 4 \operatorname{Re}(Y_{\pm 1}) \operatorname{Re}(Y_0) |Z_{0\,\pm 1}|^2. \tag{3.4}$$

This result holds for coupled mixer gain from mth port to the output if we replace the subscript "± 1" with "m". It is instructive to express the coupled gain $G_{\pm 1}$ in Eq. (3.4) as a product of the available mixer gain $G^0_{\pm 1}$ and the IF coupling coefficient C_{IF},

$$G_{\pm 1} = G^0_{\pm 1} C_{\text{IF}}, \tag{3.5}$$

where

$$G^0_{\pm 1} = R_D \operatorname{Re}(Y_{\pm 1}) \left| \frac{Z'_{0\,\pm 1}}{Z'_{00}} \right|^2. \tag{3.6}$$

Here the dynamic resistance R_D is dV/dI from the pumped $I\text{--}V$ curve, and

$$C_{\text{IF}} = 1 - \left| \frac{1/Z'_{00} - Y^*_0}{1/Z'_{00} + Y_0} \right|^2. \tag{3.7}$$

In Eqs. (3.6) and (3.7), Z' is the Z-matrix calculated with an open output $Y_0 = 0$. Equation (3.6) suggests that the available mixer gain is approximately proportional to the dynamic resistance R_D of the pumped junction. The available gain can become infinite for infinite or negative R_D. The input admittance of an SIS mixer is of order R_N^{-1}. When the gain is large, the output admittance $(Z'_{00})^{-1} \approx R_D^{-1}$ is $\ll R_N^{-1}$. Large coupled gain is observed, therefore, only when a transformer is used to provide an IF load impedance significantly larger than 50 Ω [9].

Many of these predictions regarding mixer gain can be understood from a simple picture. The mixing between the RF signal and the LO can be viewed as a small modulation of the LO amplitude at ω_{IF}. This modulation produces a current I_{IF} in the junction and provides an available IF power of $I^2_{\text{IF}} R_D/4$, resulting in a linear dependence of the mixer gain on R_D. Since R_D is a maximum at the center of each photon-assisted tunneling step, the gain is expected to have peaks at these bias points. These features are clear in Fig. 3, which shows the IF output of an SIS mixer as a function of bias voltage.

5. QUASIPARTICLE MIXERS AND DETECTORS

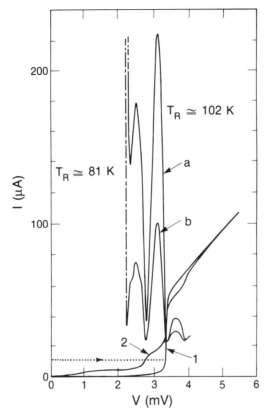

Fig. 3. Curves describing the performance of an SIS quasiparticle mixer [10]. The line labeled 1 is the unpumped $I-V$ curve. Line 2 is the $I-V$ curve when pumped with a LO at 150 GHz. Line a is the IF output power with a 300 K load at the input and line b is the output power for a 77 K load. The mixer is unstable for bias $V < 2.2$ mV because of Josephson effects. Measured values of T_R(DSB) were 102 K on the first photon peak and 81 K on the second photon peak below $2\Delta/e$.

4. Mixer Noise

The contribution of shot noise in the tunneling current of the pumped junction to mixer noise was first calculated using classical radiation fields [4]. The components of the broad-band shot noise that appear at the frequencies ω_m, $|m| \geq 1$, are downconverted and superimposed on the shot noise in the IF band. Because these contributions to the output noise are correlated, destructive interference can occur. It can be shown [11,12,13] that for a properly tuned SIS mixer, the mixer noise temperature due to shot noise can be as low as $\hbar\omega/2k$ for a SSB mixer, and zero for a DSB mixer or a mixer pumped with two local oscillators [14].

This noise theory is incomplete, however, because it does not include the effects of photon shot noise in the input circuit. General arguments related to the Heisenberg uncertainty relation $\Delta n \Delta \phi > \frac{1}{2}$ for photon number n and phase ϕ show that for signals in a coherent state, any phase preserving linear amplifier will add a noise power $\hbar\omega B/2$ for each input port [15]. Since frequency downconversion with unity power gain corresponds to a large gain in photon number, these results are valid for heterodyne mixers. An exception to this general result can occur when most of the uncertainties are squeezed into one variable, leaving the fluctuations in the other variable relatively small [16]. Calculations have shown [14] that an SIS mixer pumped by two local oscillators can be operated in a phase-sensitive mode and can thus be used to detect squeezed state signals.

When the noise power per unit bandwidth at the output of a mixer is calculated from a complete quantum mechanical treatment [12], it has the form

$$\frac{P_N}{B} = \text{Re}(Y_0) \sum_{mm'} Z_{0m} Z^*_{0m'} H_{mm'} + \sum_m G_m \left(\frac{1}{2}\hbar\omega_{m'}\right), \quad (4.1)$$

where G_m is the conversion gain from the mth port to the output as defined in Eq. (3.4), and

$$H_{mm'} = e \sum_{n,n'=-\infty}^{\infty} J_n(\alpha) J_{n'}(\alpha) \delta_{m-m', n-n'} \left\{ \coth\left[\frac{\beta(eV_0 + n'\hbar\omega + \hbar\omega_{m'})}{2}\right] \right.$$
$$\times I_{dc}\left(V_0 + \frac{n'\hbar\omega}{e} + \frac{\hbar\omega_{m'}}{e}\right) + \coth\left[\frac{\beta(eV_0 + n\hbar\omega - \hbar\omega_{m'})}{2}\right] \quad (4.2)$$
$$\left. \times I_{dc}\left(V_0 + \frac{n\hbar\omega}{e} - \frac{\hbar\omega_{m'}}{e}\right) \right\}.$$

The mixer noise temperature T_M referred to the input is defined as

$$T_M(\text{SSB}) = \frac{P_N}{kG_1 B},$$

and (4.3)

$$T_M(\text{DSB}) = \frac{P_N}{k(G_1 + G_{-1})B}.$$

The first term in (4.1) is the contribution from shot noise in the junction current. The second term arises from quantization of the external circuit. It has a minimum noise power of $\hbar\omega/2$ per bandwidth at each of the mixer input ports. Assuming that the input ports of the mixer are terminated by a blackbody with physical temperature T, then the minimum total power at each of the input ports is

$$\frac{P_N}{B} = \frac{1}{2}\coth\left(\frac{\hbar\omega}{2kT}\right)\hbar\omega. \qquad (4.4)$$

At $T = 0$, Eq. (4.4) gives a minimum noise power $\hbar\omega B/2$ at each port of the mixer input. This minimum noise power equals the radiated power from a blackbody at temperature $T_Q = \hbar\omega/k \ln 3$. Some authors define a noise temperature $T_{M'}$, which is linear in noise power, by setting $\hbar\omega B/2$ equal to $kT_{M'}B$. For this case, $T_{Q'} = \hbar\omega/2k$. Mixers have been built at 36 and 114 GHz that have measured noise temperatures within a factor of 2 of T_Q [17,18].

In addition to the intrinsic noise discussed above, SIS mixers can suffer from limitations associated with the Josephson effect. To avoid instabilities caused by the Josephson current at low bias voltages, an SIS mixer has to be biased above a threshold voltage,

$$V_T = V_{LO} + K\left(\frac{2\hbar\omega\Delta}{e^2\omega R_N C}\right)^{1/2}, \qquad (4.5)$$

where K is a constant close to unity [4]. Below this threshold, the noise is high and the bias point is unstable as is seen in Fig. 3. Because an SIS junction is usually biased at $\sim \hbar\omega/2e$ below the gap, this threshold is a problem for SIS mixers at high frequencies. To avoid this instability, a magnetic field is often used to quench the Josephson currents. Other possibilities include using SIS junctions with magnetic impurities in the tunnel barrier to suppress the pair tunneling, and using SIN junctions which do not have pair tunneling.

5. Imbedding Admittance and Computer Modeling

The importance of the imbedding admittance can be understood with the help of the equivalent circuit of an SIS mixer shown in Fig. 4. In this circuit, the signal source is represented by an RF current source in parallel with a source admittance Y_S. The imbedding admittance is represented by a series admittance Y_L and a parallel admittance Y. The SIS junction is represented with its RF admittance Y_{RF} in parallel with its geometric capacitance C. The RF coupling coefficient C_{RF} characterizes the impedance matching between the

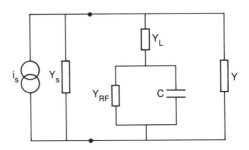

Fig. 4. Equivalent circuit of an SIS mixer. The signal source is represented by a current source in parallel with its admittance Y_S. The SIS junction is represented by its quasiparticle RF admittance Y_{RF} in parallel with its geometric capacitance C. The parasitic inductance Y_L arises mainly from the inductance of the leads to the junction, and Y is the total admittance provided by the tuning structure.

source and the mixer, and it can be calculated from the formula

$$C_{RF} = 1 - \left|\frac{Y_S - Y_J^*}{Y_S + Y_J}\right|^2, \tag{5.1}$$

where Y_J is the total admittance on the right side of Fig. 4. The mixer gain referred to the signal source is approximately proportional to C_{RF}, and the mixer noise temperature is inversely proportional to C_{RF}. Thus, C_{RF} is a key parameter for the design of SIS mixers. Since the invention of SIS mixers, much experimental work has been focused on optimizing the RF imbedding admittances.

For optimum SSB gain, the imbedding susceptance at ω_S should resonate the susceptance of the pumped junction, which arises from both the nonlinear quantum susceptance and the geometrical junction capacitance. Also, the imbedding conductance at ω_S should match the RF conductance. For DSB mixers, these conditions must be met at both ω_S and ω_I. Waveguide mixers generally have $\omega_S R_N C \gtrsim 2$ so only three mixer ports are important, and one or more mechanical tuning elements are used to obtain good RF coupling. Planar lithographed quasioptical mixers are sometimes operated with $\omega_S R_N C \sim 1$ to provide good coupling to a resistive RF source. In this case, harmonic response can be important. Alternatively, they operate with $\omega_S R_N C \geq 2$ and are provided with lithographed RF matching structures.

The quantum theory of mixing [4] can be used to compute the performance of an SIS mixer if the $I-V$ curve, the dc and LO bias, and the imbedding admittances are known. The unpumped $I-V$ curve, dc bias, and available LO power can be measured directly. The theory includes a prescription for calculating V_{LO} from the available LO power. As with classical mixers, however, it is difficult to obtain adequate information about the IF and RF imbedding admittances.

Waveguide mixers for millimeter waves are often designed using lower frequency (3–12 GHz) measurements on large scale models. Scaled modeling has not yet been used for planar lithographed mixers that use RF matching structures that rely on the properties of superconductors. The geometries used in these structures, however, are often selected to facilitate direct calculations of the RF imbedding admittances.

The dependence of the shape of the pumped I–V curve on the LO source admittance shown in Fig. 1b,c can be used to deduce values for the RF admittance. The original technique [19] used the available pump power and the dc current to obtain allowed values of Y_{LO} in the form of circles in the admittance plane. If the input data are very precise, Y_{LO} can be obtained from the intersection of several circles. More recently, considerable success has been obtained with an automated computer search for the value of Y_{LO} that produces the best fit to an experimental pumped I–V curve, as is shown in Fig. 1c [8].

A number of attempts have been made to compare calculations of mixer performance from quantum mixer theory with direct measurements of gain and noise. All of the qualitative effects predicted by theory have been observed. Predictions of gain have been quite successful [20] when the onset of quasiparticle tunneling is not very sharp on the voltage scale $\hbar\omega/e$. Predictions of noise are less successful, but still frequently agree within a factor of 2. Substantial disagreements between theory and experiment are often found when the I–V curve is very sharp, especially with regard to the conditions under which infinite gain is available [17]. It is possible that comparisons in the quantum limit are particularly sensitive to errors in the imbedding admittances.

6. LO Power and Saturation

An estimate of the V_{LO} required to pump a mixer biased on the nth photon step can be obtained from the value of $\alpha_n = eV_{LO}/\hbar\omega$, which corresponds to the first maximum of $J_n(\alpha)$. Since the input impedance of an SIS mixer is of the order of its normal resistance R_N,

$$P_{LO} \approx \frac{(N\hbar\omega\alpha_n/e)^2}{2R_N}. \tag{6.1}$$

This equation includes the case of a mixer that uses a series array of N junctions with total normal resistance R_N. It is found to account for the observed P_{LO} to within 2 dB for experimental values that range from 1 nW for single junction mixers to 30 μW for array mixers [4]. These low values of P_{LO} are a great convenience for SIS mixers, especially at submillimeter wave frequencies. It appears possible that the Josephson local oscillator can be used to pump an SIS mixer [21].

Because of the small values of P_{LO} required, quasiparticle mixers saturate at relatively low signal levels. For this reason, SIS mixers are limited to small signal applications. Saturation first occurs in the mixer output because of the rapid dependence of mixer gain on bias voltage shown in Fig. 3. The IF response of the mixer can be viewed as a modulation of the instantaneous bias point at frequency ω_{IF}. When the amplitude V_{IF} of this IF voltage swing reaches some fraction γ of the width $N\hbar\omega/e$ of the gain peak, the average gain is suppressed. If the mixer is matched at the IF output, the input RF power that will cause saturation can be written

$$P_{sat} = \frac{(\gamma N\hbar\omega/e)^2}{2GR_D}. \qquad (6.2)$$

For a single junction mixer with $R_D = 50\ \Omega$, $G = 3$ dB, and $\gamma = 0.1$ (which corresponds to 0.2 dB gain compression), this expression gives $P_{sat} = 2$ pW, in agreement with a measured value of 1.5 pW [22]. For a quantum noise limited mixer with utility gain and a 500 MHz IF bandwidth, the dynamic range is 20 dB at 36 GHz. This is large enough for most astronomical applications. However, it might be insufficient for radar and communication systems.

Although problems with saturation from 300 K noise can occur in broad-band SIS mixers, such extreme problems can be avoided by several techniques. The use of a series array of N junctions will increase P_{sat} by N^2. Also, if the coupled RF bandwidth is larger than ω_{IF}, then the V_{IF} that comes from a broad-band signal can be reduced by the use of a low-pass filter that shorts the mixer output for frequencies above ω_{IF} [23].

7. Series Arrays of Junctions

Some freedom is introduced into the design of quasiparticle mixers by the possibility of using arrays of junctions in series. If the RF currents have the same phase in N identical junctions, then the equivalent circuit of the N-junction array can be reduced to that of a single effective junction with normal state resistance and series inductance increased by the factor N and capacitance decreased by the same factor. The voltage scale of the I–V curve is increased by the factor N. If junctions with the same tunnel barrier are used, the response time $R_N C$ is unaffected. In order to retain impedance matching at RF and IF, the junctions in the array should have areas N times larger than that for the single junction mixer. Measurements of the performance of array mixers scaled in this way show that mixer gain and noise can be independent of N up to at least $N = 25$ [24].

Advantages of array mixers include a saturation level and dynamic range that scale as N^2, and junction areas that scale as N. The relaxation of fabrication requirements for the larger junctions is partly offset, however, by the need for nearly identical junctions. One disadvantage of array mixers is that the series inductance L of the array scales as N. Limits to the operating frequency ω_s of SIS mixers set by $(LC)^{-1/2}$ or R_N/L can be troublesome when arrays are used at high frequencies.

8. Types of Junctions

The SIS quasiparticle mixer depends on the availability of tunnel junctions with a well-defined onset of quasiparticle current at $V = 2\Delta/e$, values of resistance $20 \leq R_N \leq 100\ \Omega$ that can be matched at RF and IF frequencies, and small enough capacitance that the relaxation time $R_N C$ can meet the criterion $1 \leq \omega_s R_N C \leq 10$. When $\omega R_N C$ is held fixed, the Josephson critical current density, which is an exponential function of barrier thickness, scales directly with frequency. This easily measured parameter is $\sim 500\ \text{A/cm}^2$ at 100 GHz. Since the spread of useful barrier thicknesses is only 10%, the specific capacitance depends only on the type of barrier used. The values of $40\ \text{fF}/\mu\text{m}^2$ for the oxides of Pb-In alloys [25] and $45\ \text{fF}/\mu\text{m}^2$ for Al_2O_3 [26] and the value of $140\ \text{fF}/\mu\text{m}^2$ for the higher dielectric constant oxides of Nb [25] and Ta [17] can be used for design purposes. The junction areas required scale inversely with operating frequency and are typically 1–$4\ \mu\text{m}^2$ at 100 GHz.

A variety of approaches have been used to fabricate the small junction areas required. These include photoresist lift-off to produce window junctions with areas of 1–$4\ \mu\text{m}^2$, photoresist bridge masks to produce overlap junctions with areas of 0.5–$2\ \mu\text{m}^2$, and edge techniques for areas less than $0.5\ \mu\text{m}^2$.

Most early mixer experiments made use of Pb-alloy junctions with In-oxide barriers. These junctions usually degrade gradually when stored at room temperature. Junctions made from Nb/Nb-oxide/Pb-alloy are more durable, but suffer from the higher dielectric constant of the Nb-oxide. All-Nb junctions with artificial barriers such as Al_2O_3, MgO, and α-Si are becoming available that combine ruggedness with a low barrier dielectric constant. A few experiments have been done with Sn/Sn-oxide/Sn and Ta/Ta-oxide/Pb-alloy junctions which have very low leakage current and a very sharp onset of quasiparticle tunneling.

The requirement of small current flow at voltages below $2\Delta/e$ sets an upper limit of $\sim T_c/2$ on the operating temperatures that can be used for SIS mixers. Since there are significant conveniences to operation with unpumped liquid He at 4.2 K, or with mechanical refrigerators (which achieve temperatures below ~ 4.5 K only with difficulty), there are benefits from the use of Nb junctions with $T_c = 9$ K compared with ~ 7 K for the Pb alloys. In the future, the cryogenic problem will be eased by the availability of NbN junctions with $T_c = 15$ K.

It is interesting to speculate on the usefulness of SIS quasiparticle mixers made from the new superconductors with much higher values of T_c. In the radio astronomy applications, higher operating temperatures would be an advantage only if the noise temperature does not also increase. Since noise temperatures of receivers that use cooled Schottky diode mixers are less than 10 times those of the best SIS receivers at W-band, a millimeter wave high-T_c mixer would have competition if T_M is degraded significantly. The energy gap limitation to the operating frequency of SIS mixers could be significantly relaxed by the use of high-T_c superconductors. Operation at frequencies above one THz, however, will require extremely small junctions with area $<0.1\ \mu m^2$ and very high current densities $\sim 10^5$ A/cm^2. At present, there is no appropriate high-T_c SIS junction technology.

9. Quasiparticle Mixer Measurements

Measurements of the performance of SIS receivers are generally made by coupling in signals from hot and cold loads at ~ 300 K and 77 K, and by measuring the output of the cold IF amplifier on a spectrum analyzer. Coherent sources are used to test the relative gains for the signal and image ports. A bi-directional coupler is frequently introduced at the output of the mixer to measure the impedance mismatch at the output, which is important for receiver optimization. A coherent IF signal from an external source can then be reflected from the mixer output to evaluate the IF coupling, and signals can be introduced to measure the gain and bandwidth of the IF amplifier. The complications of cryogenic operation make it difficult to obtain the accurate measurements of the performance of the isolated mixer that are required to test quantum mixer theory. Special techniques such as cryogenic hot–cold loads at both the RF and IF ports have been developed for this purpose [27].

9.1. Performance of Waveguide SIS Mixers and Receivers

Soon after the first mixing experiments were reported in 1979 [1,28], SIS quasiparticle mixers began to replace Schottky diode mixers in coherent receivers for molecular line radio astronomy. These receivers are now in daily use on millimeter wave telescopes and interferometers in at least six observatories. Portable line receivers for submillimeter wavelengths are being developed for use on mountain top and airborne telescopes. Other applications include atmospheric line measurements and radiometers for measurements of the anisotropy of the cosmic microwave background [29]. The SIS quasiparticle heterodyne mixer is now the technology of choice for sensitive coherent receivers from the ~ 40 GHz upper limit of high electron mobility transistor (HEMT) amplifiers [30] to more than 800 GHz.

More than one hundred papers have been published describing the performance of SIS quasiparticle mixers and receivers. A few highlights of these developments will be described here, starting with the waveguide mixers that are typically used at frequencies below ~ 400 GHz.

Good coupling to both the real and imaginary parts of the RF mixer admittance is most easily obtained by the use of a waveguide mount with two mechanical adjustments. Early evaluations of the potential of SIS mixers done in this way gave significantly better performance than was obtained from early mixer blocks with one mechanical adjustment.

Much attention has been given to careful optimization of W-band mixers (75–110 GHz) to achieve broad instantaneous bandwidth, broad tuning range and optimum termination of the image and harmonic ports [31,32,18]. Current practice often makes use of integrated tuning elements lithographed on the junction substrate, which can take the form of lumped or distributed circuit elements [32,33,34]. An example of such a mixer with two mechanical tuning elements [18] is shown in Fig. 5. A broad tuning range with good instantaneous bandwidth has also been obtained with only one mechanical adjustment by the use of suitable lithographed tuning elements [32].

The lowest noise and highest gain thus far obtained from SIS mixers have come from waveguide devides. One experiment in K_A band at 36 GHz with very high quality Ta junctions and two experiments at 100 GHz with Pb-alloy junctions gave noise temperatures of $T_M(\text{SSB}) = 3.6$ K [17], $T_M(\text{DSB}) = 5.6$ K [18], and 6.6 K [32], respectively, which are within a factor of 2 of the quantum limits for these frequencies. Measurements of a W-band mixer with a small IF load admittance gave values of coupled gain as large as 12.5 dB [9]. The observation of such large coupled gain is an interesting confirmation of

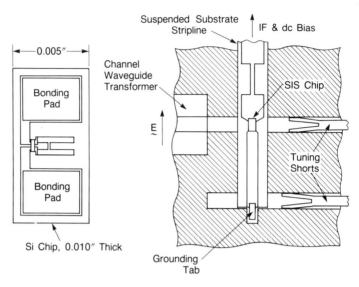

Fig. 5. Cross section of a W-band SIS quasiparticle mixer with two mechanical adjustments for RF matching [18]. The SIS junction with a lithographed RF matching structure is deposited on a Si chip that is bonded to the suspended stripline used to provide dc bias and IF output.

quantum mixer theory. Because of the low noise available from HEMT IF amplifiers, however, gains of order unity are more appropriate for practical receivers.

Despite the progress that has been made in optimizing W-band SIS quasiparticle receivers, the noise temperatures of the receivers on telescopes are a factor of 10 or more above the quantum limit. There is room for improvement before the sky temperature limit is reached. A summary [35] of some of the best reported results is shown in Fig. 6.

Waveguide SIS mixers have been constructed at frequencies up to 345 GHz by several groups [36,37,38]. Because of increased waveguide loss and increased difficulty of fabricating precise structures for these frequencies, simpler mixer blocks are often used with a single mechanical adjustment and a circular waveguide as is shown in Fig. 7. Many of the best results have been obtained by operating at $\omega R_N C \sim 1$ with submicron junctions and no lithographed tuning elements. Although the noise temperatures of these receivers shown in Fig. 6 do not approach the quantum limit as closely as do the W-band receivers, their performance is good enough to produce very valuable astronomical data.

5. QUASIPARTICLE MIXERS AND DETECTORS

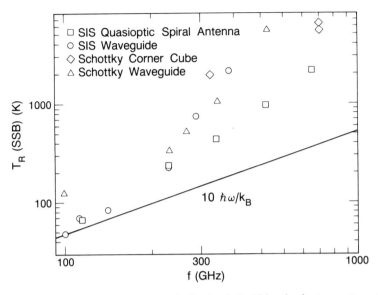

Fig. 6. A summary of some of the best results for the single sideband noise temperature of SIS quasiparticle and Schottky diode heterodyne receivers [35].

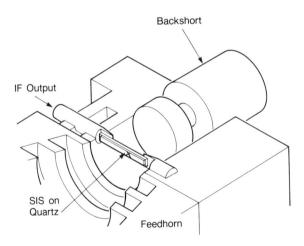

Fig. 7. Scalar feed horn and mixer block used for an SIS quasiparticle mixer from 85 to 115 GHz [31]. This mixer uses a circular waveguide and a single mechanical tuning element for RF matching.

Interference from Josephson tunneling phenomena becomes increasingly troublesome as the operating frequency is increased. Even in a magnetic field, the value of P_{LO} must sometimes be limited so that the instability described in Eq. (4.5) does not interfere with operation on the first photon step below the gap. An encouraging noise temperature of $T_M(DSB) \approx 200$ K has been obtained at 230 GHz with an SIN mixer [39] that avoids this problem.

9.2. *Quasioptical SIS Receivers*

Thin-film SIS tunnel junctions are compatible with other lithographed superconducting receiver components such as planar antennas, transmission lines, and filters. It is therefore attractive to use optical lithography to make integrated planar quasioptical receivers at high frequencies so as to avoid the fabrication problems associated with waveguide structures. Since a planar antenna located on a dielectric surface radiates primarily into the dielectric, the RF signals are introduced through the back surface of the dielectric, which is curved to form a lens as is shown in Fig. 8. Early work on planar integrated SIS mixers began with bow-tie antennas, but attention has shifted to log-periodic and spiral antennas, which have more symmetrical central lobes and so can couple more efficiently to telescopes. All three are self-complementary and so have real impedances of 120 Ω when deposited on quartz [40].

As is shown in Fig. 6, very good performance has been obtained over the extremely broad bandwidth from 100 to 760 GHz with a planar quasioptical SIS receiver [35]. This mixer used a single Pb-alloy junction with $\omega_S R_N C = 1$

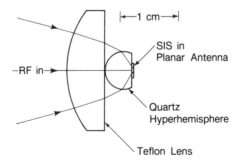

Fig. 8. Cross-section of the optical system used with a planar lithographed SIS quasiparticle mixer [23]. The junction and antenna are located on a quartz substrate attached to the back surface of a quartz lens.

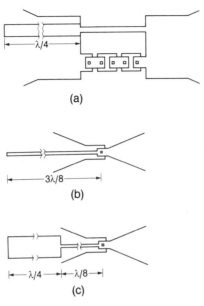

Fig. 9. Layout of window junctions and lithographed RF matching structures used in quasi-optical SIS quasiparticle mixers from 90 to 270 GHz [42]. Diagram (a) shows a series array of five junctions with a parallel wire inductor terminated in an open-ended $\lambda/4$ microstrip stub. (b) Shows a single junction with an inductive open-ended $3\lambda/8$ stub. (c) Shows a single junction with an inductive $\lambda/8$ stub whose end is RF shorted by an open-ended $\lambda/4$ stub.

at 300 GHz and a spiral antenna. Saturation on 300 K noise was avoided by shorting the mixer output for frequencies above ω_{IF}.

The future appears very bright for planar quasioptical SIS mixers for frequencies up to and beyond 1 THz, especially if junctions can be made from superconductors such as NbN with small enough areas to match the antenna resistance. These severe requirements on junction fabrication can be eased by the use of lithographed matching structures such as those shown in Fig. 9. Such structures have been used on planar quasioptical mixers with bow-tie and log-periodic antennas at frequencies from 90 to 270 GHz [41,42]. They are used to resonate the junction capacitance over RF bandwidths of 5–25% and thus to permit the efficient use of junctions with $\omega_s R_N C \geq 10$. Since conventional microwave test apparatus is not available at such high frequencies, special techniques are used to evaluate the RF coupling provided by such structures. These have included using a Fourier transform far-infrared spectrometer as a sweeper and the mixer junction a direct detector [43].

10. Quasiparticle Direct Detector

10.1. Theory

The quasiparticle direct detector, also called a video or square-law detector, uses the nonlinearity of the quasiparticle $I-V$ curve of an SIS junction to rectify the coupled RF signal. The current responsivity $S_I = \Delta I_{dc}/P_S$ of such a detector is defined as the induced change in the static current ΔI_{dc} divided by the available signal power P_S. In the quantum theory [4], the current responsivity is obtained from the small signal limit of the theory of the pumped $I-V$ curve discussed in Section 2,

$$S_I = \frac{e}{\hbar\omega} \left| \frac{I_{dc}(V_0 + \hbar\omega/e) - 2I_{dc}(V_0) + I_{dc}(V_0 - \hbar\omega/e)}{I_{dc}(V_0 + \hbar\omega/e) - I_{dc}(V_0 - \hbar\omega/e)} \right|. \tag{10.1}$$

If the RF source is not matched to the detector, the current responsivity will be reduced from (10.1) by the RF coupling coefficient C_{RF} defined in Eq. (5.1). In the small signal limit, the RF conductance and susceptance of an SIS junction can be calculated analytically,

$$G_{RF} = \frac{e}{2\hbar\omega} \left[I_{dc}\left(V_0 + \frac{\hbar\omega}{e}\right) - I_{dc}\left(V_0 - \frac{\hbar\omega}{e}\right) \right], \tag{10.2}$$

$$B_{RF} = \frac{e}{2\hbar\omega} \left[I_{KK}\left(V_0 + \frac{\hbar\omega}{e}\right) - 2I_{KK}(V_0) + I_{KK}\left(V_0 - \frac{\hbar\omega}{e}\right) \right], \tag{10.3}$$

where I_{KK} is the Kramers–Kronig transform of the dc current defined in Eq. (2.4).

The quantity in the square brackets in Eq. (10.1) is the second difference of the unpumped $I-V$ curve computed for the three points $V = V_0$ and $V_0 \pm \hbar\omega/e$, divided by the first difference computed between $V = V_0 \pm \hbar\omega/e$. In the classical limit, where the current changes slowly on the voltage scale $\hbar\omega/e$, the differential approximation gives the usual result for a diode detector $S_I = (d^2I/dV^2)/2(dI/dV)$. If the $I-V$ curve is sharp enough that the current rise at 2Δ occurs within the voltage scale $\hbar\omega/e$ and if the bias voltage V_0 is just below $2\Delta/e$, then Eq. (10.1) becomes $S_I = e/\hbar\omega$. This quantum limit to the responsivity corresponds to one extra tunneling electron for each coupled photon. The SIS direct detector makes a continuous transition between the classical energy detector and the quantum photon detector.

Since direct detectors do not preserve phase, there is no quantum limit to the detector noise analogous to that for the mixer. The intrinsic noise limit of the quasiparticle direct detector is the shot noise $\langle I_N^2 \rangle = 2eI_{dc}(V_0)B$ in the dark

5. QUASIPARTICLE MIXERS AND DETECTORS 191

current I at the bias point. Here B is the post-detection bandwidth. The noise equivalent power (NEP) in W Hz$^{-1/2}$ of an RF-matched SIS direct detector is then

$$\text{NEP} = \frac{\langle I_N^2 \rangle^{1/2}}{S_I B^{1/2}} = \frac{[2eI_{\text{dc}}(V_0)]^{1/2}}{S_I}. \tag{10.4}$$

In the quantum limit, the NEP increases linearly with signal frequency ω_S.

When the signal power is increased, the responsivity of an SIS dtector falls [44] as $1-P_S/P_{\text{sat}}$, where

$$P_{\text{sat}} \approx 16 \left[\frac{(\hbar\omega/e)^2}{2R_{\text{RF}}} \right]. \tag{10.5}$$

This expression for saturation power is $16G/\gamma^2 N^2$ times the corresponding Eq. (6.2) for an SIS mixer. For a single-junction mixer, this factor is $\sim 10^3$. It can be significantly smaller for mixers made with many junctions in series. Series arrays of junctions are not useful for SIS detectors because the responsivity is reduced and the NEP increased by the factor N.

10.2. *Detector Performance*

The first experimental test of an SIS direct detector [45] showed excellent agreement with the quantum theory. The current responsivity of 3.6×10^3 A/W was within a factor of 2 of the quantum-limited value $e/\hbar\omega$ at 36 GHz. Figure 10 shows the measured and calculated current responsivity S_I as a function of bias voltage. Similar results were reported at the W-band [44]. In these experiments, the shot noise was measured with amplifiers at 50 MHz and 1.4 GHz that were designed for use as IF amplifiers for heterodyne mixers and found to agree with theory. The NEP was deduced to be 2.6×10^{-16} WHz$^{-1/2}$ at 36 GHz [45], which is essentially equal to the performance of a millimeter wave astronomical radiometer based on the ^{3}He-cooled composite bolometer. In usual radiometric applications, the signal is modulated to some low frequency $1 < f < 100$ Hz. A receiver for such signals is sensitive to $1/f$ noise at the frequency f, which is commonly observed in tunnel junctions. Also, an amplifier at frequency f must be used that does not contribute significant excess noise for a source resistance of a few hundred ohms. Even if these problems are solved, an astronomical radiometer based on the SIS direct detector will not be significantly more sensitive at millimeter wavelengths than the SIS heterodyne radiometer that uses a Schottky diode detector at the output of the IF amplifier [46]. It will be less sensitive at submillimeter wavelengths than the ^{3}He-cooled bolometric radiometer.

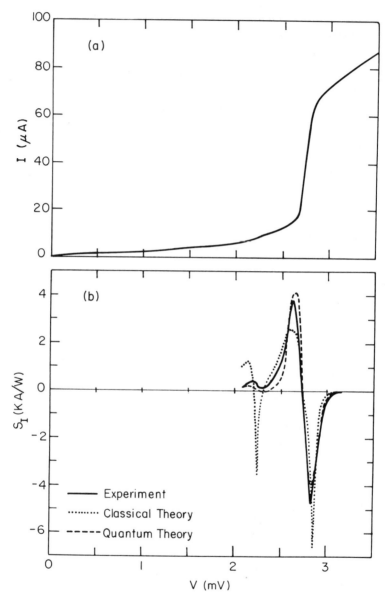

Fig. 10. (a) Measured dc $I-V$ curve of an $\sim 4\,\mu\text{m}^2$ Pb-In-Au alloy SIS tunnel junction at 1.4 K. (b) Measured and calculated responsivities of an SIS quasiparticle direct detector made from the above junction as a function of bias voltage.

The potential remains for applications of the SIS direct detector at near-millimeter and submillimeter wavelengths which benefit from its higher operating temperature and faster speed when compared with the composite bolometer. It is easier to make in-planar arrays than either the composite bolometer or the SIS mixer.

A novel radiometer configuration has been suggested [47] that uses one SIS junction pumped with an LO as a heterodyne downconverter followed by a second SIS junction used as a photon detector. From one viewpoint, the SIS mixer is a preamplifier for the photon detector. It amplifies the photon rate by the product of its power gain and the frequency downconversion ratio. The SIS photon detector is used at the relatively low IF frequency where its NEP is better than at the RF. From another viewpoint, this system is a simplified version of a conventional heterodyne radiometer. Because of the conversion gain of the SIS mixer and the excellent performance of the SIS direct detector, there is no need for an IF amplifier. This configuration appears to have higher sensitivity than the SIS direct detector while retaining its high speed and operating temperature. It appears to be simple enough that planar-integrated arrays of detectors are possible.

Acknowledgments

The authors are grateful to C.A. Mears for many helpful discussions. This work was supported in part by the Director, Office of Energy Research, Office of Basic Energy Sciences, Materials Sciences Division of the U.S. Department of Energy under Contract No. DE-AC03-SF00098, and by the Department of Defense.

References

1. Richards, P.L., Shen, T.-M., Harris, R.E., and Lloyd, F.L. Quasiparticle heterodyne mixing in SIS tunnel junctions. *Appl. Phys. Lett.* **34**, 345–347 (1979).
2. Tucker, J.R. Quantum limited detection in tunnel junction mixers. *IEEE J. Quantum Electron.* **15**, 1234–1258 (1979).
3. Phillips, T.G., and Woody, D.P. Millimeter- and submillimeter-wave receivers. *Ann. Rev. Astron. Astrophys.* **20**, 285–321 (1982).
4. Tucker, J.R., and Feldman, M.J. Quantum detection at millimeter wavelengths. *Rev. Mod. Phys.* **57**, 1055–1113 (1985).
5. Richards, P.L. The Josephson junction as a detector of microwave and far-infrared radiation. *In* "Semiconductors and Semimetals" (Willardson and Beer, eds.), Vol. 12, pp. 395–440. Academic Press, New York, 1977.

6. Richards, P.L., and Hu, Q. Superconducting components for infrared and millimeter-wave receivers. *IEEE Proceeding*, **77**, 1233–1246 (1989).
7. Richards, P.L., and Greenberg, L.T. Infrared detectors for low-background astronomy: Incoherent and coherent devices from one micrometer to one millimeter. *Infrared and Mm Waves* **6**, 149–207 (1982).
8. Mears, C.A., Hu, Q., and Richards, P.L. Numerical simulation on experimental data from planar SIS mixers with integrated tuning elements. *IEEE Trans. Magn.* **25**, 1050–1053 (1989).
9. Räisänen, A.V., Crété, D.G., Richards, P.L., and Lloyd, F.L. A 100 GHz SIS mixer with 10 dB coupled gain. *In* "IEEE MTT-S Digest," pp. 929–930, 1987.
10. Ibruegger, J., Okuyama, K., Blundell, R., Gundlach, K.H., and Blum, E.J. *In* "Proceedings of LT-17," pp. 937–938. Elsevier, 1984.
11. Feldman, M.J. Quantum noise in the quantum theory of mixing. *IEEE Trans. Magn.* **23**, 1054–1057 (1987).
12. Wengler, M.J., and Woody, D.P. Quantum noise in heterodyne detection. *IEEE J. Quantum Electron.* **23**, 613–622. (1987).
13. Devyatov, I.A., Kuzmin, L.S., Likharev, K.K., Migulin, V.V., and Zorin, A.B. Quantum statistical theory of microwave detection using superconducting tunnel junctions. *J. Appl. Phys.* **66**, 1808–1828. (1986).
14. Wengler, M.J., and Bocko, M.F. Beating the quantum limit in SIS mixers. *IEEE Trans. Magn.* **25**, 1376–1379 (1989).
15. Caves, C.M. Quantum limits on noise in linear amplifiers. *Phys. Rev. D* **26**, 1817–1839 (1982).
16. Yurke, B., Kaminsky, P.G., Miller, R.E., Shittaker, E.A., Smith, A.D., Silver, A.H., and Simon, R.W. Observation of 4.2-K equilibrium-noise squeezing via a Josephson-parametric amplifier. *Phys. Rev. Lett.* **60**, 764–767 (1988).
17. McGrath, W.R., Richards, P.L., Face, D.W., Prober, D.E., and Lloyd, F.L. Accurate experimental and theoretical comparisons between superconductor–insulator–superconductor mixers showing weak and strong quantum effects. *J. Appl. Phys.* **63**, 2479–2491 (1988).
18. Pan, S.-K., Kerr, A.R., Feldman, M.J., Kleinsasser, A.W., Stasiak, J., Sandstrom, R.L., and Gallagher, W.J. An 85–116 GHz SIS receiver using inductively shunted edge-junctions. *IEEE Trans. Microwave Theory Tech.*, **37**, 580–592 (1989).
19. Shen, T.-M. Conversion gain in millimeter wave quasiparticle heterodyne mixers. *IEEE J. Quantum Electron.* **17**, 1151–1165 (1981).
20. Feldman, M.J. and Rudner, S. Mixing with SIS arrays. *In* "Infrared and Millimeter Waves" (K.J. Button, ed.), Vol. 1, pp. 47–75. Plenum, New York, 1983.
21. Lukens, J.E., Jain, A.K., and Wan, K.L. Application of Josephson effect arrays for submillimeter sources. Presented at NATO Advanced Study Institute on Superconducting Electronics, 1988.
22. Smith, A.D., and Richards, P.L. Analytic solutions to superconductor–insulator–superconductor quantum mixer theory. *J. Appl. Phys.* **53**, 3806–3812 (1982).
23. Wengler, M.J., Woody, D.P., Miller, R.E., Phillips, T.G. A low noise receiver for submillimeter astronomy. *Proc. SPIE* **598**, 27–32 (1985).

24. Crété, D.G., McGrath, W.R., Richards, P.L., and Lloyd, F.L. Performance of arrays of SIS junctions in heterodyne mixers. *IEEE Trans. Microwave Theory Tech.* **35**, 435–440 (1987).
25. Magerlein, J.H. Specific capacitance of Josephson tunnel junctions. *IEEE Trans. Magn.* **17**, 286–289 (1981).
26. Gurvitch, M., Washington, M.A., and Huggins, H.A. High quality refractory Josephson tunnel junctions using thin aluminium layers. *Appl. Phys. Lett.* **42**, 472–474 (1983).
27. McGrath, W.R., Räisänen, A.V., and Richards, P.L. Variable temperature loads for use in accurate noise measurements of cryogenically cooled microwave amplifiers and mixers. *Int. J. Infrared and Mm Waves* **7**, 543–553 (1986).
28. Dolan, G.J., Phillips, T.G., and Woody, D.P. Low-noise 115 GHz mixing in superconducting oxide barrier tunnel junctions. *Appl. Phys. Lett.* **34**, 347–349 (1979).
29. Timbie, P.T., and Wilkinson, D.T. Low-noise interferometer for microwave radiometry. *Rev. Sci. Instrum.* **59**, 914–920 (1988).
30. Pospieszalski, M.W., Weinreb, S., Norrod, R.O., and Harris, R. FET's and HEMT's at cryogenic temperatures—Their properties and use in low-noise amplifiers. *IEEE Trans. Microwave Theory Tech.* **36**, 552–560 (1988).
31. Woody, D.P., Miller, R.E., and Wengler, M.J. 85 to 115 GHz receivers for radio astronomy. *IEEE Trans. Microwave Theory Tech.* **33**, 90–95 (1985).
32. Räisänen, A.V., Crété, D.G., Richards, P.L., and Lloyd, F.L. Wide-band low noise mm wave SIS mixers with a single tuning element. *Int. J. Infrared and Mm Waves* **7**, 1835–1851 (1986).
33. D'Addario, L.R. An SIS mixer for 90–120 GHz with gain and wide bandwidth. *Int. J. Infrared and Mm Waves* **5**, 1419–1442 (1984).
34. Kerr, A.R., Pan, S.-K., and Feldman, M.J. Integrated tuning elements for SIS mixers. *Int. J. Infrared and Mm Waves* **9**, 203–212 (1988).
35. Büttgenbach, T.H., Miller, R.E., Wengler, M.J., Watson, D.M., and Phillips, T.G. *IEEE Trans. Microwave Theory Tech.*, **36**, 1720–1726 (1988).
36. Ibruegger, J., Carter, M., and Blundell, R. A low noise broadband 125–175 GHz SIS receiver. *Int. J. Infrared and Mm Waves* **8**, 595–607 (1987).
37. Blundell, R., Carter, M., and Gundlach, K.H. A low noise SIS receiver covering the frequency range 215–250 GHz. *Int. J. Infrared and Mm Waves*, **9**, 361–370 (1988).
38. Sutton, E.C. Private communication (1988).
39. Blundell, R., and Gundlach, K.H. A quasiparticle SIN mixer for the 230 GHz frequency range. *Int. J. Infrared and Mm Waves* **8**, 1573–1579 (1987).
40. Rutledge, D.B., Neikirk, D.P., and Kasilingam, D.P. Integrated-circuit antennas. In "Infrared and Mm Waves" (K.G. Button, ed.), Vol. 10, pp. 1–90. Academic Press, New York, 1983.
41. Li, X., Richards, P.L., Lloyd, F.L. SIS quasiparticle mixers with bow-tie antennas. *Int. J. Infrared and Mm Waves* **9**, 101–133 (1988).
42. Hu, Q., Mears, C.A., Richards, P.L., and Lloyd, F.L. *IEEE Trans. Magn.* **25**, 1380–1383 (1989).

43. Hu, Q., Mears, C.A., Richards, P.L., and Lloyd, F.L. Measurement of integrated tuning elements for SIS mixers with a fourier transform spectrometer. *Int. J. Infrared and Mm Waves* **9**, 303–320 (1988).
44. Feldman, M.J., and D'Addario, L.R. Saturation of the SIS direct detector and the SIS mixer. *IEEE Trans. Magn.* **23**, 1254–1258 (1987).
45. Richards, P.L., Shen, T.-M., Harris, R.E., and Lloyd, F.L. Superconductor–insulator–superconductor quasiparticle junctions as microwave photon detectors. *Appl. Phys. Lett.* **36**, 480–482 (1980).
46. Weinreb, S. Feasibility of millimeter-wave SIS direct detectors. *Memorandum*, National Radio Astronomy Observatory, Charlottesville, Virginia, 1986.
47. Richards, P.L. A novel superconducting radiometer. To be published (1989).

CHAPTER 6

Digital Signal Processing

THEODORE VAN DUZER

*Department of Electrical Engineering and Computer Sciences
and the Electronics Research Laboratory
University of California
Berkeley, California
and*

GREGORY LEE

*TRW Space and Technology Group
Redondo Beach, California*[†]

1. Introduction	197
2. Analog-to-Digital Converters	199
2.1. Counting A/D Converters	199
2.2. Flash-Type A/D Converters	207
3. Shift Registers	219
References	223

1. Introduction

The very high switching speeds achievable with superconductive circuits suggests that a natural application would be wideband signal processing. The intrinsic switching speed of a Josephson junction has been shown to be $h/2\Delta$, where h is the reduced Planck's constant and 2Δ is the superconductor gap. This is 0.22 ps for a niobium junction. The record speed of switching circuits of 1.5 ps [1] is about an order of magnitude greater than the intrinsic limit. This larger value is a result of parasitic circuit elements such as the resistances and capacitances of the junctions. An additional important feature of superconductive circuits is the availability of nearly dispersion-free matched transmission lines.

The work on digital superconductive devices dates back to the invention of the cryotron [2] in which a superconducting wire was switched into the normal state by the magnetic field produced by another superconducting wire.

[†] Present address: Hewlett Packard Laboratories, Palo Alto, California

This was followed by thin-film cryotron work in several laboratories until it was discovered in the 1960s that the switching speed of the cryotron was quite limited and would be exceeded by the emerging semiconductor transistor.

Meanwhile, the Josephson junction, which was predicted theoretically in 1962 [3] and demonstrated experimentally in the following year [4], was known to be sensitive to magnetic fields and predicted to be a possible switching device. Work was begun at IBM in 1964 to study the potential usefulness of the Josephson junction as a switching device for digital circuit applications. Subnanosecond switching was demonstrated [5] and a flip–flop was made [6]. This work led to the formation of a small group that grew by the early 1980s to over 150 researchers. A small effort was started in Japan under the sponsorship of the Ministry of International Trade and Industry (MITI) in the late 1970s. In 1982, MITI initiated a major 8-year supercomputer project aimed at evaluating various approaches to the next generation of high-speed computers. These included GaAs, Josephson devices, and silicon software. Even after 1983, when IBM terminated its effort to develop a superconductive computer based on the Josephson junction, the work in the MITI project continued. This work is still in progress and its completion is planned for March 1990. New fabrication technology was introduced and perfected, and now quite large (LSI) circuits can be made. Many new types of circuits and increasing speed have been reported; singly loaded OR gates of several families have shown switching speeds under 5 ps.

The main difficulty in the Josephson digital technology lies in the random access memory (RAM), mainly because of the need to tightly control circuit parameters. As of this writing, no fully functional 1 K-bit RAM has been made, although such a memory unit has been partially functional and has been demonstrated to have an access time as low as 570 ps [7]. A 4 K-bit RAM is also under development.

The last part of the MITI project will be used for the demonstration of data processors of various configurations. A four-bit processor was made as a copy of one made in silicon and GaAs technologies; the GaAs device had a clocking speed of 72 MHz with a power dissipation of 2.2 W. The Josephson processor was clocked at 770 MHz under worst-case conditions with a power dissipation of 5 mW [8]. It is reasonable to expect that data processors will have clocking speeds limited to about 1 GHz for some time to come.

There is interest in superconductive signal processing circuits that would make possible very wide bandwidth signal channels. It now seems possible to make analog-to-digital (A/D) converters of limited dynamic range (say, four bits) that can operate with bandwidths as large as 10 GHz. This implies a clock frequency of at least 20 GHz. Such a data rate greatly exceeds even the very fast processors mentioned above. However, the data can be handled by shift registers that can be used to effectively reduce the rate; simulations indicate that Josephson-based shift registers can work at clocking speeds up to over 60 GHz. These circuits will be discussed below.

Another realm of application of Josephson signal processing is the slower, wider dynamic range for such purposes as digitizing the output of infrared detectors. Two different approaches to this task will be discussed below.

2. Analog-to-Digital Converters

The first proposal for a Josephson A/D converter was made by Klein [9]; this was a successive-approximation device and was limited to a sampling rate of 62.5 MHz. Shortly thereafter, Zappe [10] suggested an A/D converter that made use of the periodicity of multi-junction SQUIDs so that, for example, only four comparators would be required for a 4-bit converter. This was followed by experiments to demonstrate the possibility of multi-gigahertz sampling rates and then by refinements to achieve the highest possible bandwidth. Some designs have appeared for A/D converters that are configured with one input comparator for each digital level, as in semiconductor A/D converters. These circuits will be discussed below. A different type of superconductive A/D converter has been devised in which the key to operation is very high speed counting. This is of special importance for high dynamic range circuits. We will start the discussion with this type.

2.1. Counting A/D Converters

This intriguing class of superconducting A/D converter uses the principle of binary counting to establish the digital signal. Counting A/D converters also exist in semiconductor electronics, although they are generally quite slow. In superconductor electronics, this class of A/D converter enjoys the flexibility of application to either lower accuracy and higher speed or higher accuracy and lower speed. Most of the work has focused on the latter option.

There are several proposed types of counting A/D converters, but one can understand how all of them work by referring to the simplified "generic" counting A/D converter shown in Fig. 1. The quantizer is used to generate discrete digital pulses in response to the analog signal. It may put out a pulse corresponding to a change by a discrete amount in the analog signal, or it may produce a train of pulses at a rate proportional to the signal amplitude—we will discuss these choices of A/D conversion below. Once the pulses are

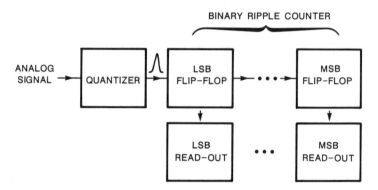

Fig. 1. Simplified block diagram of a counting A/D converter.

present, the information is in digital form and what remains to be done to get a binary equivalent is to count the pulses with a binary counter. This task is performed by the flip–flops in Fig. 1. The present high-speed Josephson devices are asynchronous ripple counters based on nonlatching logic. Since the state of these flip–flops is not recorded as a latched voltage, the flip–flops must be read by latching read-out devices at the end of each sampling interval.

2.1.1. Quantizers

The classes of superconducting counting A/D converters are distinguished by their quantizer. One class uses voltage-to-frequency (V/f) conversion to represent the analog signal. This type of converter is being pursued by the National Institute of Standards and Technology. The registered count is proportional to the average frequency of pulsing during the sampling interval. If the quantizer converts a voltage to a train of pulses at a proportionate frequency, then the counter will record a number proportional to the signal voltage, averaged over the sampling interval. The ac Josephson effect in a single junction affords a nearly perfect V/f converter since the Josephson relation gives $f = 2eV/h$, or, in other words, the frequency is 483.6 MHz per microvolt. The circuit shown in Fig. 2 is the practical realization of a V/f quantizer. Because a Josephson junction has a very low impedance, one needs a bias resistor R_B on the order of a milliohm in order for the analog signal to voltage-bias the junction. The inductance L is needed to block the produced pulses from draining entirely into R_B. Instead, the pulses pass to the counter through R_0, which damps the junction. This type of A/D converter is capable

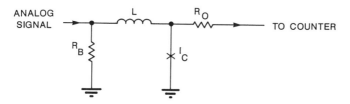

Fig. 2. Voltage-to-frequency quantizer.

of measuring a dc voltage. The resolution increases the longer one counts, i.e., the more flip–flop stages one has and the longer the sampling period. The accuracy cannot increase indefinitely, however, due to the finite linewidth of the Josephson frequency. This linewidth is given by

$$\Delta f = \left(\frac{2e}{h}\right)^2 (4\pi R_B k_B T) \tag{2.1}$$

and is approximately 200 kHz for $R_B = 1$ mΩ at 4 K. The maximum bit accuracy is $\log_2(F/\Delta f)$, where F is the maximum counting rate. The V/f type of A/D converter has an extreme nonlinearity near zero signal due to the inevitable nonlinearity in the junction's I–V characteristic.

The second class of superconducting counting A/D converter quantizes the changes in the analog signal rather than the signal itself and is thus called an incremental converter. (It is unable to convert a true dc signal.) An exemplary incremental quantizer circuit is shown in Fig. 3a. It is a two-junction SQUID into which the analog signal is magnetically coupled. The periodic nature of the SQUID's mode diagram, shown in Fig. 3b, is exploited to establish all quantization levels. Whenever the signal increases sufficiently to cross the right-hand side of a mode boundary, the phase of J_+ changes by 2π and a pulse is sent out the "+" side. Likewise, whenever the signal decreases sufficiently to cross the left-hand side of a mode boundary, J_- sends a pulse out the "−" side. In general, the signal can slew up and down, and the quantizer simply tracks its motion by recording increments and decrements. The counting required is a little more sophisticated. For example, one design uses separate "+" and "−" counters as in Fig. 4, the outputs of which are subtracted from each other at the end of each sampling interval. The incremental A/D converter theoretically has a very high degree of linearity, although in practice the linearity and dynamic range will be limited by stray flux suppressing the critical currents of the individual junctions.

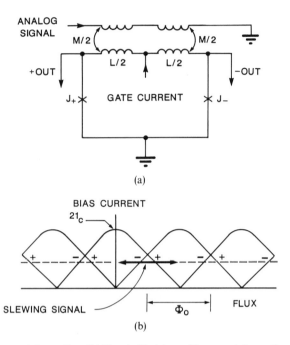

Fig. 3. (a) Incremental quantizer. (b) Threshold picture of incremental quantizer operation.

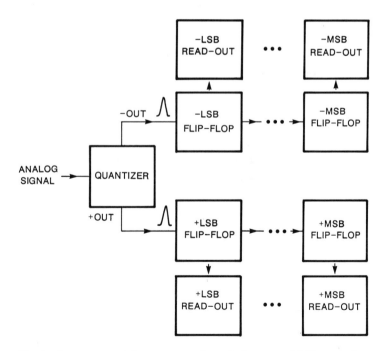

Fig. 4. Separate up and down counter scheme for incremental A/D conversion.

2.1.2. Flip–Flops

The key circuit element that makes all of this possible at high speed is the Josephson "D" flip–flop. This device, shown in Fig. 5a, is also based on a two-junction SQUID and was proposed by Hurrell, Pridmore-Brown, and Silver [11]. The critical currents of both junctions are nominally I_c, and $\beta_L = 2\pi L I_c/\Phi_0$ is approximately π. The two states are the counterclockwise circulation "0" state and the clockwise circulation "1" state. A control line is used to flux-bias the SQUID at the half-flux-quantum point where it rests symmetrically between the two states, as shown in the mode diagram of Fig. 5b. Input current arrives symmetrically to the two junctions in the form of pulses which toggle the flip–flop. As mentioned above, the junctions do not

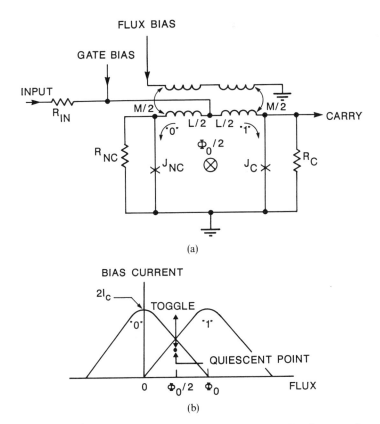

Fig. 5. (a) SQUID counting flip–flop. (b) Threshold picture of flip–flop operation.

latch but simply pulse upon each event such that the pulsed junction's phase changes by approximately 2π. There is one-to-one correspondence between the junctions and the possible flip–flop transitions. The left junction J_{NC} pulses upon a state change "0" to "1" since it is the one carrying more current in the initial "0" state; hence, it is the "No Carry" junction. Likewise, the right junction J_C pulses upon a state change "1" to "0"; hence, it is the "Carry" junction. To prevent latching, the McCumber parameter $\beta_c = 2\pi I_c R^2 C/\Phi_0$ for each junction is chosen to be less than unity. The damping of J_{NC} is by R_{NC}, but since J_C drives the next stage and is partially damped by it, one chooses $R_C > R_{NC}$.

These nonlatching Josephson flip–flops are faster than any other kind in electronics. Hamilton and Lloyd [12] have demonstrated over 100 GHz in counting rate and Silver, Phillips, and Sandell [13] have demonstrated 40 GHz. At a critical current density of about 2×10^4 A/cm^2, the counting rate should approach 1 THz. This brings us to the question of how these ultra-fast devices are tested. One well-known property of any binary frequency counter is that each stage puts out a frequency that is half of its input. Furthermore, an average voltage develops across the SQUID that is proportional to the frequency. Therefore, by observing that the low-frequency voltages of successive stages scaled in the ratio 2:1, they were able to infer extremely high-frequency counting. Direct observation of counting was also made at low (kilohertz) frequencies by using clocked latching SQUIDs to magnetically read the nonlatching flip–flops [12]. The first of a chain of three flip–flops was toggled at the clock rate and the read-out proved that the subsequent flip–flops toggled correctly.

2.1.3. Aperture

Thus far, we have presented the basic components for performing A/D conversion by counting. What we have not yet addressed is the clocking problem that arises in trying to establish the sampling intervals. This problem, which we will call the "collision" problem, is almost inherent to the counting technique and must be circumvented in order to achieve high performance in the A/D converter. The speed and accuracy of any A/D converter is rated by its aperture $\tau = 1/(2^n \pi f_{max})$, where n is the number of bits and f_{max} is the maximum signal frequency to be converted. For a counting A/D converter, one obvious limit to τ is the pulsewidth of the counted pulses, since counting can occur accurately only up to the rate at which pulses are distinguishable. However, there is a less obvious condition on both τ and the accuracy n that

6. DIGITAL SIGNAL PROCESSING

comes from the dynamics at the sampling time. The analog signal is asynchronous with respect to sampling; therefore, so are the emitted quantizer pulses. Unless some trick is used, when the clock rises at sampling time, there is a high probability of "collision"—one or more pulses may be arriving while sampling occurs and they will likely be lost. What makes the situation worse is that in order to obtain the next sample, the A/D converter must also process all counts in the following sampling period, which begins *immediately* afterwards.

Hamilton [14] has studied the impact of the collision problem on the V/f converter. His conclusion is that although one loses some pulses at the beginning and end of each sampling period by temporarily shutting down the counter to read, the averaging of the signal that occurs over the counting period saves the accuracy at dc and yields a gradual roll-off in accuracy at high signal frequency.

For the incremental class of converter, there are two proposed methods for handling the collision problem. Each method works by building the switching needed to form aperture into the quantizer itself. The first scheme uses a device with four identical junctions called the SQUAD, standing for Superconducting QUADruplet [15]. The SQUAD quantizer is shown in Fig. 6; it can be thought of as two SQUID quantizers built into one loop. It has two sides, A and B, and its output can be directed to either side. Bias current is supplied to a symmetric point; if more of it flows to node A, junctions J_1 and J_4 are selected and the B-side bipolar counter receives the pulses. By steering the bias current back and forth by means of a simple driver device, one is able to toggle back and forth between the A and B counters between successive sampling intervals. The data are read in an interleaved fashion; while the active counter is working, the inactive counter can be read to yield the last sample. Even if a collision occurs, information cannot be lost because flux quantization in the SQUAD loop guarantees that one and only one pulse is emitted for each change in the analog signal amounting to Φ_0 and that one and only one side will pick up the pulse.

A second aperturing method uses a two-junction SQUID quantizer as in Fig. 3a and is called TRAP, standing for Time Release after Aperturing of Pulses [16]. The salient feature of TRAP is that $\beta_L \gg \pi$. Because β_L is so large, there is extensive overlap of modes in the mode diagram, shown in Fig. 7. Although the overlap of modes is large, the hysteresis in the quantizer's response to the analog signal can be small if the bias current is raised nearly to $2I_c$. This is the case for the majority of the sampling interval. At the sampling time, however, the bias current is forced low so that the hysteresis suddenly

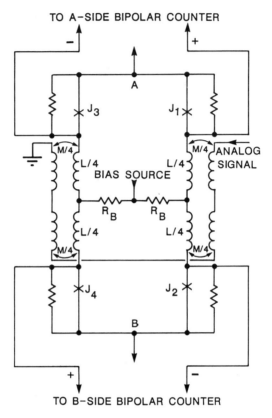

Fig. 6. SQUAD aperturing quantizer used in a toggling scheme for incremental A/D conversion.

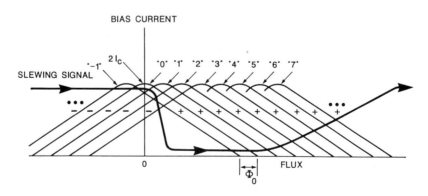

Fig. 7. Threshold picture of TRAP aperturing principle.

becomes large. The SQUID stops pulsing for a period of time, even if the analog signal is rapidly slewing. At this point, information is not lost but merely "trapped" in the form of stored flux. The counter(s) can be read quickly and safely during the nonpulsing (window) period. Eventually the quantizer begins to pulse again as the signal finally exits the mode in which it was temporarily trapped. The bias current is gradually restored in order to record all of the signal increments for the next sampling period, including the ones that were initially stored immediately after aperture. The sequence then repeats. The TRAP was simulated, and it was found that there is, typically, a reading window of 30 ps after aperture.

2.1.4. Target Performance

Since counting A/D converters have the property that resolution increases with counting time, they enjoy flexibility in application to either lower accuracy, higher speed or higher accuracy, lower speed systems. They are especially attractive for the latter purpose because a single quantizer device is used to establish all levels, hence there is no need for high-precision matching of components. The National Institute of Standards and Technology has built and tested a 30-bit counter and is developing a 16-bit, 5 MHz-bandwidth A/D converter based on V/f conversion. At TRW work is in progress on a 16-bit, 5 MHz-bandwidth A/D converter using the TRAP principle and also on developing an 8–10 bit, 500 MHz-bandwidth converter using the SQUAD quantizer.

2.2. Flash-Type A/D Converters

Our concentration in this section is the achievement of A/D conversion with the highest possible bandwidth. This is done at the sacrifice of dynamic range in a so-called flash-type A/D converter.

For many types of A/D converters, the conversion of the analog signal to a train of binary words can be understood in terms of Fig. 8. Samples are taken at evenly spaced intervals T_s. The time required to take the sample is called the aperture time. Each sample amplitude must be converted into a binary word during the time between samples. There are two major criteria that limit the bandwidth. The Nyquist criterion requires that samples be taken at twice the highest frequency f_{max} of the analog signal. Thus, $f_{max} = 2/T_s$. Also, the aperture time must be short enough that the signal does not change by as much as one least significant bit while the sample is being taken. Using the maximum rate of change of a sine wave of frequency f_{max}, it is easy to see that the

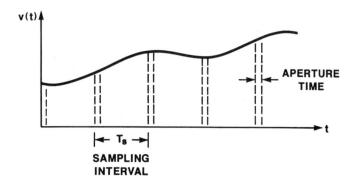

Fig. 8. Times of importance in A/D conversion.

aperture time $\tau = (\pi 2^n f_{max})^{-1}$, where n is the number of bits. In what follows, we will assume that it is always desired to have a bandwidth given by the Nyquist frequency so that the sampling frequency f_s is twice the bandwidth f_{max}. It should be noted that the aperture time requirement is quite severe for large bandwidths. As examples, for four bits, $\tau = 20$ ps if $f_{max} = 1.0$ GHz and $\tau = 2$ ps if $f_{max} = 10$ GHz. A switching time of 2 ps is pressing the limits of any kind of electronic circuit. Other factors that affect the accuracy of converter A/D performance include noise, aperture jitter, and parameter spread in fabrication.

2.2.1. Bit-Parallel Flash-Type A/D Converters

As was mentioned earlier, Zappe first suggested the use of the SQUID periodicity to make possible A/D conversion with one input comparator circuit for each bit [10]. The basic idea is illustrated in Fig. 9, which shows a set of four threshold characteristics for the SQUIDs of the comparators of a four-bit A/D converter. The periodicity differs by a factor of two between each successive comparator; this can be achieved in various ways, as will be seen below. The most significant bit is on the top line in Fig. 9 and the least, on the bottom line. The "0"s and "1"s superimposed on the chart indicate the binary levels associated with the various positions along the horizontal axis (analog signal strength). The coding is Gray code in which only one of the bits changes value on crossing from one digital level to the next. In natural binary, if one of the comparators were to change at a different point from the others, as a result of circuit imperfections, large errors in the binary word value would result. In Gray code, if that happens, the maximum error is one least significant bit. The

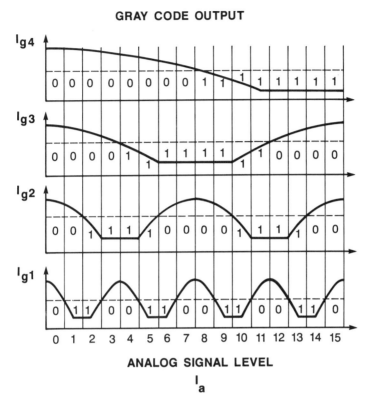

Fig. 9. The basic concept of bit-parallel A/D converters, which require devices with periodic dependence on analog signal level.

broken lines on the threshold curves are set to give equal division of "1"s and "0"s. The method of setting this threshold depends on the choice of circuit realization.

The first experiments on this type of A/D converter [17] used one SQUID for each bit, with the various sensitivities to analog signal effected by changing the coupling to the SQUIDs for successive bits by a factor of two. This was achieved with accuracy sufficient also to make a 6-bit converter [18]. The comparator SQUIDs for two of the bits are shown in Fig. 10. An ac trapezoidal clock signal is applied to the gates of all SQUIDs simultaneously, and the analog signal is applied to the control lines. The analog signal determines the position on the horizontal axis in Fig. 9. Then the gate current rises to the level of the broken line. If located in a region marked "0" on the graph, the SQUID does not switch and if in a region marked "1", it crosses the

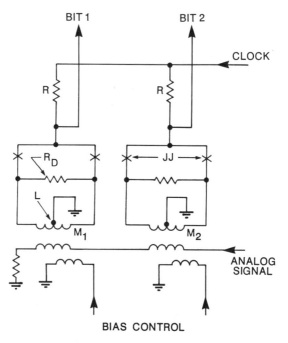

Fig. 10. Two comparator stages in a bit-parallel A/D converter, in which factor-of-two sensitivity difference is achieved by a difference of mutual coupling factors M_1 and M_2 [17,18].

threshold characteristic and switches to the voltage state. Thus, for example, for the four-bit converter depicted in Fig. 9, a digital level "8" would be represented by Gray code word of 1100. It was shown that this arrangement could convert a low-frequency analog signal into 6-bit Gray code words at a rate of four gigasamples per second. However, these simple SQUID comparators had no way to achieve the small aperture required for large analog bandwidth.

In a subsequent work, an A/D converter with the required short aperture time was studied [19]. The comparator circuit is an edge-triggered latch, the state of which is determined by a race in the input circuit that occurs during the rise of the clock. It was adapted from use as a computer circuit where it was called a Self-Gating AND gate [20]. A block diagram of the comparator is shown in Fig. 11a. Figure 11b shows the control characteristic applicable to the SQUIDs in both S_1 and S_2. A bias adjustment is made on S_2 such that it switches at the broken line in Fig. 11b (which corresponds to the broken lines in Fig. 9). The SQUID S_1 switches at the threshold characteristic, which

6. DIGITAL SIGNAL PROCESSING

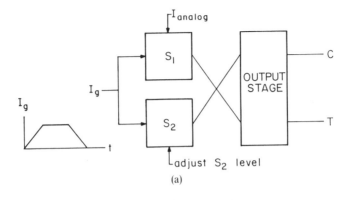

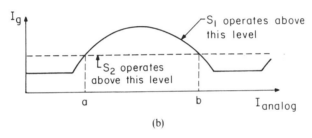

Fig. 11. (a) Structure of an edge-triggered latch used as an A/D comparator. The input circuits S_1 and S_2 are SQUIDs, both of which have the threshold characteristic shown in (b).

depends on the value of the analog current. As the clock rises, either S_1 or S_2 will switch first, depending on I_a. If I_a is between a and b in Fig. 11b, S_2 switches first and the output stage latches into a state with $C = 1$ and $T = 0$. When I_a is outside that region, the opposite result obtains. By having a fast rising clock, the decision is based on the value of the analog signal in a very short aperture time.

This circuit has been demonstrated experimentally in a 4-bit A/D converter [21]; see Fig. 12. The four comparators, which correspond to the four threshold characteristics in Fig. 9 comprise the left column of circuits in Fig. 12. These were clocked at 1 GHz so that the Gray code data coming out of the comparators were at too high a rate for the room-temperature test circuits. Therefore, a second set of latches (circuit is identical to the comparator) was included on the chip to reduce the data rate. The second column of latches was clocked at 1/32 of the 1 GHz clock rate so the data taken off of the chip is at about 31 MHz. By using a beat-frequency test scheme in which the frequency of the analog signal to be evaluated is slightly different from

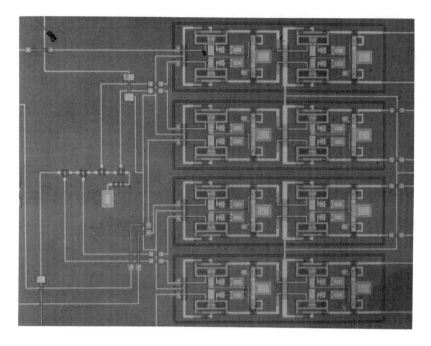

Fig. 12. Realization of a 4-bit A/D converter based on the edge-triggered latch in Fig. 11. The four latching comparators in the left column are fed by a binary resistor divider and are clocked at 1 GHz. The right column of latches is subharmonically clocked to select samples from the comparators for data-rate reduction [21].

a subharmonic of the clock frequency, the complete analog sine wave is mapped out. The data were fed into a minicomputer and the sine wave was reconstructed. The result was that a 500 MHz sine wave was converted with 3-bit accuracy.

This type of circuit has been evaluated theoretically and in simulation to determine the factors limiting its performance. The circuit involves a transfer of current, as illustrated by the crossed lines in Fig. 11a, when one of the input SQUIDs switches upon winning the race during the rise of the clock. This transfer of current is the limiting factor. The circuit can be improved within the same general structure, and simulations indicate that four bits at $f_{max} = 500$ MHz is about the best that can be done [22].

Other circuits have been proposed that eliminate the limitation in the above-described edge-triggered latch. In one of these, the comparator consists of two SQUIDs connected in series [23]. One is biased with a fixed control

6. DIGITAL SIGNAL PROCESSING

current and the other has the analog signal as control current. The operation is similar to the race of the circuits in Fig. 11 except that here the circuits are in series. This comparator eliminates the delay involved in the current transfer discussed above in connection with the circuit in Fig. 11. The aperture time is a fraction of the clock-rise time here also. In another circuit, the periodic threshold of a two-junction SQUID is still used, but a short aperture time is achieved by switching a one-junction SQUID on the rising edge of the clock; the aperture time is 0.5–1.0 of the rise time of the clock [24]. The dominant limitation on the two types of A/D converters discussed in this paragraph is the distortion of the threshold characteristics of the least significant bits that results from the rapid changes of analog control currents in the SQUIDs. It has been estimated that, for any 4-bit flash-type A/D converter that employs the periodic threshold characteristic of multi-junction SQUIDs, the analog bandwidth will be limited to about 2.0 GHz [22].

A recently reported comparator circuit employs the periodic characteristic of a one-junction SQUID as basis for a bit-parallel A/D converter [25,26]. The one-junction SQUID in Fig. 13a has a periodic relationship between the junction current I_j and the SQUID current I_a. The relationship is single-valued as in Fig. 13b if the product $\beta_L = 2\pi L I_c/\Phi_0 \leq 1.0$, where I_c is the junction critical current and Φ_0 is the flux quantum, but is multi-valued for larger β_L. The design of this comparator circuit depends on the I_j–I_a relation being single-valued. The comparator circuit actually uses a quasi-one-junction SQUID in which the leg with the junction in Fig. 13a has two junctions in series, but with one having such a larger critical current than the other that the circuit behaves essentially as a one-junction SQUID. The comparator for one bit is shown in Fig. 14. The inductor L along with J_0 and J_s comprise the quasi-one-junction SQUID. The magnitude of the analog current determines whether the junction current from that source is positive or negative, as in Fig. 13b. The clocked reference current is adjusted so that, in the absence of an analog current in J_s, the pulse from the SQUID at the top of the circuit will switch J_s 50% of the time in the presence of noise. If I_a is such that I_j is positive, J_s goes into the voltage state and remains there until the reference clock current falls back to zero. If the value of I_a is such that I_j in Fig. 13b is negative, the junction J_s does not switch and the output is zero. Thus, "1"s and "0"s can be associated with the positive and negative regions in Fig. 13b. A set of four such comparators fed by a binary resistor divider (as used in the circuit of Fig. 12) will give a Gray code output, as introduced in connection with Fig. 9. The pulse generator at the top of the comparator circuit in Fig. 14 is shared among the four comparators in order to synchronize the sampling.

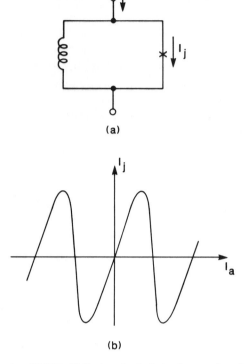

Fig. 13. (a) A one-junction SQUID. (b) Single-valued relation between junction current and total SQUID current that is obtained if $\beta_L = 2\pi L I_c/\Phi_0 \leq 1.0$.

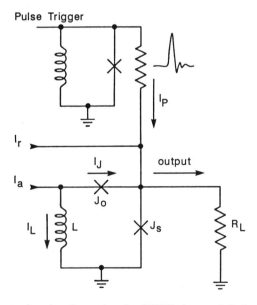

Fig. 14. A comparator based on the one-junction SQUID characteristic shown in Fig. 13b. The key element is the quasi-one-junction SQUID involving L, J_0, and J_s [25,26].

The pulse width sets the aperture time. Simulations showed successful 4-bit conversion of a 10 GHz analog sine wave using a 20 GHz clock. The higher bandwidth compared with the circuits using multi-junction SQUIDs results from the better dynamic behavior of the one-junction SQUID.

2.2.2. Fully Parallel Flash-Type A/D Converters

The architecture of the fully parallel flash-type A/D converter is shown in Fig. 15. The set of comparators acts as a digital "thermometer" for the analog signal amplitude, which is fed to all of the comparators in parallel and is there compared with references that represent the $2^n - 1$ digital levels (15 for a 4-bit converter). The outputs of the comparators must then be combined in an encoder to give the binary word output. We examine in this section two different circuits that employ this architecture.

The circuit shown in Fig. 16 is being studied as a comparator for a fully parallel A/D converter [27,28]. The central element of the comparator is the bridge containing two junctions and two resistors. The circuit is clocked with fast rise-time (≈ 5 ps) supply. If no current I_{net} were present, the clock current would pass through the superconducting junctions J_1 and J_2 to ground. The current I_{net} is proportional to the difference between the analog and reference inputs. If I_{net} is positive, it adds to the clock current in J_2, causing it to switch, and it subtracts from the current in J_1, which does not switch. The result is that the clock current circulates through L_1 in a counterclockwise direction,

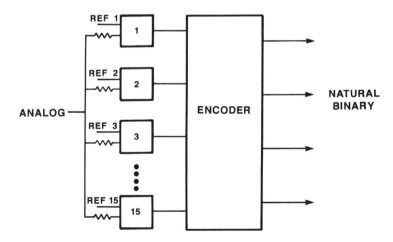

Fig. 15.   Structure of a fully parallel A/D converter.

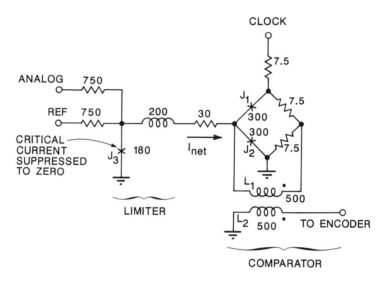

Fig. 16. Comparator with input limiter for an experimental fully parallel A/D converter. Resistor values are in ohms, inductances in picohenrys, and junction critical currents in microamperes [28,29].

passing through J_1 and the lower 7.5 Ω resistor. On the other hand, if I_{net} is negative, it adds to the clock current in J_1, which then switches, and it subtracts from the clock current in J_2 so that it does not switch. The clock current then circulates clockwise through the top 7.5 Ω resistor in the bridge and through L_1 and J_2. The current in L_1 induces current in the secondary circuit to the encoder. Circuit simulations showed a 3.5 GHz bandwidth for a 4-bit A/D converter. Dynamics of the comparator were verified experimentally. It was found that the dynamic range of the bridge is insufficient to make a 4-bit A/D converter that has adequate design margins. The limiter circuit preceding the bridge is included to increase the dynamic range. The critical current of the limiter junction is suppressed to zero and the quasi-particle part of the I–V characteristic is used to make a soft limiter. The conductance is very low in the sub-gap region and then rises rapidly and shunts the input to ground for larger voltages. The junction capacitance and the inductor form a low-pass filter to remove high-frequency inference [29].

The encoder is a set of four multiple-input two-junction SQUIDs [27]. The fifteen lines are used in various combinations for the SQUIDs to convert the digital levels to natural binary code. In each SQUID, the various control lines cause transitions back and forth across one threshold so the problem of the

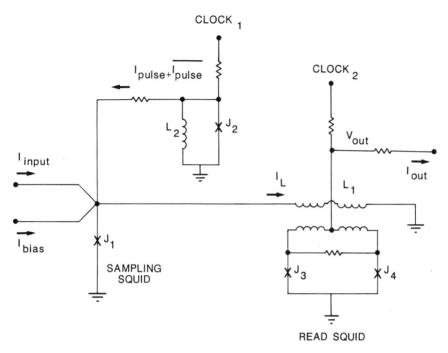

Fig. 17. Comparator based on a one-junction SQUID for a fully parallel A/D converter [24].

threshold distortion discussed in Section 2.2.1 does not occur. Work is still in progress on this converter.

Another circuit that is designed for use in a fully parallel A/D converter is based on the current latching property of a one-junction SQUID [24]. The comparator circuit is shown in Fig. 17. The central element in the comparator is the one-junction SQUID consisting of J_1 and L_1. It is fed the analog current input, a dc bias current, and pulses generated in the one-junction SQUID at the top of the drawing. The power supply CLOCK$_1$ has a trapesoidal form so that, when applied to the J_2–L_2 SQUID, a sharp positive pulse is produced on the leading edge and a negative pulse on the trailing edge. The multi-valued relation between current in the inductor L_1 and the current applied to the node above J_1 is shown in Fig. 18. Shown there is a bais current adjusted so that the positive pulses would drive past the critical point on the I_L vs. I_{external} curve, and the SQUID would switch up to the higher I_L if any positive analog signal were present. When the negative pulse arrives at the end of CLOCK$_1$, the circuit is reset. The two-junction SQUID is switched to the voltage state if the larger value of I_L passes through L_1 while CLOCK$_2$ is high. The aperture time

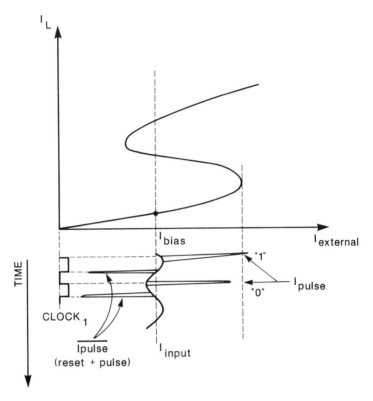

Fig. 18. Characteristic of the one-junction SQUID for the circuit of Fig. 17, shown to illustrate the principle of operation.

of the comparator is less than the width of the positive pulses produced by J_2-L_2, which can be a few picoseconds. Thus, for a 4-bit A/D converter, the bandwidth should be in excess of 5 GHz. The circuit is very sensitive to the signal; in circuit simulations, it was possible to discriminate signal level differences of 1 μA at an input frequency of 5 GHz while sampling at the Nyquist frequency. The experimental sensitivity will be limited by noise.

A pipelined encoder can be made with circuits identical with the comparator since it can perform the AND-OR function. For a complete converter, four clocks with appropriate phase shifts are required.

Some other flash-type A/D converters, these with comparators based on the quantum flux parametron (QFP), have recently been reported [30,31]. Some of this work also suggests the possibility of multigaghertz operation.

3. Shift Registers

A number of different circuits using Josephson junctions have been proposed as shift registers. A few of them have been evaluated by simulation and low clock-rate experiments, and one has been reported with high clock-rate test results. The first report was by Yao and Herrell [32]. We concentrate here on the most recent work.

There was a recent report of an 8-bit shift register that was powered by a three-phase sinusoidal clock with a frequency of 2.3 GHz [33]. The gates used in the circuit are modified variable threshold logic (MVTL) gates, circuits devised at Fujitsu Laboratories. The shift-register circuit diagram is shown in Fig. 19 for one bit. The output for the 1-bit shift register is given by the logical function $S \cdot DS + L \cdot DL + H \cdot T_{\phi_3}$, where S, L, and H represent the control signals for SHIFT, LOAD, and HOLD, respectively, DS and DL represent the data for SHIFT and LOAD operations, respectively, and T_{ϕ_3} is the output. The 8-bit shift register used a circuit area of 1.1×2.1 mm^2 and contained 328 Josephson junctions and 516 resistors. It was fabricated using niobium technology with Nb-AlO$_x$-Nb junctions having critical current density of 1700 A/cm^2.

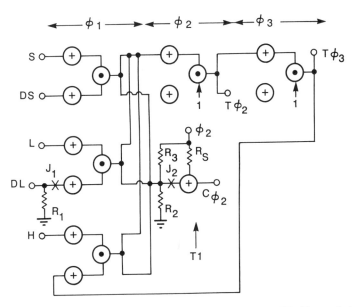

Fig. 19. One stage of a shift register with a clocking frequency of 2.3 GHz. The symbols $\oplus$ and $\odot$ are OR and AND gates, respectively [33].

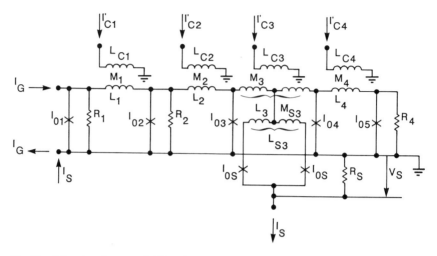

Fig. 20. A flux-transfer type of shift register. Loops 1–3 form one stage. Read-out is by means of the two-junction SQUID containing L_3 and M_{S3}. The fourth loop is the first one of the next stage [34,35].

Another circuit that has received some experimental evaluation at low clock rate, as well as extensive simulation, is shown in Fig. 20. The clocking is three-phase: I'_{c1}, I'_{c2}, and I'_{c3} are currents with the three phases. A flux quantum is shifted from one loop to the next by application of the clocks. Simulations indicated that the clocking can be done successfully at 55 GHz. The circuits do not go into a voltage state except momentarily in order to transfer the flux. To read out the data, a latching two-junction SQUID is connected in the third loop. The loop on the right in Fig. 20 is the first stage of the next bit [34,35].

Several other circuit designs that involve flux transfer and memory by circulating currents have appeared. Work is in progress on a circuit that involves shifting currents between arms of a superconducting loop to represent the logic states [36] (see Fig. 21). Every memory location consists of two superconducting loops, each of which contains two-junction SQUIDs for redirecting a dc current from one leg of a loop to the other, as seen in Fig. 21. Logical "1" and logical "0" are represented by the paths taken by the currents. If the data signal D is high when CLOCK comes up, the SQUID S_L switches to the voltage state and diverts the portion of I_{dc} that was flowing through it to the right side, as shown by the curved arrow in Fig. 21. The SQUID S_L then quickly resets to the zero-voltage state. When CLOCK falls and $\overline{\text{CLOCK}}$ rises, the current is forced to the left side of the second loop because of the

6. DIGITAL SIGNAL PROCESSING

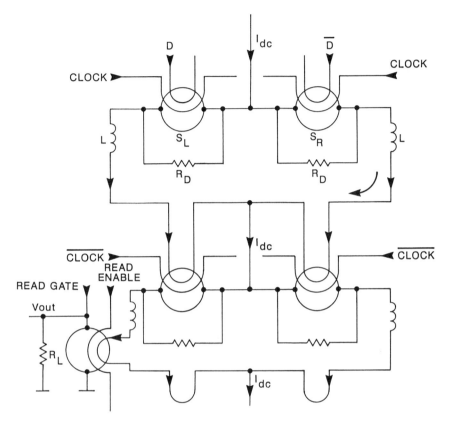

Fig. 21. One stage of a shift register based on current steering [36].

switching of the SQUID on the right side of that loop. The current latched into the left side of the loop can be read by the READ GATE without interrupting the shifting operation. The power dissipation is extremely low at 0.1 μW/bit with worst-case switching at 62.5 GHz.

Low-frequency tests have verified operation of another flux-transfer circuit that employs a two-phase sinusoidal clock C_{L0} and $C_{L\pi}$ [37]. Each bit-storage location requires two SQUIDs, as shown in Fig. 22; Q_s is dc-biased and Q_g has an offset-sinusoid clock. Suppose the stage is initially storing a logical "0" and is fed with a "1" on the data line D while the clock C_{L0} of the previous stage is high. The SQUID Q_s has been dc-biased at a point in the overlap of modes of zero- and one-stored flux-quantum and contains zero; the addition of the control current D causes a transition to the one-flux quantum condition. A

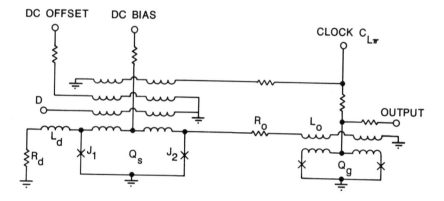

Fig. 22. One stage of a shift register in which a datum is stored as a circulating current in Q_s [37].

short voltage pulse is generated across J_1 and it is dissipated in R_d. The SQUID Q_s is now in the stable one-flux-quantum state. When the offset-sinusoid clock of this stage $C_{L\pi}$ comes up, it adds a control current to Q_s (D is now removed as C_{L0} of the preceding stage goes to zero), which causes Q_s to reset to the zero-flux-quantum state. This generates a pulse across J_2 that produces a control current to Q_g, which switches into the voltage state and transfers its gate current to the next stage, causing its input SQUID to switch to the "1" state. Thus, the input "1" is transferred to the next stage. If the input D had been "0", nothing would have been transferred. This circuit has been shown to work correctly at low speeds; simulations indicate correct operation for a clock frequency of at least 60 GHz with current density of 5 kA/cm^2 in

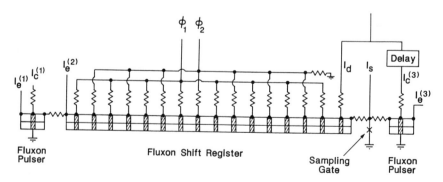

Fig. 23. One stage of a shift register based on fluxon transfer along a long Josephson junction with energy-well storage location [40].

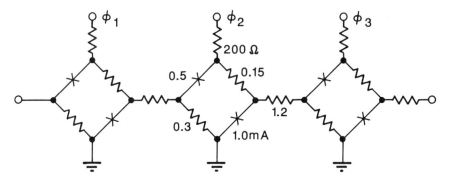

Fig. 24. One stage of a shift register with a three-phase clock. Resistor values are in ohms and junction critical currents are in milliamperes [39,40].

the junctions. The circuit has the advantage of not requiring the complement of the data D input.

There is also a report of a long-junction flux-shuttle type of shift register illustrated in Fig. 23 [38]. It employs a two-phase clock and transfers a flux quantum between low-potential-energy locations along a long Josephson junction. Low-frequency tests have shown correct operation and sampler measurements indicate single-flux-quantum bits.

Other circuits have been simulated and show promise for multi-gigahertz clocking. Some of these employ resistor-junction circuits and AND gates; three such gates with three-phase clocking can form one bit storage. The circuit shown in Fig. 24 has been simulated and showed correct operation at 25 GHz [39]. Low-speed tests have been successfully completed [40].

References

1. Kotani, S., Imamura, T., and Hasuo, S. A 1.5 ps Josephson OR gate. *In* "Technical Digest of the International Electron Devices Meeting, Washington, D.C.," pp. 884–885, 1988.
2. Buck, D.A. The cryotron—A superconductive component. *Proc. IRE* **44**, 482–493 (1956).
3. Josephson, B.D. Possible new effects in superconductive tunneling. *Phys. Lett.* **1**, 251–253 (1962).
4. Anderson, P.W., and Rowell, J.M. Probable observation of the Josephson superconducting tunneling effect. *Phys. Rev. Lett.* **10**, 230–232 (1963).
5. Matisoo, J. Subnanosecond pair-tunneling to single-particle tunneling transitions in Josephson junctions. *Appl. Phys. Lett.* **9**, 167–168 (1966).

6. Matisoo, J. Measurement of current transfer time in a tunneling cryotron flip-flop. *Proc. IEEE* **55**, 2052–2053 (1967).
7. Wada, Y., Nagasawa, S., Ishida, I., Hidaka, M., Tsuge, H., and Tahara, S. A 570 ps, 13 mW Josephson 1 kb RAM. *In* "Extended Abstracts, 1988 International Solid-State Circuits Conference, San Francisco, February 1988," pp. 84–85, 310–311, 1988.
8. Kotani, S., Fujimaki, N., Imamura, T., and Hasuo, S. A Josephson 4b microprocessor. *In* "Extended Abstracts, 1988 Int. Solid-State Circuits Conference, San Francisco, February 1988." pp. 150–151, 1988.
9. Klein, M. Analog to digital converter using Josephson junctions. *In* "Digest of Tech. Papers, Int. Solid-State Circuits Conference," Vol. XX. pp. 202–203, 1977.
10. Zappe, H.H. Ultrasensitive analog-to-digital converter using Josephson junctions. *IBM Tech. Disclosure Bull.* **17**, 3053–3054 (1975).
11. Hurrell, J.P., Pridmore-Brown, D.C., and Silver, A.H. Analog-to-digital conversion with unlatched SQUIDs. *IEEE Trans. Electron Devices* **27**, 1887–1896 (1980).
12. Hamilton, C.A., and Lloyd, F.L. 100 GHz binary counter based on DC SQUIDs. *IEEE Electron Device Lett.* **3**, 335–338 (1982).
13. Silver, A.H., Phillips, R.R., and Sandell, R.D. High speed nonlatching SQUID binary ripple counter. *IEEE Trans. Magn.* **21**, 204–207 (1985).
14. Hamilton, C.A. Private communication, 1988.
15. Lee, G.S. SQUAD: Superconducting QUADruplett. A superconducting 4-junction loop as a bidirectional bipolar incrementing quantizer. Submitted for publication (1989).
16. Lee, G.S. A variable hysteresis aperturing method for superconducting counting A/D conversion. *IEEE Trans. Magn.* **25**, 830–833 (1989).
17. Harris, R.E., Hamilton, C.A., and Lloyd, F.L. Multiple-quantum interference superconducting analog-to-digital converter. *Appl. Phys. Lett.* **35**, 720–721 (1979).
18. Hamilton, C.A., and Lloyd, F.L. A superconducting 6-bit analog-to-digital converter with operation to 2×10^9 samples/second. *IEEE Electron Device Lett.* **1**, 92–94 (1980).
19. Dhong, S.H., Jewett, R.E., and Van Duzer, T. Josephson analog-to-digital converter using self-gating-AND circuits as comparators. *IEEE Trans. Magn.* **19**, 1282–1285 (1983).
20. Davidson, A. A Josephson latch. *IEEE J. Solid-State Circuits* **13**, 583–587 (1978).
21. Petersen, D.A., Ko, H., Jewett, R.E., Nakajima, K., Nandakumar, V., Spargo, J.W., and Van Duzer, T. A high-speed analog-to-digital converter using Josephson self-gating-AND comparators. *IEEE Trans. Magn.* **21**, 200–203 (1985).
22. Fang, E.S., and Van Duzer, T. Speed-limiting factors in flash-type Josephson A/D converters. *IEEE Trans. Magn.* **25**, 822–825 (1989).
23. Hamilton, C.A., Lloyd, F.L., and Kautz, R.L. Superconducting A/D converters using latching comparators. *IEEE Trans. Magn.* **21**, 197–199 (1985).
24. Fang, E., Nandakumar, V., Petersen, D.A., and Van Duzer, T. High-speed A/D converters and shift registers. *In* "Extended Abstracts of the 1987 International

Superconductivity Electronics Conference (ISEC'87), August 28–29, 1987, Tokyo," pp. 325–328, 1987.
25. Ko, H., and Van Duzer, T. A new high-speed periodic-threshold comparator for use in a Josephson A/D converter. *IEEE J. Solid-State Circuits* **23**, 1017–1021 (1988).
26. Ko, H. A flash Josephson A/D converter constructed with one-junction SQUIDs. *IEEE Trans. Magn.* **25**, 826–829 (1989).
27. Petersen, D.A., Ko, H., and Van Duzer, T. Dynamic behavior of a Josephson latching comparator for use in a high-speed analog-to-digital converter. *IEEE Trans. Magn.* **23**, 891–894 (1987).
28. Petersen, D.A., Nandakumar, V., Fang, E., Hebert, D.F., and Van Duzer, T. Flash-type A/D converters and shift registers. *In* "Proceedings of SPIE Sensing, Discrimination, and Signal Processing and Superconducting Materials and Instrumentation, Los Angeles," Vol. 879, pp. 76–80, 1988.
29. Petersen, D.A., Hebert, D., and Van Duzer, T. A Josephson analog limiter circuit. *IEEE Trans. Magn.* **25**, 818–821 (1989).
30. Askerzade, I.N., Korncv, V.K., Semenov, V.K., and Schedrin, V.D. Josephson-junction balanced comparators: Dynamics and fluctuations. 1988 Applied Superconductivity Conference, San Francisco, CA, Aug. 21–25, 1988.
31. Shimizu, N., Harada, Y., Miyamoto, N., and Goto, E. New A/D converter with quantum flux parametron. *IEEE Trans. Magn.* **25**, 865–868 (1989).
32. Yao, Y.L., and Herrell, D.J. An experimental Josephson junction shift register element. *In* "Tech. Digest of Int. Electron Devices Meeting, Washington, D.C., December 9–11, 1974," pp. 145–148, 1974.
33. Fujimaki, N., Kotani, S., Imamura, T., and Hasuo, S. Josephson 8-bit shift register. *IEEE Trans. Solid-State Circuits* **22**, 886–891 (1987).
34. Beha, H., Jutzi, W., and Mischke, G. Margins of a 16-ps/bit interferometer shift register. *IEEE Trans. Electron Devices* **27**, 1882–1887 (1980).
35. Jutzi, W, Crocoll, E., Herwig, R., Kratz, H., Neuhaus, M., Sadorf, H., and Wunsch, J. Experimental SFQ interferometer shift register prototype with Josephson junctions. *IEEE Electron Device Lett.* **4**, 49–50 (1983).
36. Nandakumar, V., and Van Duzer, T. Design of a fast variable-frequency shift register. *IEEE Trans. Circuits and Systems* **35**, 1172–1174 (1988).
37. Kuo, F., Whiteley, S.R., and Faris, S.M. A fast Josephson SFQ shift register. *IEEE Trans. Magn.* **25**, 841–844 (1989).
38. Sakai, S. Fluxon shift register. Private communication, 1988.
39. Przybysz, J.X., and Blaugher, R.D. Josephson data latch for frequency agile shift registers. *IEEE Trans. Magn.* **23**, 777–780 (1987).
40. Przybysz, J.X., and Blaugher, R.D. Josephson shift register design and layout. *IEEE Trans. Magn.* **25**, 837–840 (1989).

CHAPTER 7

Wideband Analog Signal Processing*

RICHARD S. WITHERS

Analog Device Technology Group
MIT Lincoln Laboratory
Lexington, Massachusetts

1.	Functional Overview of Signal Processing	228
	1.1. Definition of Signal Processing	228
	1.2. Example: Signal Processing in a Radar System	228
	1.3. Signal-Processing Device Architectures	231
	1.4. Spectral Analysis	234
	1.5. Resonators	235
	1.6. Data Acquisition and Buffering	236
2.	Mature Signal-Processing Device Technologies	236
	2.1. Digital Integrated Circuits	237
	2.2. Charge-Coupled Devices	237
	2.3. Surface-Acoustic-Wave Devices	238
3.	Superconductive Devices	238
	3.1. Motivation	238
	3.2. Circuit Elements	238
	3.3. Families of Devices	239
4.	Fabrication Technology	255
	4.1. Delay Lines	255
	4.2. Mixers	260
	4.3. Lumped Elements	260
	4.4. Logic	261
5.	System Integration	261
	5.1. Spectrum Analyzer	261
	5.2. Pseudonoise Radar	261
6.	Comparisons and Conclusions	264
	6.1. Comparison with SAW Devices	264
	6.2. Analog and Digital Processing	267
	6.3. Use of High-T_c Superconductors	268
	Acknowledgments	269
	References	269

*This work is supported by the Departments of the Air Force, Army, Navy, and DARPA.

In many emerging applications, such as pulse-compression radar, spread-spectrum communications, and electronic warfare, real-time signal processing is stressed, and the required computational rate is of the order of 10^{12} arithmetic operations per second and the required instantaneous bandwidths approach 10 GHz. These exceed by nearly three orders of magnitude the capabilities projected for digital systems in the near future, and even exceed that of recently developed analog technologies, such as surface-acoustic-wave (SAW) signal-processing devices. To meet future system needs, superconductive analog signal-processing components with bandwidths exceeding 2 GHz have recently been developed [1,2,3] and the technology is being developed for the realization of 10-GHz bandwidths.

1. Functional Overview of Signal Processing

1.1. Definition of Signal Processing

The task of the signal processor is to accept signals from our analog world and extract the essential information. More specifically, the purpose behind the processing of signals is the improvement of the signal-to-noise or signal-to-interference ratio to an acceptable level at which to make detection and other decisions with high accuracy. Signal processing is usually accomplished by a combination of pre-processing, which involves linear techniques, and post-processing, which involves nonlinear or conditional techniques (e.g., threshold detection). This is indicated in Fig. 1.

1.2. Example: Signal Processing in a Radar System

A radar signal processor, depicted schematically in Fig. 2, is described as an example of a demanding signal-processing application [4]. The purpose of

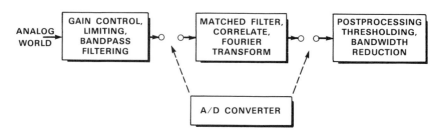

Fig. 1. Canonic signal processor. The A/D converter is placed in the left position if the computationally intensive operations (matched filtering, correlation, Fourier transformation) are performed digitally and in the right position if these operations are carried out in the analog domain.

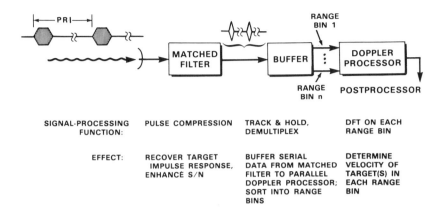

Fig. 2. Radar signal processing.

this processor is to produce a map of target intensity versus range and velocity. In order to radiate sufficient power, the transmitted pulses are typically much longer than the desired receiver temporal resolution. Target reflections of the periodically emitted pulses are first time-compressed to recover this range resolution to that limited only by the pulse bandwidth. For fixed waveforms, tapped delay lines can perform this function. For agile or nonrepeating waveforms, a programmable matched filter such as a convolver or time-integrating correlator is needed. The matched-filter output is buffered to reduce its bandwidth to that accepted by the Doppler processor. The data is reformatted into range bins, each range bin being a resolvable time sample at a fixed delay from the beginning of the waveform. The Doppler processor performs a discrete Fourier transform on the successive returns within each range bin and thereby determines the target velocity within each bin.

The buffering operation has been performed by high-speed semiconducting and optoelectronic devices and is a candidate for superconductive circuits. Doppler processing has been performed by CCDs and by digital circuits. As will be shown, all three functions illustrated in Fig. 2 can be performed by superconductive time-integrating correlators.

The motivation for the use of pulse-compression matched filters (e.g., dispersive delay lines) in radar is illustrated in Fig. 3. A short CW pulse has good range resolution but suffers from low average power and hence poor signal-to-noise ratio (S/N). A long CW pulse has higher average power but poor range resolution. This tradeoff is avoided by using a chirp waveform, which is simultaneously long (hence high power) and wideband (hence

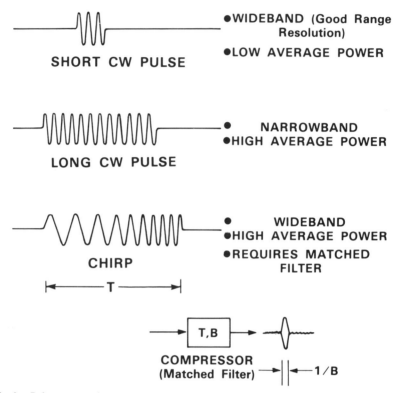

Fig. 3. Pulse compression radar. The chirp waveform provides both wide bandwidth (hence good range resolution) and high average power but requires a matched filter for pulse compression.

potentially good resolution). To recover the range resolution, a matched filter (pulse compressor) is required.

Figure 4 illustrates more generally how a pre-processor can enhance the signal-to-noise ratio (S/N) with a matched filter. The filter is chosen with an impulse response equal to the time-reverse of the signal to be processed. Consider equal signal and noise powers input to this filter. The noise, which does not match the impulse response, suffers the CW insertion loss of the filter as it passes to the output. The matched signal, however, has its energy collapsed coherently and an impulse is generated at the output. Functionally, the filter forms the autocorrelation of the signal. The width of the central lobe of the correlation is proportional to the bandwidth (B) of the waveform. The correlation peak power rises to a level above the noise at the output in direct proportion to the number of information cycles of the waveform gathered

7. WIDEBAND ANALOG SIGNAL PROCESSING

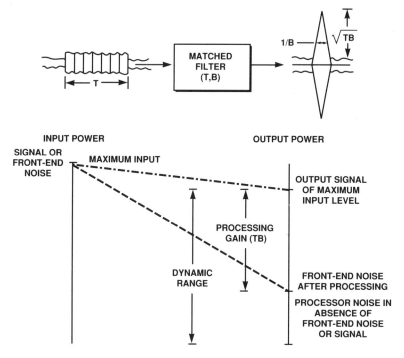

Fig. 4. Matched filter operation. The signal-to-noise power ratio at the output is increased by TB over that at the input. The lower part shows an input $S/N = 1$.

coherently in the filter. This improvement in S/N is termed processing gain, and, in an ideal filter, is given by the time–bandwidth (TB) product of the waveform. It is typically in the range of 20 to 30 dB, or 100 to 1000 cycles of information combined.

1.3. Signal-Processing Device Architectures

Most of the analog signal-processing structures are variations of the transversal filter indicated schematically in Fig. 5. A tapped delay line provides temporary storage and sampling of the input signal at the intervals τ. The samples are multiplied by weights w and coherently combined in space or time. Spatial summation is used for matched filtering and time integration for correlation.

Another important device in signal-processing systems is a stable frequency source. This is typically realized by a feedback oscillator stabilized by a high-Q resonator. Such resonators are made in the form of a lightly coupled section of

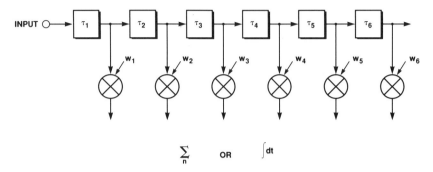

Fig. 5. Transversal filter architecture.

low-loss transmission line (analogous to a Fabry–Perot resonator in optics) or a delay line with periodically spaced taps or reflectors (analogous to a grating).

1.3.1. Fixed-Waveform Matched Filter

Figure 6 illustrates one of the classic analog signal-processing components and the functions it performs. The component is a dispersive delay line (DDL) and is most commonly designed with geometrically fixed weights and sampling intervals to produce a linear frequency-vs.-time characteristic. The impulse response is often referred to as an upchirp or downchirp. The most widespread application of such filters is currently at 10–500-MHz bandwidths (using acoustic devices; see [5]) in pulse-compression radars. The pulse expander is excited with an impulse to generate the waveform for transmission; the finite extent of the waveform permits adequate energy for target detection to be transmitted. Upon reception of the target return, the pulse compressor (of opposite chirp slope) provides processing gain. The wider the chirp bandwidth, the sharper the compressed pulse and the better the range resolution. Weighting accuracy is important to achieving low side lobes and an ability to discern weak scatterers adjacent to strong scatterers.

1.3.2. Programmable Matched Filter

Jammers that mimic the signal can negate the noise suppression of the matched filter. An efficient solution is modern systems is to use rapidly reprogrammable matched filters. An example of such a device, a convolver, is shown in Fig. 7. The device structure combines the elements of delay, distributed mixing, and spatial integration. The impulse response of the filter

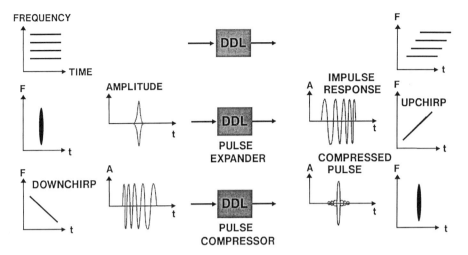

Fig. 6. Dispersive delay line pulse expansion/compression. The frequency plots illustrate instantaneous frequency $d\varphi/dt$ versus time.

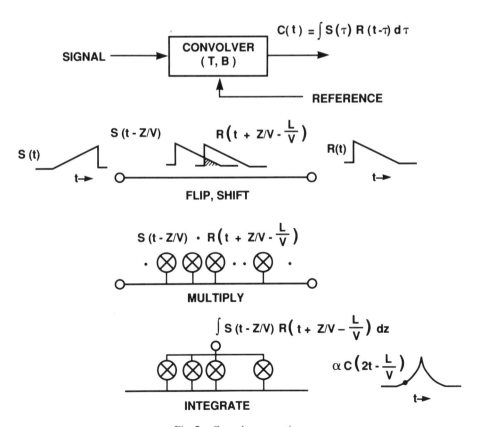

Fig. 7. Convolver operation.

is set by the reference that shifts past the incoming signal. The convolver is usually used as a programmable matched filter (within the time window set by the delay line length). This requires that the reference be the time-reversed replica (as indicated in Fig. 7) of the signal to be processed. Programmable matched filters using stationary weights have also been constructed [6]; these do not require constant regeneration of the reference waveform.

1.3.3. Correlator

Correlation is the integration over time of the product of a signal and a reference waveform. If the signal and reference are correlated and in synchronism, the correlation product will be large. The tapped-delay-line architecture of Fig. 5 can perform multiple correlations, each with a different time offset, if the multiplying elements can use the rapidly varying reference for the weights w_n, and if time-integrating circuits are provided at the output of each multiplier. In a radar system, the reference is a replica of the transmitted waveform. The output of each correlation cell is a function of the target cross-section in one range bin. With respect to a matched filter, the correlator has the advantages of an integration time (and hence TB product) not limited by the delay-line length and of parallel outputs of the range-bin data (lessening the demands on the bandwidth of the subsequent processor). The disadvantage is that the range extent that is processed is limited by the delay-line length.

1.4. Spectral Analysis

Fourier analysis is certainly an important signal processing function. Here, too, the constituent elements of the signal are coherently combined by means of the Fourier algorithm for enhanced S/N. Applications include Doppler analysis in moving-target and imaging radar and power spectral analysis in frequency-shift-keyed communication and radar-warning receivers. The standard Fourier integral can be written as

$$H(\mu\tau) = \exp(-j\mu\tau^2/2) \int h(t)\exp\left(-\frac{j\mu t^2}{2}\right)\exp\left(\frac{j\mu(t-\tau)^2}{2}\right)dt. \quad (1.1)$$

The quadratic phase terms are linear-frequency chirps with slope $\pm\mu$. It is seen that the time function $h(t)$ is Fourier analyzed by the following steps: multiply with a chirp, convolve with a chirp of opposite slope, and post-multiply with the original chirp. The analog hardware realization of this transform is shown

7. WIDEBAND ANALOG SIGNAL PROCESSING

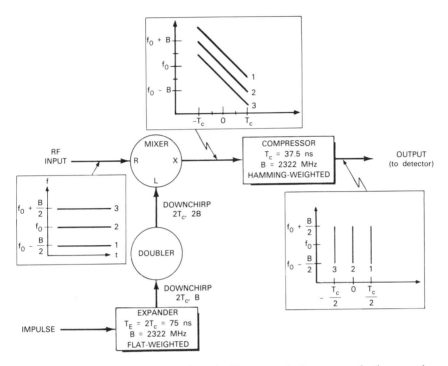

Fig. 8. Chirp-transform spectrum analyzer, in this case employing superconductive expander and compressor. Frequency–time plots are shown for various points in the signal path. The RF signal input shown in the figure is an example of the possible signals within the $f_0 - B/2$ to $f_0 + B/2$ analyzer input band; in this case, three tones are shown, one at f_0, a second at $f_0 - B/2$, and a third at $f_0 + B/2$. These are mapped by the analyzer into time slots spaced by $T_c/2$.

schematically in Fig. 8. Three chirp lines are required for full amplitude and phase analysis; the post-multiply chirp is neglected for power spectral analysis. Essentially the transformer maps the spectral channels into time bins. The analysis bandwidth is equal to the convolving compressor bandwidth and the number of channels is the TB product of that device. The pre-multiply chirp is required to be twice the T and B of the compressor.

1.5. Resonators

High-Q resonators require a low-loss transmission medium. A lightly coupled section of line of length l will resonate at frequencies $v/2l$, i.e., at frequencies for which it is an integral number of half-wavelengths long. The maximum quality factor is reciprocally proportional to the loss per wavelength in the medium, being exactly 27.3 divided by the loss in dB per wavelength. Such resonators are often used to stabilize low-noise oscillators, a critical function in Doppler radar and communication systems.

1.6. Data Acquisition and Buffering

Some analog processors, such as the time-integrating correlator, provide outputs that have much lower bandwidths than the input signals and which can be directly applied to much slower conventional circuitry for further processing. However, other processors, such as a radar pulse compressor or the real-time spectrum analyzer described above, produce outputs whose bandwidths exceed the input capabilities of post-processors. The solution of this interface problem typically requires sampling circuits and either data-reduction or demultiplexing circuits.

Sampling circuits must (1) have an input bandwidth at least that of the processor output and (2) take samples at a rate exceeding the Nyquist criterion. The first criterion is necessary to preserve the accuracy of the samples taken, while the second ensures that no information is lost. The first criterion requires fast circuit elements, in particular the active time-varying elements (e.g., diode switches) that set the sampling aperture. The second criterion is often met by interleaving a bank of sampling circuits, an architecture that also provides a demultiplexing (serial-to-parallel) function.

The second criterion is often less important than might appear on the surface. This occurs when the signal to be sampled is repetitive, as is often the case in laboratory situations where the measurement can be repeated. In this case, a few samples of the waveform to be measured can be taken during each measurement interval, and a set of samples sufficient to reconstruct the waveform is assembled after a number of intervals. This "sampling-oscilloscope" mode of operation also effects a time-buffering operation.

2. Mature Signal-Processing Device Technologies

Signal-processing functions are currently realized in a number of technologies [7]. Relatively off-the-shelf implementations are available in the form of digital integrated circuits [8], charge-coupled devices (CCDs) [9,10], and surface-acoustic-wave (SAW) devices [11,12]. These devices are compared in Fig. 9 in terms of their two most important parameters, bandwidth and processing time.

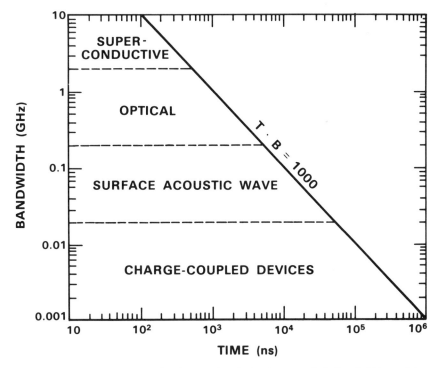

Fig. 9. Performance of analog signal-processing devices in terms of the bandwidths and time durations of signals.

2.1. Digital Integrated Circuits

Digital circuits offer the most flexible implementation but are limited in effective bandwidth for real-time operations because of the massive number of computations required for signal processing. Power and weight requirements become onerous at bandwidths of more than a few tens of megahertz.

2.2. Charge-Coupled Devices

CCDs are analog sampled-data devices in which signal samples are represented by charge packets that can be transported and operated on within the device. Specifically, tapped delay lines can be realized using this technology. CCDs are often fully programmable, low-power, and can operate at bandwidths of several tens of MHz. Sampling rates of 50 MHz are achieved using relatively modest silicon fabrication processes, and rates of a few hundred MHz are expected to result from more aggressive bipolar/CMOS technology [9] and from acoustically clocked GaAs devices [13].

2.3. Surface-Acoustic-Wave Devices

Surface-acoustic-wave (SAW) devices were developed through the 1970s and have secured a place in systems requiring high-Q filters [14] and dispersive delay [5]. Bandwidths of a few hundred megahertz, delays of tens of microseconds, and quality factors of 10^4 are readily achieved in compact, accurate SAW devices [12].

3. Superconductive Devices

3.1. Motivation

In order to obtain useful delay at still higher bandwidth and high quality factors at higher frequencies, devices based on superconductive electromagnetic delay lines have been under development. As will be seen in subsequent sections, it is the low microwave surface resistance of superconductors that makes their application to wideband signal processing so attractive. Quite simply, there is no way, other than with superconducting thin-film electromagnetic transmission lines, to achieve tapped delay at microwave frequencies in a compact package.

3.2. Circuit Elements

Devices for analog signal processing clearly must simultaneously perform multiple functions, such as delay, multiplication, and summation, all with adequate dynamic range. Superconductive components can perform all the required functions. A list of the functions, the most stressing device requirements, and a superconductive component that can provide each function is given in Table 1 [15]. Individual superconductive components (delay lines, couplers, and mixers) have been developed and are being integrated in configurations that will yield extremely wide bandwith devices for pulse compression, convolution, matched filtering, and correlation. Using these elements, dispersive delay lines with $T = 88$ ns and $B = 2.6$ GHz have been produced, as have convolvers with $T = 14$ ns and $B = 2$ GHz. Extension of the dispersive delay lines to greater dispersion, the development of a signal correlator based on the convolver technology, and the investigation of programmable tap weights for a transversal filter are underway, as

Table 1. Functions Required for Analog Signal Processing

Function	Requirement	Superconductive component
Delay	Low dispersion, low loss, compactness	Stripline
Tapping	Accurate weights	Proximity coupler
Multiplication	Adequate dynamic range	Tunnel-junction mixer
Spatial summation	Phase coherence	Microstrip
Time integration	Adequate storage time	LC or stripline resonator

is the ongoing development of materials technology needed for extending the operating temperature range of these devices.

3.3. Families of Devices

All of these devices exploit a single unique property of superconducting thin films: low RF surface resistance. This property permits the realization of long delay lines or high-Q resonators in very compact planar structures. This section of this chapter presents the dependence of the loss of planar transmission lines on surface resistance R_s and the actual values of R_s for several superconductors of interest. The following section presents the design, fabrication, and performance of superconducting chirp filters, and the subsequent section discusses superconductive resonators and oscillators.

3.3.1. Superconducting Transmission Lines

Planar superconducting transmission lines have been fabricated in the forms familiar to microwave engineers—stripline, microstrip, and coplanar—by relatively straightforward substitution of superconducting films for the standard copper or gold. However, as will be shown in this section, the low R_s of superconductors permits a great reduction in the cross-sectional dimensions of these lines and a corresponding miniaturization.

Consider the simplest of all planar transmission lines, the parallel-plate line shown in Fig. 10. Assume that we are in the limit $w \gg d$. The amplitude of a wave propagating down this line will diminish as

$$V(z) = V_0 \exp(-\alpha z), \quad (3.1)$$

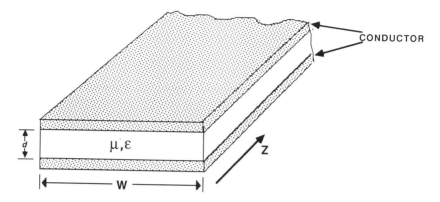

Fig. 10. Planar transmission line consisting of a dielectric of thickness d sandwiched between conducting planes of width w.

where α is the attenuation constant given by

$$\alpha = \frac{R_{\text{series}}}{2Z_0} + \frac{\pi \tan \delta}{\lambda}. \tag{3.2}$$

Here R_{series} is the series resistance per unit length, Z_0 is the characteristic impedance, $\tan \delta$ is the loss tangent of the dielectric, and λ is the wavelength on the line. The series resistance may be expressed in terms of the surface resistance as

$$R_{\text{series}} = \frac{2R_s}{w}. \tag{3.3}$$

With the characteristic impedance given by

$$Z_0 = \left(\frac{\mu}{\varepsilon}\right)^{1/2} \frac{d}{w}, \tag{3.4}$$

we find that the part of the attenuation constant attributable to the conductor loss is independent of the line width w and is

$$\alpha_c = \left(\frac{\varepsilon}{\mu}\right)^{1/2} \frac{R_s}{d}. \tag{3.5}$$

We may further define a conductor quality factor Q_c that is inversely proportional to the conductor loss per wavelength and is, in fact, the quality factor displayed by a lightly loaded resonant section of line with a lossless dielectric. This quality factor is

$$Q_c = \frac{\pi}{\alpha_c \lambda} \quad (3.6)$$

$$= \frac{\pi \mu f d}{R_s}, \quad (3.7)$$

where f is the frequency of excitation. With d expressed in micrometers, f in GHz, R_s in ohms, and $\mu = \mu_0$, the conductor quality factor is, numerically,

$$Q_c = 3.95 \times 10^{-3} \frac{fd}{R_s}. \quad (3.8)$$

The significance of this expression will be seen by examining the surface resistances of various materials. Figure 11 is a plot of measured and predicted

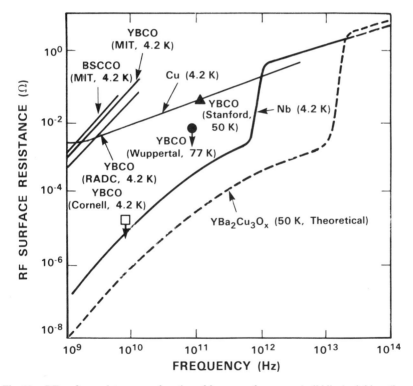

Fig. 11. RF surface resistance as a function of frequency for copper (solid line), niobium (heavy solid line), YBa$_2$Cu$_3$O$_x$, and Bi$_2$Sr$_2$Ca$_1$Cu$_2$O$_y$ (measured data points and dashed theoretical curve) at cryogenic temperatures. The downward-pointing arrows indicate that the measured point is an upper bound.

values of R_s for copper, niobium, and the high-T_c superconductor yttrium-barium-copper oxide (YBCO) as a function of frequency. (Copper, as all normal conductors, exhibits an $f^{1/2}$ or $f^{2/3}$ dependence of R_s, depending on whether it is in the normal or anomalous skin-depth regime. The superconductors exhibit a characteristic f^2 dependence of R_s.) At 10 GHz, cold copper has a resistance of about 10^{-2} ohms and therefore, with a 100-micrometer-thick dielectric, could not be used for a resonator with a Q exceeding 400. Also, since from Equations (3.1) and (3.6), we can see that the loss in a line n wavelengths long is 8.69 $\pi n/Q_c$ dB, copper could not be used for a delay line longer than 44 wavelengths if no more than 3 dB of insertion loss can be tolerated. This imposes a severe limit on the available time-bandwidth product of tapped-delay-line signal-processing devices.

On the other hand, it is clear that niobium, with a surface resistance at 10 GHz that is three orders of magnitude less, is not so constrained. Resonators with quality factors of several hundred thousand can (and have) been built, and tapped-delay-line signal processors with time-bandwidth products of many thousands are possible.

Another aspect of this analysis becomes evident upon studying the two neighboring striplines of Fig. 12. Symmetric stripline is typically used for high-density tapped-delay-line circuits because the upper ground plane provides maximum shielding between neighboring lines. Also note that, to conserve space as well as to maintain a reasonably high impedance (>10 ohms), the lines are not constructed in the parallel-plate limit of $w \gg h$. This results in somewhat higher losses than in the parallel limit because of current crowding.

To maintain reasonable isolation between the lines (>40 dB over a quarter wavelength), the striplines must be separated by a distance s of several substrate thicknesses h. It is this separation that in fact limits the density of the microwave circuit—for a delay line, it limits the number of nanoseconds of

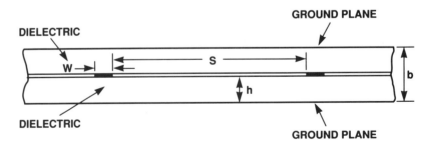

Fig. 12. Cross-section of the coupled stripline structure.

7. WIDEBAND ANALOG SIGNAL PROCESSING

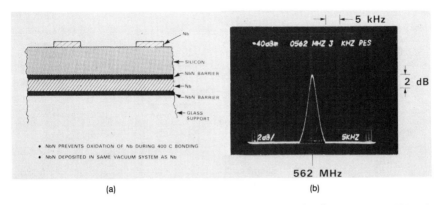

Fig. 13. Resonator consisting of a niobium center conductor, thinned epitaxial silicon dielectric, and composite niobium/niobium nitride ground plane bonded to a glass support wafer. [16] (a) Cross-section; (b) amplitude-vs.-frequency response near resonance.

delay per square meter of substrate. Superconductors permit the substrate thickness to be reduced—and the lines packed commensurately closer together and the delay per unit area increased—without incurring excessive losses. Niobium at 4.2 K can support devices with thousands of wavelengths on 10-micrometer-thick substrates. Figure 13 is proof of this, showing a resonator with a quality factor of 2.9×10^5 at 562 MHz on a 10-micrometer silicon substrate [16]. (Lightly doped silicon is an excellent dielectric at 4.2 K.) This structure was fabricated using a substrate bonding and thinning process developed by Anderson et al. [17] and is discussed in a later section.

Another advantage of superconducting lines is that they are dispersionless. This is because the skin depth is essentially a material property called the penetration depth and is constant with frequency to very high (submillimeter wave) frequencies. Normal conductors exhibit a skin depth that varies as $f^{-1/2}$ or $f^{-2/3}$, and thus exhibit a frequency-dependent inductance and consequent dispersion.

3.3.2. Tapped Delay Lines

Fixed-Weight Tapped Delay Lines. The basic device cross-section is shown in Fig. 12. A coupled pair of superconductive striplines of width w is separated by a variable distance s. The striplines are surrounded by two layers of low-loss dielectric of thickness h that are coated with superconductive ground planes.

The two lines are coupled by a cascaded array of backward-wave couplers

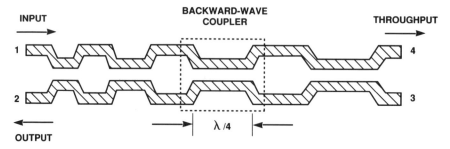

Fig. 14. Chirp filter formed by cascading backward-wave couplers between adjacent electromagnetic delay lines. A typical device actually contains hundreds of such couplers.

as shown in the schematic of Fig. 14. Each coupler has a peak response at frequencies for which it is an odd number of quarter-wavelengths long. If the length of the couplers is proportional to the reciprocal of distance down the line, then the resulting structure has a local resonant frequency that is a linear function of delay. In other words, it is a chirp filter with a linear group-delay-vs.-frequency (or quadratic phase) relationship. The strength of each coupler is controlled by varying the line spacing s to give a desired amplitude weighting to the response.

Design and Analysis. The design and analysis of stripline tapped-delay-line chirp filters was described previously [18]. The designer specifies the desired center frequency, bandwidth, insertion loss at center frequency, and weighting function. A grating function (coupling vs. spatial coordinate) is generated that includes a correction for amplitude distortions caused by strong coupling and the consequent input wave depletion. The design is analyzed using a coupling-of-modes solution; predistortions of the grating phase can be introduced at this point to compensate for phase distortions caused by strong coupling and the procedure iterated [19]. From the grating function, an electromagnetic analysis is used to specify the line spacing function, and a photomask is generated.

In typical systems employing chirp filters, one filter has a flat amplitude response, the full amplitude weighting function being carried by the second (quasi-matched) filter. Unlike most chirp filters, however, the superconductive tapped delay lines are four-port devices and, provided that propagation losses are small and the grating coupling is not too strong, the coupled responses (S_{12} and S_{34} in Fig. 14) are phase-conjugate; that is, the impulse response from one end is the time-reverse of that from the other. One filter can thus serve as its own matched filter. By weighting the filter with the square root of the desired weighting function, the overall response of a pulse-compression system will be the desired one. In some applications (e.g., spectrum analysis), however, this partitioning of weighting functions is disadvantageous and separate flat and fully weighted filters must be used.

Filters with characteristic impedances of less than 50 ohms offer several advantages: (1) greater width of the lines (for example, on silicon, 25-ohm lines are 4.5 times the width of 50-ohm lines) reduces the sensitivity of the impedance to linewidth variations, thereby reducing spurious reflections; (2) conductor losses become asymptotically small as the parallel-plate limit is approached; and (3) the perturbation in effective dielectric constant caused by voids between the dielectric layers is minimized in the parallel-plate limit. The first point is especially important as we scale down linewidths for thinner substrates. There is, however, some increase in sensitivity of impedance to substrate thickness.

Provision for impedances other than 50 ohms is included in the design routines [1]. Integral tapered-line transformers that present 50-ohm impedances at the device ports are included. Devices with 25-ohm lines have been fabricated; the results are presented in a following section.

Fabrication and Packaging. The delay line shown schematically in Fig. 14 would be difficult to fabricate, being typically 1 mm wide and 3 m long. To utilize round substrates with a minimum number of small-radius bends, the delay line is wound up into a quadruple-spiral configuration, with all four electrical ports brought to the wafer edge. The lower wafer (cf. Fig. 12) carries a niobium ground plane on its lower surface and the niobium stripline pattern on the upper surface. The upper wafer carries a single niobium ground plane.

The niobium films are deposited to a thickness of 300 nm by RF sputtering and patterned by reactive ion etching. The two substrates are 5-cm diameter, 125-μm-thick silicon wafers. The use of high-resistivity silicon as a low-loss dielectric substrate has been discussed [17]. Resonator measurements indicate that wafers with resistivities as low as 3 ohm-cm can support *TB* products of 1000.

In the package, the second wafer with its ground plane is held in place against the patterned wafer by an array of springs, preventing air gaps that would cause spurious electrical responses. The four RF terminals, brought to bonding pads at the wafer edge, are connected with multiple short wire bonds to SMA connectors brought through the back of the package. Although other laboratories use the wedge bonding of aluminum wires directly to niobium, we find it convenient to evaporate Ti/Au (20 nm/500 nm) films on top of the niobium pads, to which gold wires can be ball-bonded. These connections are reliable, can be removed for wafer cleaning, repackaging, or other processing without destroying the underlying metallization, and, provided that residual niobium oxides are stripped by a brief dilute HF dip prior to Ti/Au deposition, exhibit little resistance (0.2 ohm). The Ti/Au layer, patterned by liftoff of a spot-exposed photoresist layer, is also utilized to repair defects in the Nb striplines.

Electrical Performance. Both the measured and simulated amplitude responses of a flat-weighted device are shown together in Fig. 15a [20]. The simulations were made with the first-principles theory used to design the filter and were based on the physical design parameters of the device. Further details concerning the theory and simulations can be found elsewhere [18, 19]. The agreement between theory and experiment is excellent, even to the amplitude oscillations (Fresnel ripple).

The phase of a linear chirp filter is designed to be a quadratic function of frequency, the quadratic coefficient being proportional to the "chirp slope" of the device T/B. The measured best-fit quadratic to the phase-vs.-frequency response yields a chirp slope that is within 1.3% of the designed chirp slope. The remaining phase error (deviation from quadratic) has a rms value of 6.0°, which is very close to the predicted rms phase error of 5.9° and is an artifact of the Fresnel ripple rather than a device error that would degrade side-lobe performance.

The measured and simulated amplitude responses of a Hamming-weighted filter are shown in Fig. 15b [20]. Again the agreement is excellent. The measured chirp slope differs from the design value by 0.6%. The measured rms phase error has been weighted by the Hamming function to yield a weighted rms phase error of 3.1°, which compares favorably with the predicted weighted rms phase error of 3.5°.

The compressed pulse response of a matched pair of chirp filters was obtained by applying a 7-V video impulse of 85-ps width to the input of a flat-weighted filter [20]. The expanded upchirp response was then directly applied to the downchirp input port of the Hamming-weighted filter. The resulting compressed pulse had a 4-dB full-width of 0.6 ns, consistent with the 2.6-GHz bandwidth and Hamming weighting (which broadens the compressed pulse). The largest side lobe lies at -32 dB relative to the central peak.

Devices have been weighted with the square root of the Hamming function to give an overall Hamming response when using a single line as both

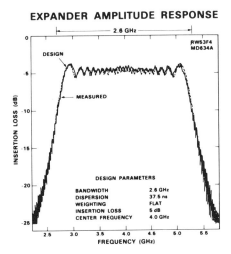

Fig. 15. (a) Predicted and measured insertion loss vs. frequency of a flat-weighted filter. (b) Predicted and measured insertion loss vs. frequency of a Hamming-weighted filter.

expander and compressor in pulse-compression tests. Side-lobe levels of −28 dB and phase deviations of 3.4° rms across the central 1.5 GHz of bandwidth have been obtained.

Tapped delay lines with 25-ohm characteristic impedance were made by using 173-μm-wide lines on 125-μm silicon [1]. Tapered transformers of 1-ns length were incorporated to bring the impedance to 50 ohms at all four ports.

The available dispersion on 5-cm-diameter wafers was reduced to 27 ns by the wider lines and the transformers. A 25.8-ohm impedance was measured by time-domain reflectometry, and a VSWR of less than 1.5 was observed. The 2.6-GHz-bandwidth root-Hamming-weighted devices exhibited a 7° rms phase error over the central 2 GHz of bandwidth and -27-dB side lobes.

Programmable Tapped Delay Lines. If the tap weights w_n in Fig. 5 are made programmable and the output, as in the dispersive delay line, is formed by summing the weighted samples, the device is an electrically programmable transversal filter. MOSFETs fabricated on the same silicon substrate as a niobium delay line have been used as programmable weighting elements [21]. The programming voltages, which can be input at rates much less than the signal bandwidth of the device, are applied to the gates of the FETs and thereby control the channel conductances and thus the RF transmission from source to drain. Similar devices have been used as weighting elements in SAW transversal filters operating in the UHF frequency range [6]. MOSFETs, operating on a frozen-out and hence insulating substrate and with the gate RF-bypassed to ground, can be used to control signals at microwave frequencies.

3.3.3. Convolvers

Operation of the superconductive convolver [22] is shown schematically in Fig. 16. A signal $s(t)$ and a reference $r(t)$ are entered into opposite ends of a superconductive delay line. Samples (delayed replicas) of the two counter-

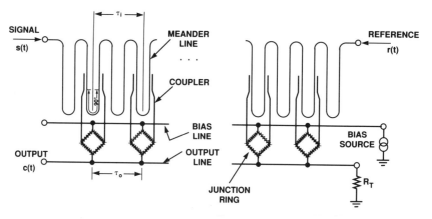

Fig. 16. Schematic of the superconductive convolver.

propagating signals are taken at discrete points by proximity taps weakly coupled to the delay line. Sampled energy is directed into junction ring mixers. The mixing products are spatially integrated by summing in multiple nodes connected to short transmission lines. The mixer ring with quadrature feed is chosen to enhance the desired cross product (signal times reference), while suppressing undesired self-products and higher order terms. Because both the $s(t)$ and $r(t)$ spatial patterns are moving, there is a halving of the time scale at the output; that is, the center frequency and bandwidth are doubled. If the reference is a time-reversed version of a selected waveform, then the convolver in effect becomes a correlator and functions as a programmable matched filter for that waveform.

Efficient mixing is obtained using the very sharp nonlinearity produced by tunneling of electrons in a superconductor–insulator–superconductor junction. The ring structure has a series array of four tunnel junctions in each of its four legs. The junctions consist of a niobium base electrode, a 3-nm thick niobium oxide tunnel barrier and a lead counterelectrode. Junction areas are defined by 4-μm-diameter windows etched in a silicon monoxide insulation layer. Unlike logic circuits that employ the Josephson current due to tunneling of superelectrons, for mixing the junctions are biased at the onset of quasiparticle (normal electron) tunneling because of the larger RF impedance available.

The convolver circuit is fabricated on a 125-μm-thin, 2.5 × 4 cm sapphire substrate with a niobium ground plane deposited on the reverse side. The central region of the device consists of a 14-ns meander delay line with a 50-ohm characteristic impedance. Twenty-five proximity tap pairs, located along opposite sides of the delay line, sample the propagating waveforms and direct the sampled energy into a corresponding number of junction-ring mixers. The resultant mixing products are collected and summed by two 15-ohm transmission lines located near opposite edges of the rectangular substrate. Each end of the output transmission line has a tapered line section that transforms the characteristic impedance of the line to a standard 50 ohms. The desired outputs from the two transmission lines are then summed externally with a microwave combiner. During device assembly, a second sapphire substrate with another niobium ground plane is placed against the delay line region of the first substrate to form a stripline circuit.

The real-time output of the convolver with CW input tones gated to a 14-ns duration and entered into signal and reference ports is shown in Fig. 17a [22]. The envelope of the convolver output has a triangular shape as expected for the convolution of two square input envelopes. The trailing side lobes are

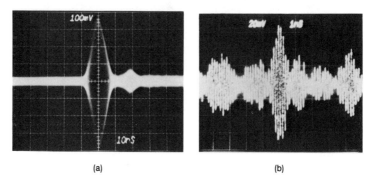

Fig. 17. Output waveform of convolver with input of (a) two gated CW input tones and (b) two linearly chirped waveforms.

associated with reflections in the measurement set and spurious signals in the device. The maximum output power level of the device is -58 dB and is set by the saturation of the mixer rings as determined by the number of junctions used in each leg.

In wideband measurements, input waveforms consisting of a flat-weighted upchirp and a complementary downchirp were applied to the signal and reference ports of the convolver. The waveforms were generated by two superconductive tapped-delay-line filters. The waveforms had chirp slopes of about 62 MHz/ns and were effectively truncated to bandwidths of about 0.85 GHz by the 14-ns-long interaction length of the convolver. The resultant output waveform shown in Fig. 17b has a bandwidth of about 1.7 GHz. Use of flat-weighted chirps should yield a $(\sin x)/x$ response with a null-to-null width of 1.2 ns and peak relative side lobes of -13 dB. A null-to-null width of 1.5 ns was observed with excessively high -7 dB side-lobe levels. These distortions are attributed primarily to mixer products produced from undesired leakage of input signal onto the output line and inadequate balance in the taps. Further engineering is required to provide adequate dynamic range.

3.3.4. Correlators

As in the convolver, the tap weights of the time-integrating correlator (TIC) (Fig. 18) [2,23] are controlled by a counterpropagating reference. However, the product signals generated by the mixers at each tap are not summed together spatially; instead, each one is integrated in time. The TIC thus forms a set of correlations between signal and reference waveforms with different time

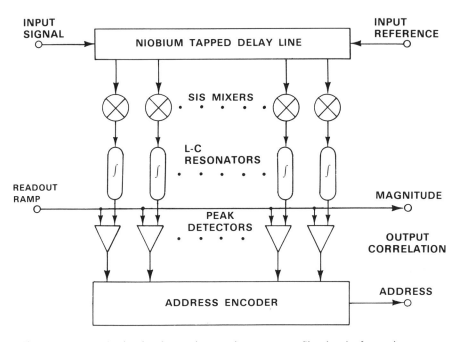

Fig. 18. Superconductive time-integrating correlator structure. Signal and reference inputs are differentially delayed by counterpropagating along the niobium line, tapped out at various points, and multiplied in the mixers. The difference-frequency products are integrated in the low-frequency LC resonators to form the correlations. By comparing these correlations with a readout ramp, the peak detectors identify the correlation cells with the largest contents and report them to the address encoder.

offsets. The TIC structure is similar to that of the convolver but requires two new elements: (1) low-frequency lumped-element resonators to provide time integration and (2) an output multiplexing circuit.

Signal and reference are offset by 20 MHz to allow AC integration in superconductive LC resonators. Anodized-niobium capacitors and meander inductors are used for this purpose. Process control of the element values is the most stringent demand because of the need for uniform resonant frequencies across the array of integrating cells.

The TIC, in contrast with the other devices described, provides an output at a much lower bandwidth than the input, making it more readily interfaced with lower-bandwidth room-temperature systems. However, the TIC provides many parallel output channels, only a few of which, at a given moment in a typical application, contain interesting output. A digital output circuit measures the signal amplitude in the resonators and reports first the addresses of those resonators with the largest stored signals. Josephson junctions embedded in the resonators serve as variable-threshold comparators, and multiple-input logic gates serve as address encoders.

3.3.5. Resonators

Structure, Design, and Fabrication. Superconducting resonators can be built in the form of TE or TM cavities [24] or as quasi-TEM transmission lines such as the planar versions discussed above. In either case, the achievable quality factor will be limited to a value inversely proportional to the RF surface resistance of the superconductor. The cavity resonators offer some advantage in having more energy stored for a given amount of surface current and thus a higher Q for a given R_s. Cavities also dispense with dielectrics and so are not limited by their loss tangent. Planar resonators, on the other hand, can be much more compact, since only one dimension must be a significant fraction of a wavelength (and this can be coiled into a compact form), and, not requiring the coating of curved surfaces, they are amenable to all thin-film deposition techniques. Planar resonators can also be integrated with other circuitry. Further discussion will be confined to the planar resonators.

Stripline, as depicted in Fig. 12, is a preferred structure, as it suffers no radiation loss and current-crowding effects are not as severe as in coplanar line. A stripline resonator is shown schematically in Fig. 19 [25]. A section of

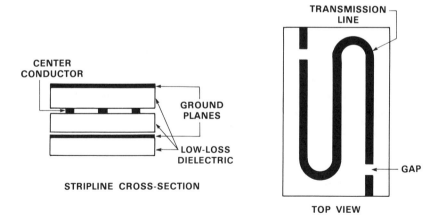

Fig. 19. Schematic view of a stripline resonator. On the left is a cross-section showing dielectrics and conductors. On the right is a top view of the center section showing the center conductor that has been patterned photolithographically. The gap shown determines the coupling to the resonant section of the line.

line is capacitively coupled at either end to input and output transmission lines. The circuit will resonate at frequencies for which the isolated line section is an integral number of half-wavelengths long.

Fabrication is conducted using the same processes employed to make tapped delay lines. However, the required line lengths are much less, so smaller substrates are used, typically 2.5 × 1.3 cm.

Electrical Performance. The quality factor of each resonance is given by

$$Q^{-1} = Q_c^{-1} + Q_d^{-1} + Q_L^{-1} = Q_{res}^{-1} + Q_L^{-1}, \qquad (3.9)$$

where Q_c is as given in Equation (3.6), $Q_d = 1/\tan \delta$, and Q_L is the loading Q given by

$$Q_L = \frac{n\pi}{4\omega^2 C^2 R_L Z_0}, \qquad (3.10)$$

where n is the mode number (the number of half-wavelengths within the resonator), ω is the angular frequency of excitation, C is the coupling capacitance at both ends of the resonator, and R_L is the source and load resistance. The insertion loss, in dB, at resonance is

$$L = 20 \log_{10} \left(1 + \frac{Q_L}{Q_{res}} \right), \qquad (3.11)$$

which provides for an independent determination of the coupling parameters.

It is evident from the surface resistance shown in Fig. 11 that Qs exceeding 10^5 can be achieved using Nb on low-loss dielectrics. This has indeed been achieved using Nb as well as NbN on substrates of silicon and sapphire.

Another important parameter is the maximum input power P. The limit imposed by the critical current of the superconductor is [25]

$$P_{in} = \frac{n\pi Z_0 I_c^2}{8 I_L (1 - I_L) Q}, \qquad (3.12)$$

where I_c is the maximum current for the center conductor, I_L the insertion loss (fractional power transmitted), and Q the loaded Q.

Oscillator Performance. Having calculated the Q and oscillator power that does not exceed the superconductor critical current, one can predict the values of phase noise to be expected from a simple feedback resonator stabilized by a superconducting resonator [25] using the standard Leeson model [26] of the phase noise of a feedback oscillator. The single sideband phase noise $L(\omega)$ is given by the following expression:

$$L(\omega) = 10\log_{10}\left[N^2\left(1 + \frac{\omega_0^2}{4Q^2\omega^2}\right)\left(\frac{GFkT}{P} + \frac{\alpha}{\omega}\right)\right], \quad (3.13)$$

where $L(\omega)$ is the noise power relative to the carrier in a one-hertz bandwidth given in dBc/Hz, P is the power at the output of the amplifier, G is the loop gain, F is the amplifier noise figure, ω the offset frequency in radian/s, Q the loaded resonator quality factor, ω_0 the basic (unmultiplied) oscillator frequency, α the empirically determined flicker-noise constant, k is Boltzmann's constant, T is the absolute temperature, and N the frequency-multiplication factor.

The flicker ($1/f$) noise can originate in either the amplifier or the resonator or both. Since the flicker noise is not well understood theoretically, it must be empirically determined. Reasonable estimates for the amplifier can be made, but, for the resonator, experimentally determined values must be used. The projections of phase noise that follow (Section 6.1.2) assume that the contribution of the resonator to the flicker noise is small compared to the amplifier contribution.

3.3.6. Bandwidth-Reducing Circuits

There are several signal-processing functions for which development of superconductive devices has only begun and which must be demonstrated in order to make it a viable technology. They all fall into what may be called bandwidth-reducing circuits. Such circuits are needed in order to interface the very wideband analog devices described earlier with lower-frequency, perhaps room-temperature electronics. The serial-in/parallel-out demultiplexer spreads data from a single input channel over several output channels by, for example, interleaved sampling. The fast-in/slow-out buffer reduces the peak data rate by spreading "bursty" data out uniformly in time. (A bank of the track-and-hold circuits developed by Go et al. [27] could be used for these functions.) Note that neither of these circuits alters the *total average* data rate. A thresholding and sorting circuit, on the other hand, does exactly this by discarding uninteresting samples, e.g., those below a certain threshold. Furthermore, it prioritizes the remaining samples by, e.g., sorting them based on amplitude. Such a thresholding and sorting circuit is incorporated in, for example, the superconducting correlator design.

4. Fabrication Technology

Although analog signal-processing devices incorporate many types of elements such as mixers, capacitors, and even logic gates, they are unique in their exploitation of delay lines. For this reason, the evolution of fabrication technology for delay lines is stressed in this section, with much less attention being given to the more conventional elements.

4.1. Delay Lines

4.1.1. Impact of Fabrication and Packaging on Performance

Two fabrication-related issues have been shown to exert a particularly strong effect on device performance [1]:

(1) *Wafer thickness variations.* The strength of 30-dB backward-wave stripline couplers varies by approximately 1 dB for every 8% change in substrate thickness of either wafer, and 60-dB couplers experience a 1-dB coupling variation for every 4% change in thickness. A uniform error in thickness causes a nearly frequency-independent error in insertion loss. Thickness errors that are nonuniform across the wafer surface, however, cause frequency-dependent amplitude errors and serious side-lobe degradation. A Hamming-weighted filter on a wedge-shaped substrate with a total thickness variation equal to 10% of its thickness was predicted, using a modified coupling-of-modes analysis, to have a 31-dB relative side-lobe level. To first order, thickness variations do not cause phase errors, as the effective dielectric constant and hence phase velocity are not affected.

(2) *Gaps between wafers.* The presence of gaps between the two dielectric wafers lowers the effective dielectric constant of the structure and consequently increases the propagation velocity. Phase distortions in chirp-filter response and frequency shifts in resonators result. Even if the gap is uniform over the wafer area, the inhomogeneous dielectric also produces differences in even- and odd-mode velocities and thus forward coupling [17] and consequent amplitude errors. In resonators, gaps are particularly troublesome if they change in response to mechanical vibrations of the package. Such microphonics are manifested as phase noise in oscillators. Gaps can result from several imperfections: (1) failure of the package to provide sufficient pressure to clamp the two wafers (8.8×10^4 N/m^2 is currently used); (2) the unavoidable step caused by the niobium thickness, possibly made larger by the nonselectivity of the RIE process and consequent etching of the silicon itself

following the removal of the niobium; (3) nonplanar wafer surfaces; and (4) particulate matter trapped between the two wafers.

The presence of gaps can be monitored by measurement of the chirp slope from which is inferred the effective dielectric constant. Numerous devices were tested which revealed effective constants as low as 10.0 instead of the design value of 11.4; poor overall performance was also noted. Investigation led to the conclusion that the third and fourth factors listed above were the cause. Single-side-polished wafers had been used as substrates. The spiral pattern was always defined on the polished surface and the deposition of the ground plane on the rough backside appeared to cause no degradation. However, the upper ground plane was deposited on the polished surface, requiring that the bare rough surface be assembled in contact with the spiral Nb pattern. These commercially etched surfaces were found to have an rms roughness of as much as 600 nm on lateral scales of 20 to 100 μm, in comparison with polished surfaces with rms roughnesses of 5 nm. The rms gap could easily exceed twice the roughness because of the peaked nature of the surface. Even in the parallel-plate limit, such a 0.5% void fraction (based on 125-μm-thick wafers) would, with a high-dielectric-constant material such as silicon, result in a 6% decrease in effective dielectric constant. With the simple corrective measure of using polished mating surfaces, effective dielectric constants of 11.5 ± 0.1 are achieved uniformly. For the elimination of trapped particulate matter between the wafers, tightened clean-room procedures during packaging are essential [20].

4.1.2. Longer Lines

Requirements. Bandwidths of several gigahertz are relatively easy to achieve with electromagnetic delay lines. Dispersion, however, is difficult to obtain because of the high propagation velocities.

The conductor loss of superconductive stripline increases with frequency and varies as the reciprocal of substrate thickness. Based on the assumption that the maximum tolerable attenuation of a dispersive delay line is 3 dB, the limit imposed by loss on the length of a niobium delay line operating at 4.2 K with a 4-GHz center frequency and 2.7-GHz bandwidth is about 13 μs on 125-μm-thick silicon and 1.0 μs on a 10-μm-thick substrate. Devices with these frequency specifications and a 370-ns dispersion ($TB = 1000$) require, assuming 55-dB isolation, areas of 200 cm^2 and 16 cm^2, respectively, on these two substrates. It is apparent that, to reach the nominal goal of $TB = 1000$, one must either: (1) use large-area (>100 cm^2) 125-μm-thick substrates; (2) electrically cascade filter sections on multiple 5-cm-diameter, 125-μm-thick substrates; or (3) use thinner 5-cm-diameter substrates.

Stacked Substrates. Two additional problems are encountered in cascading wafers [1]: (1) physically holding them together in the package and (2) making the necessary RF connections from wafer to wafer. A packaging scheme that addresses these aspects is shown in Fig. 20.

Two or more stripline assemblies (4 or more wafers) can in principle be held together by spring pressure in the same package used to house single assemblies. However, the possibility of relative motion during assembly makes the electrical bonding operations difficult. For this reason, phenolic resin (AZ 1350B photoresist) is used as an adhesive between each wafer pair to ensure the mechanical stability of the stacked structure. Measurements on a Nb-on-Si resin-bonded resonator demonstrate that a 375-nm film at the stripline interface increases the effective loss tangent of the 250-μm-thick structure by less than 2×10^{-5}. To ensure reliable, uniform contact, both mating surfaces

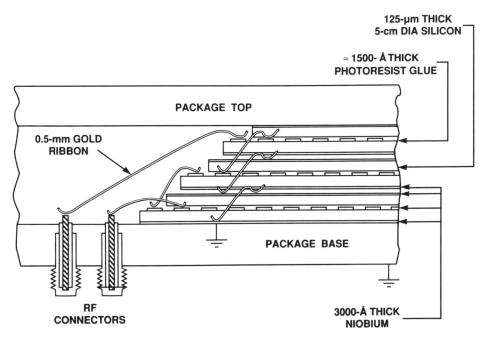

Fig. 20. Stack of three stripline wafer pairs. The connections to the ground planes and one of the two coupled lines are shown.

must be polished and coated with resin. The resin losses do not impose severe limits on devices on 125-μm-thick substrates but could limit attainable TB products on thinner substrates.

The resonator measurements also demonstrated a repeatability of resonant frequencies of approximately one part in 10,000 after each of five thermal cycles between room and liquid helium temperatures. This stability is a factor of six better than that of the same device using the identical spring-loaded package without resin. The technique therefore offers improvement for single-wafer-pair devices as well.

As suggested by Fig. 20, which for simplicity shows only half of the connections needed for a proximity-tapped delay line, low-parasitic-inductance interconnections are made with short ribbon or multiple-wire bonds. Distances are kept short by placing the bonding pads of one wafer nearly on top of the mating pads on the wafer below and slightly displacing the upper wafer laterally. Time-domain reflectometry indicates that the interlayer parasitic elements are comparable to those of the coax-to-stripline connections.

Thin Supported Wafers. Techniques for the fabrication of very thin dielectric superstrates supported by thick substrates with an intervening superconducting ground plane [17] were mentioned above. The steps in the process may be envisioned with the aid of Fig. 21 and are the following: (1) growth of a high-resistivity layer of epitaxial silicon on a heavily doped n-type Si wafer; (2) RF-sputter deposition of a Nb film onto the epitaxial layer; (3) bonding of the Si wafer to a 7740 glass (Pyrex) wafer using a low-temperature electrostatic bonding technique, with the Nb at the interface of the two wafers; (4) removal of the heavily doped silicon by a combination of mechanical lapping and nonselective and resistivity-selective chemical etching; (5) removal of a small area of the exposed epitaxial layer to allow contact to the interfacial Nb layer that will serve as the ground plane of the device that will be built on the epilayer; and (6) deposition of a Nb thin film on the silicon that can later be photolithographically patterned to form the stripline central conductor.

If the ground plane is omitted, this assembly may be viewed by transmission of white light. Red light is transmitted, and the uniformity of its intensity indicates the uniformity of the silicon thickness. The finished silicon superstrate has a thickness uniform to less than 1μm, comparable to the uniformity of the original epilayer.

To ensure uniform bonding, care must be taken that there is no epilayer

7. WIDEBAND ANALOG SIGNAL PROCESSING

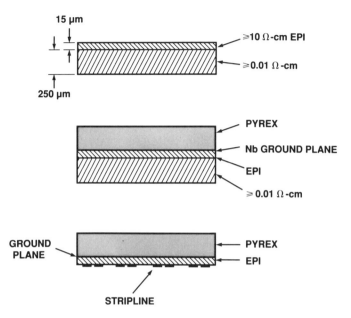

Fig. 21. Fabrication of 15-μm-thick silicon wafers by epitaxial growth, electrostatic bonding, and preferential etching. Top: starting epitaxial wafer. Middle: assembly after deposition of Nb ground plane and bonding to support wafer. Bottom: completed structure after preferential removal of substrate, niobium deposition, and patterning of the stripline.

"crowning" at the wafer edge. The possibility that a niobium/glass reaction occurs at the 400°C bonding temperature has been borne out by sputter-Auger analysis, which indicated permeation of the Nb film by oxygen. This has been circumvented by the use of reactively sputtered NbN [16], which is dense and less reactive. Sputter Auger measurements demonstrate that oxygen does not penetrate the film.

The resistivity-selective etch used to remove the heavily doped substrate is an HF: HNO_3: acetic mixture. Under optimum conditions the etch is very selective but it is inherently unstable; the generation of HNO_2 degrades the selectivity and necessitates careful monitoring. Potassium hydroxide has been used to etch holes through the epilayer without attacking the underlying niobium, permitting establishment of contact to the ground plane.

Resonator measurements have been conducted to verify the low RF surface resistance of the NbN ground plane and to ensure that recombination centers in the epitaxial silicon do not contribute to hopping conduction and increase the dielectric loss. Quality factors of 2.9×10^5 at 562 MHz and 4.2 K have been measured, as shown in Fig. 13 [16]. A 25-ohm, 220-ns-long delay line can be built on the 15-μm superstrates. Delays of up to 500 ns with a 5-dB loss at 5 GHz are expected to be acheivable on a 5-cm-diameter, 9-μm-thick superstrate.

4.2. Mixers

Mixers are used for the signal multiplication depicted in Fig. 5 as well as for the more usual frequency-translation applications. For both purposes, SIS tunnel junctions are biased to the knee of the quasiparticle branch of their $I(V)$ curve to obtain efficient low-noise mixing.

Fabrication of these mixers using Nb/Pb junctions [28] and, more recently, Nb/Al-AlO$_x$/Nb junctions [29] has been discussed. These SIS structures are similar to those used for Josephson junctions in logic circuits. The all-refractory junctions using niobium counterelectrodes offer somewhat higher voltage output. Output voltages and input and output impedances can also be increased by using series arrays of junctions [2]. NbN microbridges [30] have also been used. As with semiconductor diodes, balanced configurations can be used to suppress unwanted signal products.

4.3. Lumped Elements

Lumped-element components are needed in many signal-processing circuits: resistors for RF loads, mixer-bias supply, and logic gates; capacitors for DC blocks and low-frequency resonators; and inductors for RF blocks and low-frequency resonators.

Resistors can be realized as meander lines of thin-film tantalum [31], Ti and Ti-Pd-Au [32], Al [33], or palladium-gold [34], depending on the resistance value desired. Because of the relatively low values of resistivity available, resistors can often require considerable area.

Capacitors are made by anodizing a niobium base electrode and depositing an appropriate counterelectrode (Nb or Pb). Quality factors at 10 MHz of several hundred are obtained, although damage to the dielectric during sputtering of niobium counterelectrodes can damage the dielectric and reduce the Q [35], favoring the use of evaporated Pb. More recently, anodized tantalum dielectrics have been shown to give Q's exceeding 10^3 even with Nb counterelectrodes [36].

Inductors are readily made using niobium meander lines [34,2]. On thin-film dielectrics, it is often necessary to open holes in the ground plane to increase the inductance. Quality factors are extremely high at frequencies less than one GHz.

4.4. Logic

Digital logic has been described elsewhere in this volume and will not be discussed at length here. In analog signal-processing applications, digital circuits serve to process the output for acquisition by room-temperature electronics. For example, in the time-integrating correlator [23], resistively coupled multiple-input OR gates are used to encode the address of correlation cells whose contents have crossed a set threshold. Integrated-circuit processes have utilized $Nb/Nb_2O_5/Pb$-Au-In junctions [34], $Nb/Al\text{-}AlO_x/Nb$ [32,37], and $NbN/MgO/NbN$ [33].

5. System Integration

5.1. Spectrum Analyzer

An important application of chirp filters, in addition to radar pulse compression, is the real-time spectral analysis of wideband signals. Flat- and Hamming-weighted filters have been used as dispersive elements in a chirp-transform [38] real-time spectrum analyzer. Recently this performance has been improved in a 4-channel implementation covering 10 GHz [39].

The configuration shown in Fig. 8 is used. This arrangement is termed the multiply-convolve arrangement [40] because of the multiplication of the input signal by a chirp waveform followed by convolution of the product with a chirp waveform in the compressor.

The mixer used to form the product of the chirp and the RF input is a commercially available unit with broadband IF characteristics. All components, including amplifiers and filters not shown in Fig. 8, operate at approximately 60 K except for the expander and compressor, which operate in liquid helium at 4.2 K.

Figure 22 shows the compressor output in response to 18 sequential CW tones from 10.7 to 13.3 GHz in 150-MHz increments. The system responds to RF inputs with an amplitude uniformity of ± 0.5 dB, a resolution of 45 MHz, greater than 22 dB of channel isolation, and 45 dB of dynamic range.

5.2. Pseudonoise Radar

An imaging radar processor employing time-integrating correlators is illustrated in Fig. 23. Essentially, each TIC processes one Doppler channel, with

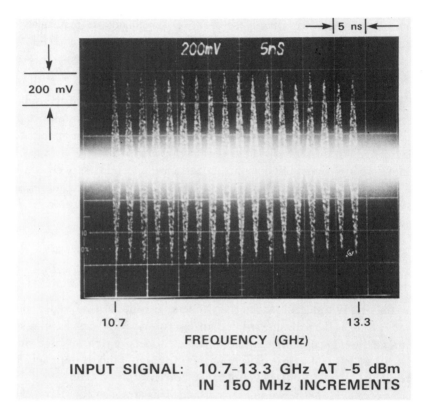

Fig. 22. Output of a chirp-transform spectrum analyzer in response to a set of tones uniformly spaced in frequency.

the Doppler frequency controlled by the LO offset of the reference waveform input and the Doppler resolution controlled by the integration time. Each cell of the correlator processes one range cell, the center of the total range window being set by the reference delay. The virtues of this architecture are the following:

(1) No high-speed range gate is needed; pulse compression and range/Doppler processing are performed in the same device. Time integration reduces the output data rate.

(2) Thresholding and sorting circuitry within the TICs reduces the volume of data. The controlling host processor can instruct the TICs to discard the data in all range/Doppler cells that do not exceed a specified threshold.

(3) The center frequency of the Doppler cells is set by the local oscillators in

the reference path, and the Doppler resolution is controlled by the integration time.

(4) The waveform is programmable and can be nonrepeating. There is no foldover in range or Doppler.

(5) If the reference waveform is literally a delayed version of the radiated waveform, rather than a regenerated replica, instabilities and inaccuracies in the RF source and modulator are for the most part compensated.

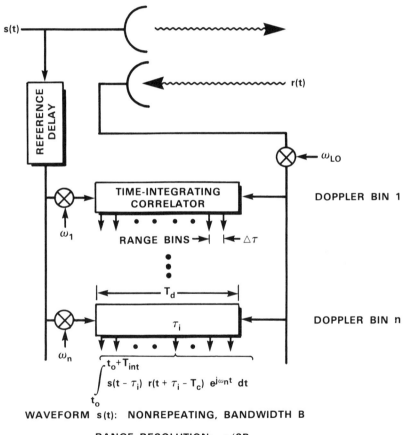

Fig. 23. Imaging radar processor with time-integrating correlators. Each correlator provides one Doppler channel with multiple range cells.

The principle drawback of the TIC-based processor is that the integration (i.e., coherence) time available in the present resonator-integrators is limited to some tens of microseconds, which is inadequate for discrimination of slowly moving objects. Longer coherence times may be obtained by higher-Q dielectrics or by developing coherent post-processors.

6. Comparisons and Conclusions

It is instructive to compare superconductivity with SAW and analog with digital processing. The next two sections compare superconductivity with SAW in the arenas of dispersive delay lines and resonators, respectively, and the third section compares analog with digital processing. The future impact of the new high-T_c superconductivity is discussed in the final section.

6.1. Comparison with SAW Devices

6.1.1. Dispersive Delay Lines

SAW DDLs have been under development for almost 20 years, and their characteristics are well known. Table 2 is an attempt to compare this rather mature technology with the fledgling technology of superconductive devices. As expected, the limits to SAW technology are relatively well known and are approached by existing devices, while the limits to superconductive technology are less well ascertained and are well beyond what has been achieved.

Table 2. Comparison of SAW and Superconductive Chirp Filters

Parameter	SAW		Superconductive	
	Achieved	Limit	Achieved	Limit
Bandwidth (MHz)	1,000	2,000	4,000	10,000
Dispersion (ns)	1.5×10^5	3×10^5	88	1,000
Time-bandwidth product	10,000	10,000	200	2,000
Phase accuracy (° rms)	1	0.2	3	1
Amplitude accuracy (dB)	0.2	0.05	0.1	0.05
Side-lobe level	45	55	33	45
Insertion loss (dB, CW)	15	15	5	2
Up- and down-chirp from same device?	no	no	yes	yes
Operating temperature (K)	300–350	—	4.2	77

(Even without the uncertainties of predicting the progress of technology, tables such as this one are full of pitfalls for the author. For example, the maximum time-bandwidth product is not the product of the maximum dispersion and maximum bandwidth, because the two cannot be achieved simultaneously. Also, side-lobe levels in SAW devices better than those listed have been achieved, but only for low-time-bandwidth devices.)

The salient advantage of superconductive devices is their bandwidth, which is already well beyond that which can be achieved by SAW DDLs. Being 4-port rather than 2-port devices with a phase-conjugate relationship between the two responses of the two terminal pairs, a lightly tapped superconductive delay line can be used for matched up- and down-chirp responses. Another significant advantage is their insertion loss; there is no transduction or diffraction loss and, in devices demonstrated to date that were limited in dispersion by the available substrate area, immeasurable propagation loss. Even the grating loss can be reduced to 2 dB by the use of coupling predistortion [19], although this precludes the use of a single device for both up- and down-chirps. Triple-transit effects are also minimal, even with low grating loss, because of the excellent match at input and output connectors.

The greatest advantage of the SAW devices is their dispersion. It is noteworthy, however, that the dispersion of superconductive devices demonstrated to date is limited not by fundamental physical phenomena such as loss but by the area and thickness of available substrates. This is a problem that will yield to further engineering effort, such as that which produced the 10-μm-thick subtrate of Fig. 13 or that which is being pursued to develop low-loss thin- and thick-film [41] dielectrics.

For either technology, there is a tradeoff between phase and amplitude accuracy (and hence side-lobe level) and time-bandwidth product, the bounds of which are related to the accuracy of the fabrication process. This is typical of analog devices. It should be noted that the phase accuracy shown in Table 2 for the superconductive devices does not account for the deviations from quadratic that are present in the design (e.g., Fresnel ripple) but do not contribute to side lobes. The *error* is thus less than that quoted.

The phase accuracy of SAW devices is somewhat better and may be further improved by *in situ* photochemical compensation [42], but superconductive devices, for which no phase-compensation technique has yet been demonstrated, may also yield to such techniques. It is entirely plausible that a modest improvement in fabrication control will be achieved, which would make compensation unnecessary.

SAW devices, of course, can operate at room temperature, a capability that would require superconductors with a transition temperature of 500 K. The highest reproducible T_c in thin films to date is 125 K, which could support operation near 77 K.

6.1.2. Resonators and Oscillators

Figure 24 shows the projected phase noise at 10 GHz for a superconducting resonator-stabilized oscillator [25]. The oscillator operates at 5 GHz and is frequency-doubled to obtain 10 GHz. The parameters are $\omega_0 = 5$ GHz, $Q = 1.5 \times 10^5$, $G = 15$ dB, $F = 6$ dB, $P = +25$ dBm (0.3 W), $\alpha = 4 \times 10^{-12}$, and $N = 2$. This value of α is that normally assumed for a GaAs FET amplifier [14]. The parameter values chosen for this curve are for illustrative purposes and should not be considered final. We have, for instance, assumed room-temperature operation for the amplifier, but when the amplifier is operated at the cryogenic temperatures, noise performance would be improved. Also shown in Fig. 24 are the phase-noise curves for quartz crystal [43] oscillators and for SAW oscillators [44]. Clearly the superconducting oscillator should provide better performance than other technologies.

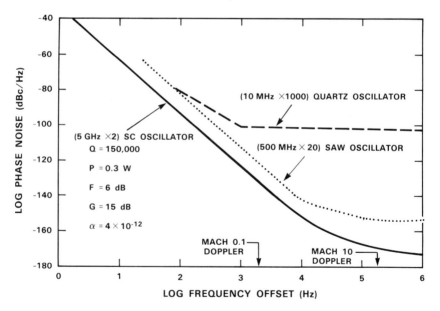

Fig. 24. Projected values of the single-sideband phase noise at 10 GHz for oscillators stabilized with resonators of three different technologies. The superconductive oscillator operates at 5 GHz and is multiplied up to 10 GHz. (See text for the detailed parameters of this oscillator.) Also shown are the phase noise for the best SAW oscillator and a production quartz crystal oscillator.

7. WIDEBAND ANALOG SIGNAL PROCESSING

Table 3. Comparison of SAW and Superconductive Resonators

		Superconductive	
Parameter	SAW	Demonstrated	Projected
Resonant frequency (GHz)	1	1.15	10
Q	5000	25,000	150,000
Maximum input power (W)	0.1	—	64
Flicker-noise constant (rad^2)	0.5×10^{-13}	10^{-11}	?
Oscillator phase noise (dBc, 10^5 Hz offset from carrier, 10-GHz oscillator)	-150	—	-170
Operating temperature (K)	300–350	4.2	77
Integrable with monolithic circuits?	no	yes	yes

The attributes of SAW and superconductive resonators are listed in Table 3. As with chirp filters, the superconductive technology permits a move to higher frequencies, which in this case lessens or eliminates the need for frequency multiplication. Substrate size, however, is not a limitation for resonators as it is for chirp filters. The superconductive devices have a clear advantage in terms of Q. Engineering refinements, such as the elimination of microphonics (caused by, for example, separation between the two dielectrics of a stripline structure), are beginning to be developed for the superconductive circuits.

6.2. Analog and Digital Processing

A reasonable basis of comparison is the Fourier transform [45]. The number of points N in the transform is equal to the TB product. The transform requires the order of N^2 operations as a direct computation. Substantial savings are possible by using the fast Fourier transform (FFT). For simplicity the radix-2 FFT is assumed. There are $(N/2) \log_2 N$ butterfly operations required for an N-point transform. Each butterfly requires 10 real arithmetic operations. Under the constraint of real-time processing, the digital rate required for the FFT is directly proportional to the signal bandwidth, i.e., at least $5 B \log_2(TB)$.

Correlation in time of a signal against a reference is equivalent to multiplying in the frequency domain. Thus, for a fixed filter response, two transforms are required, one to transform the signal into the frequency domain, where it is multiplied by a pre-computed weighting function, and a second to convert back to the time domain. For programmable filtering a forward transform of the changing reference is also required. This model for digital processing yields the equivalent computational rates of projected superconductive pulse compressors ($B = 4$ GHz, $T = 250$ ns) and convolvers ($B = 2$ GHz, $T = 110$ ns) as nearly 10^{12} operations/sec, orders of magnitude beyond existing or projected digital machines. Further, it should be noted that digital machines that are configured to push the state of art in computational speed are so streamlined in architecture that much of the flexibility typically attributed to digital technology is sacrificed. In a sense, the digital algorithms at the highest speeds take on much the same distributed nature as the analog components. The analog technologist is, of course, bound by the need to perfect a distributed structure whose physical behavior is described by equations equivalent to those of the desired signal-processing operation.

6.3. Use of High-T_c Superconductors

Since the discovery of superconductivity at high temperatures in copper-oxide based compounds by Bednorz and Mueller [46], a true revolution in the field has occurred. Previously unbelievable critical temperatures and critical fields have been demonstrated in an expanding family of compounds. However, the preparation of materials in practical forms remains a considerable challenge [47].

Progress in thin films has been more rapid than in bulk forms, a fact that encourages those who would exploit these materials in RF and other electronic applications [48]. However, a sobering fact is that practical Josephson junctions or quasiparticle tunnel junctions, necessary for digital logic circuits or RF mixers, respectively, have not yet been demonstrated.

On the other hand, the two applications highlighted in this section, as well as many others in the RF/microwave area, require just one of the unique attributes of superconductors: low RF surface resistance. As shown in Fig. 11, work is underway in this area. To date, thin films of the new materials are better than normal conductors at low microwave frequencies but worse at higher frequencies. Theoretical understanding is being developed [49], and experimental work is being directed at producing single-phase material and at controlling the orientation of grains of these anisotropic conductors and the properties of the grain boundaries.

For microwave applications, the loss of the dielectrics must also be low, with $\tan \delta < 10^{-4}$ in most applications. The otherwise preferred substrate

SrTiO$_3$ is inadequate in this respect. ZrO$_2$ and especially MgO are better prospects for low loss. Other substrates, such as LaGaO$_3$ [50] and LaAlO$_3$ [51], appear even more promising. For some of these materials, and certainly for candidates such as GaAs, which offer favorable semiconducting as well as dielectric properties, it will be necessary to lower the temperature of formation of the superconductor and/or provide diffusion barriers. Considerable work is underway.

There is consensus in the field of superconducting thin films that RF applications such as the ones described here will be one of the first major impact areas of high-T_c superconductivity. The ability to operate in monolithic or hybrid circuits with semiconducting devices, at a temperature which is near optimum for both, and the ability to be deployed in forward locations such as antenna arrays are two properties that are unique to high-T_c superconductors and not mere quantitative improvements over the older material systems. If surface resistances are driven down as we expect, a revolution in applications will ensue.

Acknowledgments

The contributions of my colleagues Alfredo Anderson, Manjul Bhushan, Maureen Delaney, Mark DiIorio, Jonathan Green, Daniel Oates, Richard Ralston, Stanley Reible, and Peter Wright, now or formerly in the Analog Device Technology Group at the MIT Lincoln Laboratory, are gratefully acknowledged.

References

1. Withers, R.S., Anderson, A.C., Green, J.B., and Reible, S.A. Superconductive delay-line technology and applications. *IEEE Trans. Magn.* **21**, 186–192 (1985).
2. Green, J.B., Smith, L.N., Anderson, A.C., Reible, S.A., and Withers, R.S. Analog signal correlator using superconductive integrated components. *IEEE Trans. Magn.* **23**, 895–898 (1987).
2. Ralston, R.W. Signal processing: opportunities for superconductive circuits. *IEEE Trans. Magn.* **21**, 181–185 (1985).
4. Cook, C.E., and Bernfeld, M. "Radar Signals." Academic Press, New York, 1967.
5. Williamson, R.C. Properties and applications of reflective-array devices. *Proc. IEEE* **64**, 702–710 (1976).
6. Oates, D.E., Smythe, D.L., Green, J.B., and Withers, R.S. SAW/FET programmable filter with varistor taps for improved performance. *In* "Proc. IEEE Ultrasonics Symposium," pp. 155–158, 1988.
7. Grant, P.M., and Withers, R.S. Recent advances in analog signal processing. *IEEE Trans. Aerospace and Elec. Sys.*, submitted (1988).

8. Brodersen, R.W., and Moscovitz, H.S. "VLSI Signal Processing III." IEEE, New York, 1988.
9. Munroe, S.C., Arsenault, D.R., Thompson, K.E., and Lattes, A.L. Programmable, four-channel, 128-sample, 40 Ms/s analog-ternary correlator. *In* "Proc. 1989 IEEE Custom Integrated Circuits Conference," 1989.
10. Barbe, D.F., ed. "Charge-Coupled Devices." Springer-Verlag, Berlin, 1980.
11. Withers, R.S. A comparison of superconductive and surface-acoustic-wave signal processing. *In* "Proc. 1988 Ultrasonics Symposium," IEEE, pp. 185–194, 1988.
12. Oates, D.E. Surface acoustic wave devices. *In* "VLSI Electronics" (N. Einspruch, ed.). Springer-Verlag, Berlin, 1984.
13. Hoskins, M.J., and Hunsinger, B.J. Recent developments in acoustic charge transport devices. *In* "Proc. IEEE Ultrasonics Symp.," pp. 439–450, 1986.
14. Parker, T. Characteristics and sources of phase noise in stable oscillators. *In* "Proc 41st Annual Frequency Control Symp.," pp. 99–110, 1987.
15. Reible, S.A. Wideband analog signal processing with superconductive circuits. *In* "Proc. 1982 IEEE Ultrasonics Symp.," pp 190–201, 1982.
16. Anderson, A.C., Marden, J.A., and Withers, R.S. Thin stripline dielectrics with passivated superconductors. *Solid State Research Quarterly* (MIT Lincoln Laboratory, Lexington, Mass.) No. 3, 29–32 (1985).
17. Anderson, A.C., Withers, R.S., Reible, S.A., and Ralston, R.W. Substrates for superconductive analog signal processing devices. *IEEE Trans. Magn.* **19**, 485–489 (1983).
18. Withers, R.S., Anderson, A.C., Wright, P.V., and Reible, S.A. Superconductive tapped delay lines for microwave analog signal processing. *IEEE Trans. Magn.* **21**, 480–484 (1983).
19. Withers, R.S., and Wright, P.V. Superconductive tapped delay lines for low-insertion-loss wideband analog signal-processing filters. *In* "Proc. 37th Annual Frequency Control Symp.," pp. 81–86, 1983.
20. DiIorio, M.S., Withers, R.S., and Anderson, A.C. Wideband superconductive chirp filters. *IEEE Trans. Microwave Theory Tech.* **37**, pp. 706–710, (1989).
21. Delaney, M.A., Withers, R.S., Anderson, A.C., Green, J.B., and Mountain, R.W. Superconductive delay line with integral MOSFET taps. *IEEE Trans. Magn.* **23**, 791–795 (1987).
22. Reible, S.A. Superconductive convolver with junction-ring mixers. *IEEE Trans. Magn.* **21**, 193–196 (1985).
23. Green, J.B., Anderson, A.C., and Withers, R.S. Superconductive wideband analog signal correlator with buffered digital output. *SPIE* **879**, 71–75 (1988).
24. Padamsee, H., Smathers, D., Marsh, R., and VanDoran, D. Advances in production of high-purity Nb for RF superconductivity. *IEEE Trans. Magn.* **23**, 1607–1616 (1987).
25. Oates, D.E., Anderson, A.C., and Steinbeck, J. Superconducting resonators and high-T_c materials. *In* "Proc. 42nd Annual Frequency Control Symposium," pp. 545–549, 1988.
26. Leeson, D.B. Short term stable microwave sources. *Microwave Journal* **13**, 59–69 (1970).

27. Go, D., Hamilton, C.A., Lloyd, F.L., DiIorio, M.S., and Withers, R.S. A superconducting analog track-and-hold circuit. *IEEE Trans. Electron. Dev.* **35**, 498–501 (1988).
28. Reible, S.A. Reactive ion etching in the fabrication of niobium tunnel junctions. *IEEE Trans. Magn.* **17**, 303–306 (1981).
29. Inatani, J., Kasuga, T., Sakamoto, A., Iwashita, H., and Kodaira, S. A 100-GHz SIS mixer of Nb/Al-AlO$_x$/Nb junctions. *IEEE Trans. Magn.* **23**, 1263–1266 (1987).
30. Hamasaki, K., Yakihara, T., Wang, Z., Yamashita, T., and Okabe, Y. High-frequency properties of all-NbN nanobridges. *IEEE Trans. Magn.* **23**, 1489–1492 (1987).
31. Anderson, A.C. Private communication, 1988.
32. Yu, L.-S., Berry, C.J., Drake, R.E., Li, K., Patt, R., Radparvar, M., Whiteley, S.R., and Faris, S.M. An all-niobium eight-level process for small and medium scale applications. *IEEE Trans. Magn.* **23**, 1476–1479 (1987).
33. Radparvar, M., Berry, M.J., Drake, R.E., Faris, S.M., Whiteley, S.R., and Yu, L.-S. Fabrication and performance of all-NbN Josephson-junction circuits. *IEEE Trans. Magn.* **23**, 1480–1483 (1987).
34. Sandstrom, R.L., Kleinsasser, A.W., Gallagher, W.J., and Raider, S.I. Josephson integrated-circuit process for scientific applications. *IEEE Trans. Magn.* **23**, 1484–1488 (1987).
35. Bhushan, M., Green, J.B., and Anderson, A.C. Low-loss lumped-element capacitors for superconductive integrated circuits. *IEEE Trans. Magn.*, **25**, 1143–1146 (1989).
36. Bhushan, M. Superconductive Nb thin-film capacitors with Ta_2O_5 dielectric. *Solid State Research Quarterly* (MIT Lincoln Laboratory, Lexington, Mass.) No. 4, 55 (1988).
37. Hasuo, S. High-speed Josephson integrated circuit technology. *IEEE Trans. Magn.*, **25**, 740–749 (1989).
38. Withers, R.S., and Reible, S.A. Superconductive chirp-transform spectrum analyzer. *IEEE Electron. Dev. Lett.* **6**, 261–263 (1985).
39. DiIorio, M. Private communication, 1987.
40. Judd, G.W., and Estrick, V.H. Applications of SAW chirp filters—An overview. *Proc. Soc. Photo-Opt. Instrum. Eng.* **239**, 220–235 (1980).
41. Wong, S., and Anderson, A.C., and Rudman, D.A. Processing of thick-film dielectrics compatible with thin-film superconductors for analog signal processing devices. *IEEE Trans. Magn.*, **25**, 1255–1257 (1989).
42. Dolat, V.S., Sedlacek, J.H.C., and Ehrlich, D.J. High-accuracy post-fabrication trimming of surface acoustic wave devices by laser photochemical processing. *Appl. Phys. Lett.* **53**, 651–653 (1988).
43. Gerber, E.A., Lukaszek, T., and Ballato, A. Advances in microwave acoustic frequency sources. *IEEE Trans. Microwave Theory Tech.* **34**, 1002–1016 (1986).
44. Montress, G.K., Parker, T.E., and Loboda, M.J. Extremely low phase noise SAW resonator oscillator design and performance. *In* "Proc. 1987 IEEE Ultrasonics Symp.," pp. 47–52, 1987.

45. Cafarella, J.H. Wideband signal processing for communication and radar. *In* "Proceedings 1983 National Telesystems Conference," pp. 55–58. IEEE, New York, 1983.
46. Bednorz, J.G., and Mueller, K.A. *Z. Physik* **B64**, 189–193 (1986).
47. Murphy, D.W., Johnson, Jr., D.W., Jin, S., and Howard, R.E. Processing for the 93 K superconductor $Ba_2YCu_3O_7$. *Science* **241**, 922–930 (1988).
48. Clarke, J. Small-scale analog applications of high-transition-temperature superconductors. *Nature* **333**, 29–35 (1988).
49. Hylton, T.L., Kapitulnik, A., Beasley, M.R., Carini, J.P., Drabeck, L., and Grüner, G. Weakly coupled grain model of high frequency losses in high-T_c superconducting thin films. *Appl. Phys. Lett.* **53**, 1343–1345 (1988).
50. Sandstrom, R.L., Giess, E.A., Gallagher, W.J., Segmüller, A., Cooper, E.I., Chisholm, M.F., Gupta, A., Shinde, S., and Laibowitz, R.B. Lanthanum gallate substrates for epitaxial high-temperature superconducting thin films. *Appl. Phys. Lett.* **53**, 1874–1876 (1988).
51. Simon, R.W., Platt, C.E., Lee, A.E., Lee, G.S., Daly, K.P., Wire, M.S., and Luine, J.A. Low-loss substrate for epitaxial growth of high-temperature superconductor thin films. *Appl. Phys. Lett.* **53**, 2677–2679 (1988).

CHAPTER 8

"MBE" Growth of Superconducting Materials

A.I. BRAGINSKI*
and
J. TALVACCHIO

Westinghouse Science & Technology Center
Pittsburgh, Pennsylvania

1. Introduction 273
2. "MBE" Apparatus for Superconductor Growth 277
   2.1. Analogy with Semiconductor MBE Systems—Requirements 277
   2.2. Sources and Deposition Rate Control 281
   2.3. Surface Analysis 282
3. Materials Systems 285
   3.1. Substrates for Epitaxial Growth 285
   3.2. Nb-Based Structures 290
   3.3. NbN-Based Structures 293
   3.4. A15 Superconductors 298
   3.5. High-T_c Oxide Superconductors 310
4. Conclusions 317
   References 317

1. Introduction

Recent attempts to fabricate superconducting devices of refractory low temperature superconductors (LTS) led to the exploratory use of film growth by "molecular beam epitaxy". We use this term and its acronym (MBE) in quotes since it represents a widely accepted but unfortunate misnomer. Molecular beams are really not employed to attain epitaxial growth of crystalline superconductors by physical vapor deposition (PVD). The ultra-high-vacuum growth apparatus and methodology are analogous to MBE as it is often used to fabricate semiconducting materials and devices. However, to evaporate refractory, high melting and boiling point materials—such as Nb, Mo, or Y—electron-beam heated sources are necessary. Sputtering and ion sources are also employed to deposit films. The analogies and distinctions

---

* Present address: ISI, Kernforschungsanlage Juelich, D-5170 Juelich, FRG.

between methods of superconductor and semiconductor growth will be reviewed in Section 2.

We assume that at this point the reader is familiar with basic concepts of superconducting electronic devices and circuits. The two most basic components of these circuits are the Josephson junction, an active nonlinear device, and a passive superconducting transmission line. By far the most widely used type of Josephson device—at least until the present time—is the tunnel junction. Practically all components and integrated circuits are now being fabricated exclusively by thin-film techniques typical of large-scale integration (LSI). As in the case of semiconductor technology, the component and circuit geometries are obtained by depositing multi-level layered film structures and, at various levels, patterning the circuit elements by photolithography. In these structures, superconducting films must interface with thin and thick insulators, normal metals, and, possibly, also semiconductors. In practical devices, the structures included, until recently, only polycrystalline superconductor films. The other materials were either amorphous or polycrystalline. The reader can find an extensive discussion of pertinent material problems in a review by Beasley and Kircher [1]. Update review articles concentrating on refractory materials have also been published [2,3].

The rationale for "MBE" growth of films of conventional superconductors is related to the length scales for superconductor perfection set by the two characteristic lengths: the superconducting coherence length ξ and the magnetic penetration depth λ. In superconductor–insulator–superconductor (SIS) tunnel junctions, the probed superconductor properties that define the junction characteristics are those within a depth of the order of the superconducting coherence length ξ from each interface with the barrier. Within this depth, the superconductor properties, i.e., the critical temperature T_c and the energy gap Δ, should be those of the bulk of the film. However, at interfaces between films of various materials some degradation of properties is usually unavoidable. Causes of the degradation can be many, for example, an interdiffusion and chemical reaction between the films or chemical reaction of one superconducting film with the ambient atmosphere prior to the deposition of a subsequent layer. Even if these effects are negligible, interfacial structural disorder will reduce T_c and Δ. The resulting depth of degradation observed in conventional polycrystalline superconductors d_c is on the order of 10 to 100 nm, in reasonably optimized processing conditions. When d_c is comparable to or greater than ξ, the junction characteristics are also degraded. The probed proximity layer in each electrode has now a lower T_c or may even be nonsuperconducting by itself. Consequently, the sumgap voltage V_g and also the critical current I_c are reduced [4], while the subgap conductance $1/R_s$ and

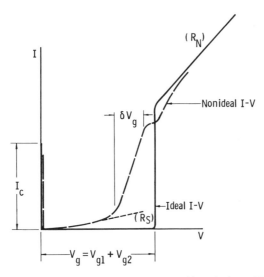

Fig. 1. Comparison of ideal tunnel junction $I-V$ curves with typical non-ideal characteristics.

the leakage current below the gap voltage are higher. The width of transition at the gap voltage δV_g (from the subgap to normal conductance $1/R_n$) broadens, and a proximity-effect knee appears. These deviations from an ideal, low temperature ($t = T/T_c < 0.5$) current–voltage ($I-V$) tunnel junction characteristic are shown in Fig. 1. The barrier non-ideality also increases $1/R_s$. Finally, the highest possible temperature of operation T_{op}, usually $T_{op} = 0.5\ T_c$, is defined by the lowest value of T_c near an interface with the barrier.

In a transmission line, surface and interface perfection of the superconductor within λ from the boundary is required to insure the lowest possible rf or microwave surface resistance and losses at a given frequency and temperature.

Junctions with Type I superconductor electrodes, having ξ of the order of 100 nm, are readily fabricated with nearly ideal $I-V$ characteristics at low t without much concern for interfacial degradation. In the case of Type II superconductors, where $\lambda \gg \xi$, it is usually much more difficult to fabricate high-quality tunnel junctions than to attain low rf surface losses in transmission lines. Table 1 includes representative examples of characteristic length scales of polycrystalline Type II superconductors listed in order of increasing T_c and decreasing ξ. Film thicknesses necessary to attain $>75\%$ of bulk T_c are also indicated. Obviously, the sensitivity of SIS junction characteristics to interfacial degradation must increase with the T_c of electrodes. In the new, highest-T_c oxide superconductors (HTS), this sensitivity must be extreme.

Table 1. Representative Parameters of Important Superconductor Films

Superconductor	Nominal transition temperature (K)	Coherence length ξ (nm) (∥Cu-O, ⊥)	Penetration depth λ (nm) (∥Cu-O, ⊥)	Required film fabrication temp. (°C)	Thickness for ≥75% T_c d_c (nm) (Non-epitaxial)	Thickness for ≥75% T_c d_c (nm) (Epitaxial)
Pb	7.2	90	40	20	3	—
Nb	9.2	40	85	20–800	25	5
Mo-Re (bcc, A15)	12, 15	20, —	70, —	20, 800	—	10
NbN	16	4	300	50–700	15	<1
Nb_3Sn	18	3	65	750–950	25	8
Nb_3Ge	23	3	90	>850	130	40
$La_{1.85}Sr_{0.15}CuO_4$	40	3.7, 0.7	80, 430	800–900	—	—
YBCO	92	3.1, 0.4	27, 180	~550–900	~200	5
Bi-Sr-Ca-Cu-O	85–110	3.8, 0.16	25, 500	870	≫100	≫100
Tl-Ba-Ca-Cu-O	108–125	⟨2.6⟩	⟨220⟩	850	≫100	≫100

Epitaxial, single crystal tunnel devices were first proposed mainly to solve the problem of stresses in soft alloy electrodes and to eliminate device failures resulting from thermal cycling [5]. Other possible advantages, however, were also mentioned by the proponents. There was, however, no further activity in this area until the early 1980s. We then realized that this approach could eliminate or minimize the degradation of T_c and Δ due to crystalline disorder near the electrode/barrier interfaces. In addition, it became clear that the *in-situ* fabrication of all or the most critical layers in an enclosed high- or ultra-high-vacuum (HV or UHV) system should minimize the contamination of free surfaces prior to the next layer deposition. These considerations led us to the exploration of "MBE" growth of refractory superconductors such as Nb, NbN, and Nb_3Sn. The next logical step was to attempt fabrication of epitaxial bi- and trilayers with insulating tunnel barriers. Both, single-crystal and polycrystalline epitaxy have proven effective. A new impetus, however, toward the use of "MBE" was provided by the advent of electronically anisotropic materials (HTS). Deposition of epitaxial films is necessary to obtain—in the plane of the film—the requisite high critical current density, $J_c = 10^5$ to 10^6 A/cm^2 at the temperature of possible utilization. Tunneling into films with parameters well defined by crystalline orientation imposed by epitaxy will probably become a necessity although at this writing technologically meaningful tunneling characteristics have not yet been demonstrated in layered film structures incorporating HTS.

2. "MBE" Apparatus for Superconductor Growth

2.1. *Analogy with Semiconductor MBE Systems—Requirements*

Considerations that led to the development of MBE growth of semiconductors [6] are also applicable to epitaxial superconductor growth. Most important are the abilities to

(1) Attain and maintain cleanliness of epitaxial substrate and film surfaces (e.g., absence of water and organic adsorbate layers) for periods of time long enough to make possible the storage between processing steps and surface characterization of the crystalline structure, composition, and contaminants prior to the subsequent processing step.

(2) Sequentially deposit epitaxial layers of similar and dissimilar materials while meeting the first requirement at all fabrication stages.

(3) Co-deposit alloys and compounds from multiple sources while maintaining a precise composition control and uniform coating thickness.

(4) Deposit at low growth rates, typically between 0.01 and 1 nm/sec to promote crystalline perfection without incorporating background impurities into the film.

(5) Maintain an acceptable throughput in the system.

To meet requirements (1) and (2), and also (4), a UHV environment must usually be employed, especially for single-crystal LTS epitaxy. Oil-free pumping methods are preferred over diffusion and turbomolecular pumps when carbon contamination is to be avoided. Cryopumping insures quick pumpdowns to 10^{-7} to 10^{-9} torr. This makes it particularly convenient for lock-chamber evacuation and degassing of sources and specimens. Pumping down further, to 10^{-10} to 10^{-11} torr, is then achieved by an ion pump, possibly assisted by a titanium sublimation pump (TSP). The UHV system should be bakeable to a temperature defined by the type of gaskets, at least 150 to 200°C. Pumps must be capable of handling the degassing load created during the bakeout.

Sequential deposition and analysis are most conveniently attained in a closed, multi-chamber system where relatively incompatible functions or fabrication methods can be grouped and separated by gate valves. Suitable manipulators and linear transfer mechanisms are installed for transporting the substrate wafers or specimens between chambers without exposure to ambient atmosphere. Prior to inserting into the system, these wafers and specimens are mounted or clamped on transfer holders or blocks that are compatible with the transfer mechanism. The manipulator in a deposition chamber is often equipped with a rotary feedthrough permitting one to rotate or oscillate the holder in the flux path of evaporant(s) to improve the deposit uniformity.

Essential for an acceptable throughput (requirement (5)) is a lock (introduction) chamber that permits one to maintain the UHV environment for many fabrication cycles without the necessity of a long-duration pumpdown and bakeout for each material batch. The throughput is also enhanced by the possibility of simultaneously carrying out different processing and analytical steps in separate, gated chambers, thus permitting the user a parallel fabrication and characterization of several batches.

Growth of epitaxial and single-crystalline layers usually requires heating the substrate to a well-defined temperature in order to clean it, attain an adequate atomic surface mobility during growth, and also carry out processes such as a bulk interdiffusion, homogenization, grain growth, chemical reactions, etc. In typical MBE systems, the maximum temperature attainable does not exceed 800°C. Higher temperatures, up to 1200°C, are necessary for

8. "MBE" GROWTH OF SUPERCONDUCTING MATERIALS

substrate cleaning and deposition or processing of A15 structure superconductors and some oxide insulators. Such high temperatures can be attained by flat strip tantalum resistive heaters. The wafers or specimens are heated by either direct thermal radiation or by contact anchoring to the thermally irradiated holder/block. Surface irradiation by halogen lamp heaters can also be employed. Small area surface heating by electron beam can be attained if temperatures in excess of 1200°C are required.

The problem of accurate temperature measurement and control on the surface of the substrate is a difficult one. Proper instrumentation of a transferable and rotatable holder is virtually impossible. The temperature of a solid block is best monitored by a thermocouple seating in a well formed in the center backside of the block. Calibration of this thermocouple to the sample surface temperature can be performed by a variety of methods including melting point observation, optical pyrometry, comparison with temporary thin film thermocouple data, etc. The uncertainty of measurement depends on the quality of thermal anchoring, with $\pm 20°C$ being a typical figure.

A block diagram of the "MBE" system used by the authors for epitaxial superconductor and layered film growth is shown in Fig. 2 [7]. Figure 3 shows

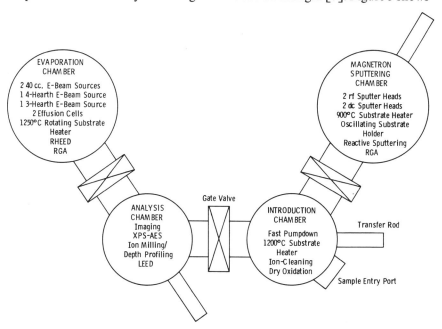

Fig. 2. Schematic of the deposition and analysis facility used at Westinghouse for "MBE" growth of superconductor films.

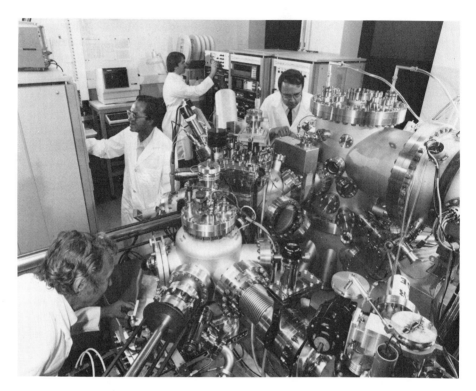

Fig. 3. Photograph of the Westinghouse deposition and analysis chambers.

the general view of the facility. This oil-free system meets satisfactorily the requirements discussed above and is equipped with most of the features reviewed in this and following sections. The advent of oxide superconductors necessitated an adaptation of the apparatus to the growth and processing in the presence of dry oxygen partial pressures ranging from 10^{-7} torr to 1 atmosphere. Obviously, the need to use UHV techniques appears questionable in this case. In our case, the first level of adaptation consisted of replacing tantalum heaters and shields with platinum, halogen lamp, and Kanthal heaters and stainless steel shields. Molybdenum transfer blocks and heater support elements were replaced with equivalents fabricated from Haynes alloy No. 230 (Haynes International, Inc., Kokomo, Indiana 46902), which resists oxidation at high temperatures. Unfortunately, it has a low thermal conductivity—detrimental to temperature uniformity of substrates clamped to the block. These changes permitted us to deposit YBCO films at

8. "MBE" GROWTH OF SUPERCONDUCTING MATERIALS

partial pressures of oxygen up to 10^{-2} torr and to *in-situ* oxidize and react (anneal) the deposits in 1 atmosphere of O_2 at temperatures up to 900°C.

2.2. Sources and Deposition Rate Control

Deposition of various materials may lead to the use of various source types and deposition rate control methods. For example, in semiconductor fabrication the co-evaporation of low-melting-point Ga, As, and Al can be conveniently performed from resistively heated Knudsen and effusion cells where a high-precision rate and deposit composition control (requirement (3)) can be attained by means of electronic temperature controllers. Evaporation of epitaxial silicon and of refractory metals or compounds usually requires an electron-beam-heated source due to the evaporants' high melting point. In this case, a rate control apparatus monitors the evaporant flux and uses a rate-proportional signal in a feedback loop to control the power and focus of a stationary electron beam or sweep amplitude of a scanning beam on the surface of an evaporant charge. The control of power results in a long response time, of the order of 1 sec. Much faster fluctuations of rate may occur in molten pools of refractory material and this necessitates using an additional feedback scheme acting on the sweep amplitude or beam focus with a time constant of much less than 0.1 sec [8]. However, even such elaborate rate control schemes are usually less precise than the resistive heater temperature control.

The choice of an e-beam gun rate control method is dictated by the vacuum environment and the type of material to be deposited. In a HV environment, rotating choppers can be used to generate an ac signal proportional to the flux of evaporant(s), which can be sensed by one or several ion gauge sensors suitably positioned in the chamber to monitor the temporal and spatial variation of flux [8]. Unfortunately, UHV-compatible chopper motors have not been readily available and, until the present, sensors different from ion gauges have been used. For certain elements, such as Si or Cu, electron impact emission spectroscopy (EIES) can be used to attain a rate control within $\pm 2\%$ [9]. For some other elements, and especially Nb, the EIES signal-to-noise ratio is too low to permit the rate control to better than $\pm 5-10\%$. Good success with short-term Nb rate control was obtained using a cross-beam mass spectrometer as a sensor [10]. Independent of the method of film deposition and growth rate control, an "MBE" growth chamber should be equipped with multiple vibrating crystal thickness monitors for absolute rate calibration. For co-depositions, the crystals should be positioned such that rate calibration is possible for individual sources and that accumulation on the substrate block can be independently measured.

The deposition of refractory materials is rarely required in semiconductor fabrication so that sources of different types are rarely grouped in the same deposition chamber. In contrast, co-evaporation of compound superconductors typically involves both high-melting-temperature T_m (Nb, V, Y) and low-T_m (Sn, Ga, Al, Ge, Ba, Sr, Bi, etc.) sources. Consequently, evaporator chambers may incorporate both e-beam guns and effusion cells. Sputtering or ion-beam deposition of epitaxial, single crystal layers is, *a priori*, less desirable than co-evaporation since the surface mobility of deposited atoms is reduced in the relatively high-pressure inert gas environment, the probability of gas impurity incorporation is augmented and bombardment by energetic particles is likely to occur, all leading to a reduced crystalline perfection. Nevertheless, crystals of superconductors and insulators incorporating gaseous components, such as NbN or oxide compounds, are conveniently grown by reactive sputtering or reactive ion beam deposition. The incorporation of the gaseous component species (N, O) is effectively done during growth, even when using a metallic source(s). The need to co-deposit or integrate dissimilar materials in one layered film structure may lead, therefore, to the incorporation of sources employing various deposition principles into either a common chamber of a UHV system or separate but interconnected chambers. This latter alternative was adopted in the "MBE" system shown in Figs. 2 and 3. A specific discussion of various deposition methods is included in Section 3.

2.3. Surface Analysis

Analytical tools that might be installed in a superconductor "MBE" system are similar to those in use for semiconductors. Only surface-sensitive techniques need to be considered. In the case of analytical tools that probe length scales on the order of typical electronic film thicknesses, 0.1 to 1.0 μm, such as x-ray diffraction, Rutherford backscattering (RBS), or electron microprobe, films can generally be removed from vacuum and transferred to a dedicated analytical machine without compromising the measurement. Similarly, surface-sensitive measurements that involve the removal of the contaminated surface layer can be performed *ex-situ*, such as secondary ion mass spectroscopy (SIMS), Auger depth profiling, or surface analysis by laser ionization (SALI) [11].

The remaining surface-sensitive techniques can be categorized by their utility for structural or chemical analysis. In the former category are reflection high-energy electron diffraction (RHEED), low-energy electron diffraction (LEED), and scanning tunneling microscopy (STM). The most common techniques in the latter category are x-ray photoelectron spectroscopy (XPS or, sometimes, ESCA) and Auger electron spectroscopy (AES). An additional technique, ellipsometry, measures optical properties determined by both surface composition and structure. All of these techniques are sensitive to surface contamination or reaction with background gases in the vacuum system. Ultra-high vacuum conditions are therefore necessary as much for *in-situ* analysis as for film growth. To allow for no more than 0.1 monolayers of contamination in the one hour time that is typically needed for analysis, one should operate in a pressure $< 10^{-10}$ torr [12].

Most analytical tools are installed in MBE systems in a chamber adjoining the growth chamber—not in the growth chamber. The exceptions are RHEED and ellipsometry [13] because the grazing angle incidence on the film and the high energy of the incident beam (10–100 keV electrons for RHEED) permit the beam source and the detector—usually a phosphor screen for RHEED—to be placed far from the sample where they do not obstruct the substrate and are not coated during deposition. The advantage of incorporating RHEED into an MBE growth chamber is that monolayer-by-monolayer growth of some semiconductor films can be monitored in real time [14]. The intensity oscillations of RHEED beams associated with monolayer growth have not been observed during superconductor growth. One reason is that most epitaxial superconducting films are composed of at least one refractory element. The black-body spectrum of the evaporating refractory source contains enough visible light to mask the RHEED pattern on a phosphor screen. Secondary electrons from the electron-beam sources that must be used with refractory materials would presumably lead to similar problems for other types of RHEED detectors. A more fundamental reason why RHEED oscillations have never been observed in superconductor film growth is that the oscillations are related to a step-propagation growth mode that may simply not occur in any of the superconducting materials. In conclusion, RHEED has generally been used to study the surface of epitaxial superconductors after film growth has stopped so there is no major advantage to having the technique available in the growth chamber.

For superconducting films, the only information gained from RHEED has been qualitative: indications of surface smoothness [7,15], evidence that some regions of the film have grown epitaxially [16,17], and the observation of reconstructed surfaces [18]. The depth scale from the surface that is probed by RHEED depends on surface roughness and varies from a few monolayers for smooth films to hundreds of nanometers for surfaces that are rough on that scale. General reviews of the merits and applications of RHEED have been published by Lagally [19] and Cohen et al. [14].

In contrast to RHEED, there is no reason why LEED cannot be used in a quantitative way with epitaxial superconductor films. Qualitatively, LEED has been used to indicate that some regions of the surface of a superconductor have a periodic structure related to the substrate [20], much in the same way as RHEED. Unlike RHEED, the depth from the surface that is probed is reliably known to be a few monolayers, independent of surface roughness. Quantitative LEED analysis involves the collection of diffracted beam intensity as a function of incident beam energy (LEED $I-V$ curves) [21]. The $I-V$ curves can be compared with a series of curves generated from structural and electronic models of the surface in a way that is widely used in other fields but has only recently been applied to superconducting film surfaces [22]. The only applications of LEED $I-V$ curves to the study of multilayer superconducting film structures have been measurements of the lattice constants of artificial tunnel barriers [23].

No other structural surface analysis techniques have been used for *in-situ* characterization of superconductors. The structure of the surfaces of superconductors have been studied by scanning tunneling microscopy (for example, [24]) and by field ion microscopy (for example, [25]). The apparatus for either of these measurements could be attached to a deposition chamber to study as-deposited surfaces.

The most common analysis techniques for obtaining *in-situ* information about surface composition and chemistry are XPS and AES. The usual advantage of AES is the small area that can be analyzed. The area on which an electron beam can be focussed is on the order of 1 μm^2, compared to typical x-ray beam diameters of several millimeters. However, for *in-situ* analysis of the surface of films that have been deposited over an entire wafer or chip that has dimensions that exceed a few millimeters, the chemical-shift information that can be obtained with XPS makes it the more useful technique. The chemical shift of XPS peak energies is due to the formal valence state of the atom from which a photon has been emitted and the atomic environment [26]. An example of its utility is the distinct separation that can be seen between the energies of photoelectrons emitted from oxidized metal overlayers used as artificial tunnel barriers, and from any part of the metallic overlayer that remained unoxidized [27]. Too many applications of the *in-situ* analysis of superconductor film structures have been published to list them here, but some examples are discussed in Section 3. For a detailed treatment of the relative merits and applications of XPS and AES, see Briggs and Seah [28].

8. "MBE" GROWTH OF SUPERCONDUCTING MATERIALS 285

3. Materials Systems

3.1. Substrates for Epitaxial Growth

Table 2 contains a complete list of substrates used for epitaxial growth of important superconductors. Some of the superconductor films listed in Table 2 are epitaxial but not single-crystal. In those cases, the symmetry of the substrate surface is higher than the symmetry of a parallel plane in the film. The film can nucleate in more than one equivalent orientation. For example, α-Al_2O_3($11\bar{2}0$) has two-fold symmetry and NbN(111) has three-fold symmetry. There are two distinct orientations in which NbN(111) will nucleate and the resulting grains will be separated by stacking faults. The references cited in Table 2 are intended to be either the first or the most complete description of a particular epitaxial relationship.

At least one of the faces of sapphire (single-crystal α-Al_2O_3) is a suitable epitaxial substrate for all of the conventional (non-oxide) superconductors listed in Table 2. It is a particularly useful substrate material since it is readily available, has good mechanical and thermal properties, and has low dielectric losses at rf frequencies. Sapphire has the most complex crystal structure of the substrates considered here, so the relative orientations of the film and substrate are not obvious and have been, therefore, listed in Table 3. Table 3 contains the growth direction of the film, the indices of the parallel planes that contain the growth direction, and the lattice mismatches in two orthogonal directions. The orientations of Mo, MgO, and Si films grown on sapphire are included for reference.

The listing of lattice mismatches in Table 3 invites predictions by numerology of when epitaxy will occur. At one time, a mismatch of $\geq 15\%$ was thought to eliminate any possibility of epitaxial growth [49]. The data in Table 3 confirms the common wisdom that a "critical lattice mismatch" is an inadequate criterion for epitaxy. Although there is no adequate set of criteria for predicting epitaxial relationships, the important issue in cases where there is a large misfit is not so much whether epitaxial growth occurs, but whether the resulting strain or dislocation density in the film proves to be detrimental to technological applications. This issue has been most thoroughly examined for the Si-on-sapphire system, and only tentatively approached for epitaxial superconductor films.

Substrate preparation is one of the most important factors in determining the structural properties of epitaxial films. The surfaces of oriented and polished substrates generally have a layer of contamination and crystalline disorder due to adsorbed gases from the air, which, additionally, may react with the surface. Absorbed carbon and other organic contamination can

Table 2. Summary of Substrates for Epitaxial Superconductor Films

Epitaxial superconductor	Single-crystal substrate	Film orientation	Reference
Nb	α-Al_2O_3(0001)	(111) (110)[a]	[29]
Nb	α-Al_2O_3(11$\bar{2}$0)	(110)	[29]
Nb	α-Al_2O_3(1$\bar{1}$02)	(100) (110)	[30]
Nb	α-Al_2O_3(10$\bar{1}$0)	(211)	[31]
Nb	α-Al_2O_3(21$\bar{1}$3)	(113)	[32]
Nb	MgO(100)	(100)	[33]
Nb	MgO(111)	(111)	[32]
Nb	GaAs(100)	(100)	[34]
Mo-Re	α-Al_2O_3(11$\bar{2}$0)	bcc-(110)	[35]
Mo-Re	α-Al_2O_3(0001)	bcc-(111)	[35]
Mo-Re	α-Al_2O_3(11$\bar{2}$0)	A15-(100)	[35]
NbN	MgO(100)	(100)	[36]
NbN	MgO(110)	(110)	[37]
NbN	MgO(111)	(111)	[37]
NbN	α-Al_2O_3(0001)	(111)	[38]
NbN	α-Al_2O_3(11$\bar{2}$0)	(111)	[38]
NbN	α-Al_2O_3(1$\bar{1}$02)	(135)	[38]
NbN	α-Al_2O_3(10$\bar{1}$0)	(110)	[38]
NbN	α-Al_2O_3(2$\bar{1}\bar{1}$3)	(113)	[32]
Nb_3Sn	α-Al_2O_3(0001)	(100)	[32]
Nb_3Sn	α-Al_2O_3(11$\bar{2}$0)	(100)	[32]
Nb_3Sn	α-Al_2O_3(1$\bar{1}$02)	(100)	[39]
Nb_3Ge	Nb_3Ir(100)	(100)	[40]
Nb_3Ge	Nb_3Ir(110)	(110)	[40]
Nb_3Ge	Nb_3Ir(111)	(111)	[40]
Nb_3Ge	Nb_3Sn(100)	(100)	[40]
Nb_3Ge	α-Al_2O_3(11$\bar{2}$0)	(100)	[41]
Nb_3Ge	α-Al_2O_3(1$\bar{1}$02)	(100)	[41]
Nb_3Ge	ZrO_2(100)[b]	(100)	[42]
Nb_3Ge	ZrO_2(111)[b]	(111)	[40]
$La_{1.85}Sr_{0.15}CuO_4$	$SrTiO_3$(100)	(100)	[16]
$La_{1.85}Sr_{0.15}CuO_4$	$SrTiO_3$(110)	(110)	[43]
$YBa_2Cu_3O_7$	$SrTiO_3$(100)	(100) (001)	[44]
$YBa_2Cu_3O_7$	$SrTiO_3$(110)	(110)	[17]
$YBa_2Cu_3O_7$	$LaAlO_3$(100)	(001)	[45]
$YBa_2Cu_3O_7$	$LaGaO_3$(100)	(001)	[46]
$YBa_2Cu_3O_7$	$KTaO_3$(100)	(001)	[47]
$YBa_2Cu_3O_7$	$LiNbO_3$(2$\bar{1}\bar{1}$)	(001)	[48]

[a] More than one orientation grown depending on substrate surface preparation, deposition temperature, or other factors.
[b] Yttria-stabilized cubic zirconia.

Table 3. Summary of Experimentally Obtained Crystallographic Orientations of Superconductor Films Deposited on Sapphire Substrates (Mo, MgO, and Si Included for Reference).[a]

Material	Structure	Lattice constant, Å	Sapphire Orientation					
			c (0001)	r ($1\bar{1}02$)	a ($11\bar{2}0$)	m ($10\bar{1}0$)	r' ($10\bar{1}2$)	n ($2\bar{1}\bar{1}3$)
Mo.65Re.35	bcc	3.13	111 $11\bar{2}\|10\bar{1}0$ 7.5% 7.5%	— — —	110 $001\|1\bar{1}02$ 11% 15%	— — —	— — —	— — —
Mo	bcc	3.15	111 (110) $11\bar{2}\|10\bar{1}0$ 6.8% 6.8%	100 $011\|11\bar{2}0$ 6.4% 6.4%	— — —	— — —	221 $1\bar{1}0\|11\bar{2}0$ 6.4% 6.4%	— — —
Nb	bcc	3.30	111 (110) $11\bar{2}\|10\bar{1}0$ 2.0% 2.0%	100 (110) $011\|11\bar{2}0$ 2.0% 9.4%	110 $001\|1\bar{1}02$ 5.4% 9.4%	211 $1\bar{1}0\|1\bar{2}10$ 1.0% 2.0%	— — —	113 $\bar{1}2\bar{1}\|01\bar{1}0$ 2.0% 9.4%
MgO	B1	4.21	— — —	100 $010\|1\bar{1}20$ 13% 21%	111 $1\bar{1}0\|0001$ 6.6% 9.2%	110 $001\|1\bar{2}10$ 13% 9.2%	— — —	— — —
NbN	B1	4.38	111 $\bar{1}10\|10\bar{1}0$ −11% −11%	135 $21\bar{1}\|\bar{1}104$ −3.9% −11%	111 $1\bar{1}0\|0001$ 2.2% 4.8%	110 $1\bar{1}0\|0001$ 4.8% −8.6%	— — —	113 $1\bar{1}0\|01\bar{1}0$ −3.8% −11%

(*continues*)

Table 3. (continued)

Material	Structure	Lattice constant, Å	Sapphire Orientation					
			c (0001)	r ($1\bar{1}02$)	a ($11\bar{2}0$)	m ($10\bar{1}0$)	r' ($10\bar{1}2$)	n ($2\bar{1}\bar{1}3$)
$Mo_{.65}Re_{.35}$	A15	4.96	poly — 18% 18%	— — —	100 (210) $010\|\|1\bar{1}00$ 11% −13%	— — —	— — —	— — —
Nb_3Ir	A15	5.14	poly — 6.9% 6.9%	100 $001\|\|11\bar{2}0$ −0.7% −7.4%	100 $010\|\|1\bar{1}00$ 6.9% −16%	poly — −7.4% −16%	— — —	— — —
Nb_3Ge	A15	5.14	poly — 6.9% 6.9%	100 $001\|\|11\bar{2}0$ −0.7% −7.4%	100 $010\|\|1\bar{1}00$ 6.9% −16%	poly — −7.4% −16%	— — —	— — —
Nb_3Sn	A15	5.29	100 $001\|\|10\bar{1}0$ 3.9% −10%	100 $001\|\|11\bar{2}0$ −3.5% −10%	100,201 $010\|\|1\bar{1}00$ 3.9% −18,10%	poly — −10% −18%	— — —	— — —
Si	diamond	5.43	111 $11\bar{2}\|\|1\bar{1}00$ 24% 24%	100 $001\|\|11\bar{2}0$ −6.0% −12%	111 $11\bar{2}\|\|1\bar{1}04$ 6.8% 13%	— — —	— — —	111 $1\bar{1}0\|\|01\bar{1}0$ 7.2% 28%

[a] The growth direction for the primary epitaxial orientation is listed in the first row of each entry and an alternative relationship, if any, is shown in parentheses. Planes in sapphire and the film are listed in the second row that are parallel to each other but perpendicular to the plane of the substrate surface. The third and fourth rows show the lattice mismatch in two orthogonal directions in the plane of the film.

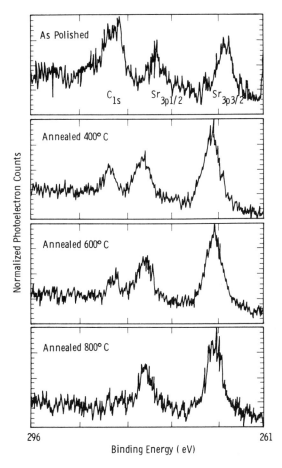

Fig. 4. *In-situ* XPS spectra centered on the energy of C_{1s} photoelectrons obtained from the surface of a $SrTiO_3(100)$ substrate after annealing at a series of temperatures in UHV.

usually be removed with an anneal in UHV just prior to film deposition. The temperatures needed for cleaning sapphire and MgO are 1250°C and 900°C, respectively. Figure 4 contains XPS data for annealed $SrTiO_3(100)$, showing that all traces of carbon disappear after an 800°C anneal.

In the case of α-$Al_2O_3(1\bar{1}02)$ substrates annealed at 800°C, Park et al. [15] found that a remaining sub-monolayer of carbon adversely affected both the structural and superconducting properties of epitaxial ultra-thin (<5 nm) Nb and V films. Argon ion-milling was effective in removing the carbon and improving the Nb and V film quality.

In some cases, the process needed to remove surface contaminants does not maintain crystalline order in the substrate's surface layer. For Nb_3Ir, a simultaneous combination of ion-milling and elevated temperature were needed to obtain clean surfaces that exhibited sharp LEED patterns [50].

3.2. Nb-Based Structures

Reports of the epitaxial growth of Nb films by vapor-phase deposition — primarily on MgO or α-Al_2O_3 single-crystal substrates — have been published for more than twenty years (see Table 2). Epitaxial Nb films have been used mainly for basic physics or materials science studies and currently do not have technological importance. The successful application of randomly oriented polycrystalline films can be attributed to superconducting length scales, ξ and λ, which are much greater than the scales of d_c, typical crystal defects, damaged surfaces due to ion-milling, etc. Possible applications where single-crystal films will be needed are in those microwave devices that are limited in performance by the surface resistance of polycrystalline Nb films.

A simple characteristic often used to compare the quality of Nb films is the resistance ratio, $RR = \rho(300\ K)/\rho(10\ K)$. For high-purity epitaxial films, a limit for RR based on an electronic mean-free path comparable to the film thickness has been inferred from a roughly linear dependence of RR on thickness with a slope of $0.5\ nm^{-1}$ for $RR \leq 170$. [51]. Since Nb is a refractory element, "MBE" growth requires the use of an electron-beam evaporation source. However, epitaxial Nb films sputtered in a UHV chamber can have equally high RR values. This section will describe the following properties of films produced by either growth technique: film/substrate epitaxial relationships, effects of film thickness, surface resistance, tunneling into single-crystal Nb films, epitaxial tunnel junctions (trilayers), and multilayer structures.

The orientation of epitaxial Nb films grown on sapphire is summarized in Table 3. As discussed by Claassen et al. [52], the sapphire/Nb registry is three dimensional. That is, the relative crystal orientation is the same regardless of which plane forms the interface between Nb and sapphire crystals. The registry is shown in Fig. 5 where stereographic projections for sapphire and Nb are overlayed corresponding to the relative crystal orientations found experimentally. Figure 5 shows that the three-dimensional orientation relationship leads to the 2.6° misorientation observed for Nb(110) grown on α-$Al_2O_3(1\bar{1}02)$ [53].

The structure of sapphire is such that there is no difference in the surface positions of oxygen ions between the $(1\bar{1}02)$ and $(10\bar{1}2)$ orientations. On the

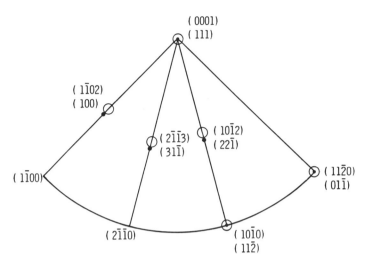

Fig. 5. Stereographic projections of α-Al$_2$O$_3$ and Nb crystals to show the relative orientation of an epitaxial film.

other hand, the two-dimensional unit cell defined by Al ions is three times larger for the latter surface. No data is available for the growth orientation of Nb deposited on α-Al$_2$O$_3$(10$\bar{1}$2). However, a Nb(22$\bar{1}$) growth orientation can be predicted based on the 3-D registry of Nb and in analogy with the data of O'Neal and Rath [54] for the epitaxial growth of Mo on sapphire (see Table 3). The combined data for Nb and Mo growth on sapphire support the model that the initial monolayer of deposited Nb (or Mo) ions occupy sites on the surface of sapphire that would be occupied by Al ions in an extension of the bulk sapphire structure. The positions of Nb ions have no simple correlation with sapphire's oxygen sites. It will be shown in Section 3.4 that a similar empirical result relating to Al ion sites holds for the orientation of A15 compounds grown on sapphire.

Niobium-based tunnel junctions are affected by the use of epitaxial films in the base electrode, the tunnel barrier, and the counterelectrode. The first reports of tunneling into single crystal Nb(110) and (111) base electrodes, and the use of the native oxide grown on a single-crystal film as a tunnel barrier, were contained in a paper by Laibowitz and Cuomo [29]. Later tunneling measurements of single-crystal Nb films were performed by Durbin et al. [55] who observed an anisotropic product of the electron-phonon coupling function $\alpha^2(\omega)$ and the phonon density of states $F(\omega)$ between $\langle 110 \rangle$ and $\langle 111 \rangle$ directions.

The thermal oxides of polycrystalline Nb have made poor tunnel barriers—even when used with Pb-alloy or other soft top electrode materials—due to the formation of conductive suboxides at the Nb/Nb_2O_5 interface [56] and to the structure of Nb_2O_5 [57]. Oxidation by an rf plasma forms a satisfactory low-leakage tunnel barrier due to the properties of a buried Nb-C-O layer [58], rather than those of the native oxide. In contrast to the polycrystalline case, the thermal oxide of epitaxial Nb(110) [59] or Nb(111) [29] formed low-leakage tunnel barriers. Another advantage of the oxidized single-crystal Nb was found by Celaschi et al. [59], who measured a reduced specific capacitance corresponding to $\varepsilon = 5-6$, compared with $\varepsilon = 29$ for oxidized polycrystalline Nb [60]. These results would seem to make the use of single-crystal Nb base electrodes preferable to polycrystalline films. However, no variation of Nb_2O_5 tunnel barriers has been found to be chemically stable during the deposition of a Nb counterelectrode. The technological state of the art employing all-Nb junctions with artificial tunnel barriers has obviated any advantages found in using the native oxide of single-crystal Nb base electrodes.

Other potential advantages of using epitaxial Nb films in tunnel junctions concern the uniformity of epitaxial artificial barriers and the gap energy of the top Nb electrode within a coherence length of the barrier/electrode interface. Epitaxial Al(111) has been grown on Nb(110) by Durbin et al. [55] and Braginski et al. [61] and subsequently oxidized. In the latter case, evaporated Nb counterelectrodes also grew epitaxially. However, the subgap leakage currents were much higher than for junctions formed with Pb top electrodes on the same type of base/barrier structure. This was interpreted as the result of non-uniform coverage by the Al layer that left pinholes in the artificial barrier filled by Nb_2O_5. The conclusion to which we are led by that singular study of epitaxial Nb/oxidized Al/Nb tunnel junctions is that randomly oriented, fine-grained Al deposited on polycrystalline Nb can provide better coverage than the epitaxial Al overlayer deposited on single-crystal Nb.

As shown in Table 1, very thin Nb films, $\leq \xi$, can have transition temperatures approaching 9 K whether the film is epitaxial [110] or not [62]. Therefore, the same contribution to the gap voltage of a tunnel junction can be obtained for a polycrystalline Nb top electrode as for one that is grown epitaxially with a suitable epitaxial tunnel barrier acting as a substrate. In contrast, Table 1 shows that NbN films thinner than ξ can have a T_c approaching that of a much thicker film only when they are grown on a suitable epitaxial substrate.

The most likely practical application for "MBE"-grown Nb films is in analog signal processing at microwave frequencies where the surface resistance of polycrystalline Nb films could limit device performance. Such devices (see, for example, [63]) have not, in general, been optimized in other ways so losses may be dominated by radiation or by dissipation in dielectrics rather than by the superconductor's surface resistance. No detailed comparisons have been made between single-crystal and polycrystalline Nb films for this application, and most devices that have been tested used polycrystalline films. In one example that is available, however, the highest-Q Nb thin-film resonator reported, which had a loaded $Q = 4 \times 10^5$ at 3 GHz and 4.2 K, was made from epitaxial films [64].

Another use for epitaxial Nb films has been in the study of metallic multilayer superlattices. The first superlattices in which three-dimensional coherence was observed were Nb(110)/Ta multilayers grown by "MBE" on α-$Al_2O_3(1\bar{1}02)$ [31]. The epitaxial Nb layers stabilized the formation of the metastable bcc Ta phase. The bcc phase is superconducting and a similar use was made [65] of an epitaxial Nb layer on sapphire to grow Ta/Ta_2O_5/Pb-Bi tunnel junctions. These junctions have a particularly sharp nonlinearity at the gap voltage that makes them useful as quasiparticle mixers for heterodyne detection of millimeter waves. Another high-quality superlattice system, Gd(0001)/Y, makes use of an epitaxial Nb(110) layer as a diffusion barrier between sapphire and Gd or Y [111].

In conclusion, the "MBE" growth of Nb has yet to make a significant impact on modern superconducting device development—despite the fact that Nb/oxidized Al/Nb tunnel junctions represent the state of the art for circuit development. The primary reason is that the coherence length of Nb is larger than the polycrystalline film thicknesses needed to obtain ~ 9 K transition temperatures. However, the relative ease of growing a single-component single-crystal film makes it likely that epitaxial Nb films will be implemented in other applications as small advantages over polycrystalline films are found.

3.3. NbN-Based Structures

The first epitaxial NbN films were grown by CVD on MgO(100) substrates [36]. Magnesium oxide and NbN both have a B1 structure and a lattice mismatch of only 4%, so MgO was a natural candidate for an epitaxial substrate. Single-crystal films were later deposited by reactive sputtering on MgO(100) [66], and on a number of different sapphire surfaces [38]. In some cases where polycrystalline substrates were used, the concept of polycrystalline epitaxy, in which each crystallite in the film aligns with one in the substrate, has been invoked to explain higher T_c's measured for NbN films grown on oxidized Mg substrates, compared with films grown on other oxide substrates [41,67].

The only all-NbN Josephson tunnel junctions reported to date with gap voltages $V_g \geq 5$ mV have used MgO tunnel barriers (for example, [67,68]). The significance of a 5 mV gap voltage at 4.2 K is the likelihood that such a junction can operate with a closed-cycle refrigerator at ~ 10 K [116]. Since the tunnel barrier is the substrate for counterelectrode formation, the junction results suggest that polycrystalline epitaxy was responsible for a substantial contribution to V_g from the energy gap of the NbN counterelectrode. The difference in the energy gap of NbN counterelectrodes formed on substrates that promote epitaxial growth and those that do not, is summarized in Table 1. Only in the case of epitaxial growth does the T_c of NbN films thinner than a coherence length approach the T_c of bulk NbN.

Despite the technological importance of polycrystalline NbN/MgO/NbN structures, the remainder of this section will focus on the "MBE growth" of single-crystal epitaxial NbN films and multilayers. In contrast to all of the other superconductor material systems described in this chapter, epitaxial NbN films have never been produced by evaporation. Typically, such films are produced by reactive sputtering of Nb in an Ar/N_2 sputtering gas which may also contain a source of carbon (e.g., CH_4, [113]) to stabilize the superconducting phase and increase T_c from ~ 15 K to 17 K. The best epitaxial films have been produced in the range of substrate temperatures between 500 and 700°C.

A unique property of epitaxial NbN films—in comparison with polycrystalline NbN—is a resistivity ratio, $\rho(300\text{ K})/\rho(100\text{ K})$, greater than unity [37]. This metallic characteristic of the normal-state resistivity is even observed for epitaxial NbN grown on sapphire. These films generally grow in a direction with a different symmetry than the substrate surface (see Table 3). Planar defects—grain boundaries, twin boundaries, or stacking faults—must form at the boundary of grains that nucleated on rotationally equivalent sites on the substrate, which are not rotationally equivalent in the film [38,20].

Low-energy electron diffraction (LEED) patterns of an α-Al_2O_3(0001) surface and a subsequently deposited NbN(111) film are shown in Figs. 6(a) and 6(b). Figures 6(c) and 6(d) contain LEED patterns obtained from an MgO(111) tunnel barrier deposited on the first NbN(111) film and from a NbN(111) counterelectrode, respectively. Figure 6 shows that an orientation relationship was maintained in the plane of the films throughout the thickness of the trilayer, and that a periodic structure was maintained in the top 1–2

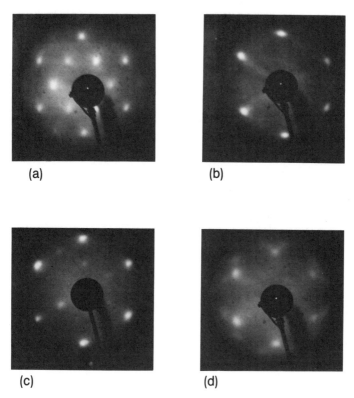

Fig. 6. LEED patterns from the surfaces of a NbN(111)/MgO/NbN epitaxial tunnel junction: (a) α-Al$_2$O$_3$(0001) substrate, electron beam voltage = 103 V; (b) 50-nm thick NbN(111) base electrode, 81 V; (c) 0.7-nm thick MgO barrier, 72 V; (d) 50-nm thick NbN top electrode, 88 V.

monolayers. The extreme surface-sensitivity of LEED was particularly important in the analysis of the structure of tunnel barriers (Fig. 6c) since, in many cases, they were only a few monolayers thick. Electrical measurements of tunnel junctions patterned from epitaxial trilayers showed that coverage by these thin MgO barriers was (nearly) complete so a pattern such as the one in Fig. 6(c) is truly a measurement of MgO and not the underlying NbN.

Typical quasiparticle I–V curves for two orientations of epitaxial NbN-based tunnel junctions are shown in Fig. 7. A complete review of epitaxial NbN junction properties has been published [20]. The junctions in Fig. 7 had resistances at 6 mV less than 10^{-4} Ω-cm^2 and critical current densities on the order of 100 A/cm^2. The I–V charactersitics were strongly influenced by the deposition temperature of the top electrode. For the junctions shown in Fig. 7,

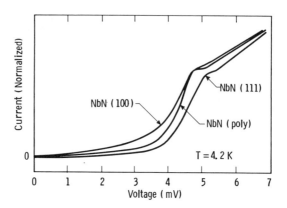

Fig. 7. Comparison of quasiparticle $I-V$ curves for NbN/MgO/NbN tunnel junctions made with single-crystal and polycrystalline electrodes.

the counterelectrode was grown at 150°C. Higher deposition temperatures, as high as 750°C, were used without significantly degrading the epitaxial barriers. Junctions with polycrystalline barriers—such as the one used for a reference curve in Fig. 7—were shorted during the process of counterelectrode deposition at such high temperatures. The epitaxial junctions formed at high temperatures had higher gap voltages, higher critical current densities, and higher subgap conductance than those completed at 150°C.

Tunnel barriers for epitaxial junctions were deposited at 700°C to obtain sharp LEED and RHEED spots indicative of good crystalline order, although low-temperature depositions provide more uniform coverage. In cases where the tunnel barriers were deposited in pure Ar, pinholes in the barrier were plugged by niobium oxide. Although Pb-alloy counterelectrodes deposited on these NbN/MgO bilayers resulted in low-leakage junctions, NbN counterelectrodes reduced the oxide in the pinholes and led to shorts through the barrier. XPS studies of NbN/MgO bilayers indicated that 1% methane added to the sputter gas during tunnel barrier deposition led to the formation of a niobium carbo-oxide in the pinholes that was more chemically stable and solved the problem [20].

A study was performed to determine whether the elimination of the 4% mismatch between NbN and MgO and, therefore, tensile stress in barriers, could lead to improved junction characteristics [20]. Figure 8 shows the lattice constant and equilibrium phase boundaries of solid solutions in the MgO-CaO pseudo-binary system as a function of CaO content. Metastable solid solutions were successfully formed in either sputtered or evaporated MgO-

8. "MBE" GROWTH OF SUPERCONDUCTING MATERIALS

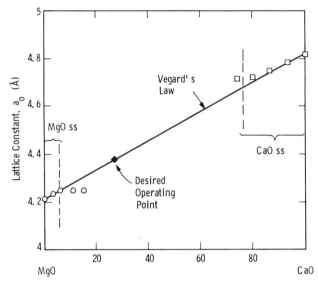

Fig. 8. Variation of lattice constants as a function of CaO content in the MgO–CaO pseudo-binary system. The desired operating point—the lattice constant of NbN—falls outside the equilibrium solid-solution phase fields.

CaO epitaxial films grown on NbN throughout the range of tensile-stress, lattice-matched, or compressive-stress solid solutions. Based on Vegard's Law, the lattice-matched composition was assumed to be 27 mole percent CaO.

Evidence that a single-phase MgO-CaO solid solution was obtained is presented in Figure 9, which shows the kind of quantitative LEED data that was discussed in Section 2.3. Ignoring dynamical diffraction effects, a lattice constant for the 7 nm thick, 50%-CaO film in Fig. 9 was found from the spacing between 00n Bragg peaks to be 0.439 nm. In contrast, reference LEED data for an MgO crystal was best fit by a lattice constant of 0.425 nm.

However, thinner MgO-CaO overlayers (∼1 nm), had a lattice constant that matched NbN regardless of CaO content. The misfit strain that was evidently present in the tunnel barriers had a negligible effect on the electrical properties of epitaxial NbN junctions. The $I-V$ curves presented in Fig. 7 were from junctions with barriers made at the lattice-matched composition, but had gap voltages and subgap conductances comparable to junctions made with pure MgO barriers.

The discovery of HTS oxides has sidetracked the development of NbN-based circuits capable of operation at 10 K, pending the outcome of efforts to make HTS junctions that could operate at much higher temperatures. As

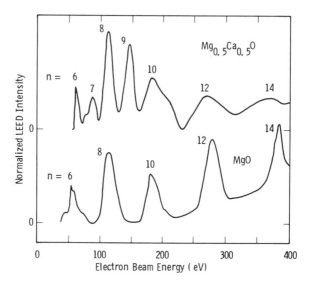

Fig. 9. LEED intensity plotted as a function of incident beam energy for the (0,0) beam: (a) a 7-nm thick $Mg_{0.5}OCa_{0.5}O(100)$ film deposited on NbN(100); (b) an MgO(100) crystal.

attention is shifted back to NbN, the most important issue in making junctions will be to obtain reproducible critical current densities. If polycrystalline NbN/MgO/NbN junctions cannot be made reproducibly, epitaxial junctions may become a practical alternative rather than simply model systems for materials studies.

An alternative to MgO or MgO-CaO epitaxial tunnel barriers, wurtzite-structure AlN(0001) grown epitaxially on NbN(111), has been used in a preliminary manner to fabricate junctions [117]. Although the junction quality was poor, the chemical compatibility of an epitaxial nitride insulator on a nitride superconductor should provide a motivation for further work. The epitaxial NbN/AlN/NbN system is also significant as a predecessor to current activity in the fabrication of the epitaxial oxide structure, $YBCO/PrBa_2Cu_3O_7/YBCO$ [115,114]. The successful formation of epitaxial NbN/MgO/NbN junctions at temperatures as high as 750°C is also an important indicator of what is possible in HTS junction development.

3.4. A15 Superconductors

3.4.1. "MBE" of A15 Superconductors

The UHV "MBE" of A15 superconductors was motivated by the interest in properties of epitaxial thin films and tunneling structures (Section 1). Prior

8. "MBE" GROWTH OF SUPERCONDUCTING MATERIALS

investigations of homo- and hetero-epitaxial effects in films of high-T_c A15's have been performed mostly in conjunction with the problem of high-T_c Nb$_3$Al, Nb$_3$Ga, Nb$_3$Ge, and Nb$_3$Si phase stabilization. In almost all cases, these films were deposited, by co-evaporation or sputtering, in HV systems having background pressures in the 10^{-5} to high 10^{-8} torr range. Below, we limit our attention to those A15's that were also grown by "MBE."

3.4.2. Mo-Re Films and Tunneling Structures

Bulk solid solution alloys of Mo-Re having the α-Mo structure are superconducting with a maximum $T_c = 12$ K at approximately 40 at.% Re, the boundary of this phase stability field [69]. A metastable A15 structure was observed in sputtered [70] and e-beam evaporated [71] polycrystalline thin films with compositions between 25 and 40 at.% Re. A maximum $T_c = 15$ K was observed in mostly A15 films containing 38 at.% Re [70] and in films having an unresolved structure and a composition Mo–(62 at.%) Re [72].

Interest in the use of Mo-Re films for tunneling structures and also microwave cavities was stimulated by the coherence length in the α-Mo phase, $\xi = 20$ nm, much longer than in other superconductors of comparable T_c, and by the low solubility of oxygen in Mo-Re alloys [73]. Superconductor surface and interface degradation could thus have a lesser effect on the tunnel junction gap voltage and subgap conductance or on the cavity losses than in the case of B1 and A15 structure films. Interest in the unresolved issue of the Mo-Re A15 phase stabilization mechanism (by impurities versus epitaxy?) also motivated the only "MBE" study by Talvacchio et al. [35], which is summarized below.

The "MBE" system of Fig. 3 was used for film and tunnel junction fabrication and analysis. Thin Mo–(25 to 40 at.%)Re films were e-beam co-evaporated on (11$\bar{2}$0) and (0001) sapphire substrates. At low deposition temperature of $T_s = 100°C$, the films were polycrystalline with α-Mo structure. At higher temperatures, however, epitaxial deposits were obtained with orientations indicated in Table 2. The crystalline quality was relatively high, with the x-ray rocking curve width of 0.4 degrees of arc. At $T_s = 800°C$, the structure was still α-Mo, while at 1000°C predominantly A15. The conditions of deposition were otherwise identical and the very low background pressure of impurities, 1×10^{-10} torr or less, made the impurity-stabilization of A15 phase very unlikely and stabilization by epitaxy probable. Due to the absence of impurities, the film T_c was independent of the deposition temperature and also of film thickness, in a marked contrast to analogous films sputtered in the presence of impurities. Figure 10 compares the T_c and residual resistivity ratio vs. thickness dependences in these two types of films.

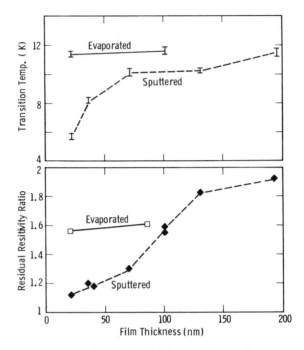

Fig. 10. Critical temperature and residual resistivity vs. thickness of sputtered and "MBE" co-evaporated Mo-Re films.

In co-evaporated films, the compositional dependence of T_c was that of the bulk α-Mo phase with a maximum T_c of 12 K, almost independent of the crystal structure. The higher $T_c = 15$ K that was obtained by sputtering was not reproduced.

The *in-situ* XPS analysis of film surfaces confirmed that Mo-Re is much less susceptible to oxidation than either Mo or Re. Due to the thinness of the native Mo-Re oxide—only 0.5 nm—a complete coverage of the Mo-Re base electrode by an artificial Al-Al$_2$O$_3$ barrier was necessary to obtain low subgap conductance tunnel junctions with Pb or Mo-Re counterelectrodes. The all-Mo-Re tunnel junctions had a low "junction T_c" temperature of only 8 K. This indicated that the counterelectrode was amorphous at the interface with crystalline Al$_2$O$_3$, in contrast to films on sapphire that were crystalline even when only one coherence length thick.

In conclusion, the available "MBE" results documented the epitaxial film growth of both α-Mo and A15 phases and the ability to fabricate all-Mo-Re tunnel junctions. The mechanism of A15 Mo-Re stabilization, however, was not revealed. It appears that epitaxy may stabilize an A15 phase with a T_c lower than that of films stabilized by impurities present in sputtering. The technological usefulness of Mo-Re for higher-T_c tunnel junctions remains in doubt since epitaxial growth of counterelectrodes was not attained.

3.4.3. Nb-Sn Films and Tunneling Structures

In the Nb-Sn binary system, the A15 structure is stable between 18 and 25 at.% Sn. A nonequilibrium phase diagram useful as a guide for Nb-Sn film deposition was proposed by Rudman *et al.* [74] and is shown in Fig. 11. Stoichiometric Nb_3Sn deposits can be conveniently obtained without a very accurate deposition rate control by co-evaporating the two constituents with some, relatively arbitrary excess of tin at substrate temperatures near 900°C. This temperature is high enough for the excess of Sn to re-evaporate as Nb_6Sn_5. The "composition-locked" field, where stoichiometric Nb_3Sn can be obtained, is shown in the phase diagram. Within the A15 stability field, some

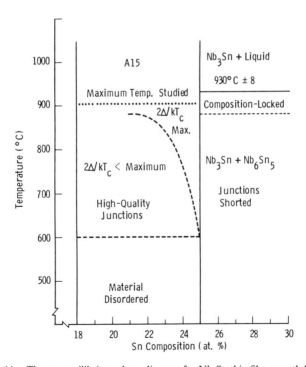

Fig. 11. The nonequilibrium phase diagram for Nb-Sn thin film growth [75].

tendency exists to segregate into two A15's differing in the tin concentration [75]. This is the cause of the observed broadening of the superconducting transition.

Epitaxial growth of (100) Nb-Sn on ($1\bar{1}02$) sapphire substrates was first observed and investigated by TEM in thick deposits that were e-beam co-evaporated in HV at substrate temperatures up to 900°C [39]. While nucleation on the substrate was preferentially (100), grains of other orientations were also nucleated so that the deposits were oriented but polycrystalline through the first 100 nm. With the increasing film thickness, random orientations became overgrown by $\langle 100 \rangle$ regions and the crystalline quality improved. However, even in films thicker than 1 micron, some domains misoriented within 5 degrees were present and separated by end-on dislocation boundaries. In addition, especially at $T_s < 900°C$, domains having another epitaxial relationship, tilted by 30 and 60 degrees with respect to the matrix, were also present. Marshall et al. [39] analyzed the observed epitaxial relationship, where the cube axes of Nb_3Sn align along the twofold symmetry orthogonal axes in the ($1\bar{1}02$) face of sapphire, and suggested that the small misorientations were related to the different symmetries of the substrate and film.

In the only "MBE" deposition study, Sn was evaporated from an effusion cell and Nb from an e-gun (Talvacchio et al. [76,7]). The main difference, when compared with Marshall et al. [39], was a much lower background pressure of impurities, $< 1 \times 10^{-10}$ torr. Films were grown in the "composition-locked" regime, simultaneously on sapphire of several orientations listed in Table 2. The deposit crystallinity was determined by in-situ RHEED and LEED. Substrate orientations where the minimum lattice parameter mismatch along one cube axis did not exceed $+4\%$ permitted the film to grow epitaxially with a (100) and (201) orientation. The epitaxial relationships are shown in Fig. 12. On sapphire ($10\bar{1}0$), where the minimum mismatch was approximately -10%, the films were polycrystalline. The LEED pattern of films only 6 to 10 nm thick is shown in Fig. 13. It indicates that "MBE" films have nucleated in highly epitaxial form. These films attained a high degree of perfection at a thickness of only 100 nm (Fig. 13), in contrast to the films grown in HV [39]. The x-ray rocking curve linewidth was only 0.4 degrees. However, the "MBE" films were not truly single crystals, as they contained two growth orientations with (100) and (201) planes tilted by about 30 degrees of arc. This was revealed by x-ray Weissenberg camera results [77] and confirmed by x-ray diffractometer data, which indicated that the two habits were present in approximately equal proportion. No TEM results were obtained. However, the narrow x-ray rocking curve suggested that the concentration of low-angle boundaries was

8. "MBE" GROWTH OF SUPERCONDUCTING MATERIALS 303

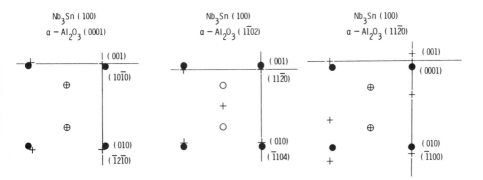

Fig. 12. Epitaxial relationships between Nb$_3$Sn(100) and three different sapphire faces showing that Nb($\bigcirc$) and Sn($\bullet$) occupy sites that would be occupied by Al($+$) if one more layer of the sapphire structure were added.

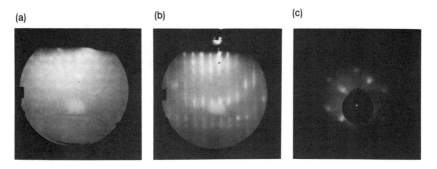

Fig. 13. Electron diffraction patterns—RHEED and LEED, respectively—of Nb$_3$Sn(100): (a) 6-nm thick film; (b), (c) 100-nm thick.

low. The polycrystalline nucleation and gradual overgrowth observed by Marshall *et al.* [39] were thus due to background impurities. The alternative growth habits in "MBE" films were different than in non-"MBE" films but could also be due to differences in the film and substrate symmetry, as suggested by Marshall *et al.*

Similarly to the case of NbN, the ability to grow reasonably high quality Nb$_3$Sn films by "MBE" made it possible to investigate growth of epitaxial Nb$_3$Sn/insulator bilayers and trilayers with Nb$_3$Sn and NbN counterelectrodes. The insulators were (1) e-beam evaporated and sputtered Al overlayer thermally or ion-beam oxidized after deposition, Al-Al$_2$O$_3$; (2) e-beam evaporated Al$_2$O$_3$; (3) thermally evaporated/sublimated CaF$_2$; and (4) thermally evaporated SrO. These overlayers were deposited at room temperature, except for SrO. The Al overlayer was epitaxial and the thin thermal

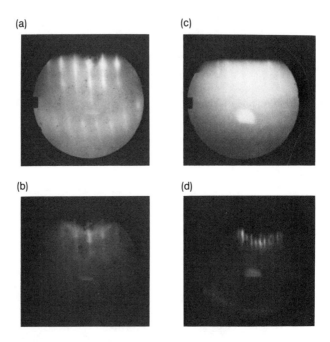

Fig. 14. RHEED patterns of (a) epitaxial Al on Nb_3Sn, (b) oxidized Al–Al_2O_3, (c) evaporated Al_2O_3 and (d) SrO tunnel barriers.

oxide was in epitaxial relationship with the metal, as shown by RHEED patterns in Fig. 14a,b. The evaporated Al_2O_3 was also epitaxial (Fig. 14c). The CaF_2 and SrO overlayers (Fig. 14d) contained random crystallites in an epitaxial matrix.

The $I-V$ characteristics of bilayers with Pb-Bi electrodes (Fig. 15a) showed low subgap conductance and high gap voltage when barrier pinholes were sealed by the native Nb-Sn oxide. This observation was first made by Rudman et al. [74]. In contrast, $I-V$ characteristics indicating microshorts in the barrier were obtained with NbN counterelectrodes deposited near room temperature. The microshorts presumably occurred at the native-oxide-sealed pinholes. Ion-beam oxidation of Nb_3Sn/Al-Al_2O_3 was reasonably effective in sealing these pinholes but a rise in conductance at the NbN gap indicated ion-beam-induced damage in Nb_3Sn (Fig. 15b). In the case of CaF_2 and SrO, barrier microshorts were always present.

Fabrication of all-Nb_3Sn junctions required depositing or annealing the counterelectrode at temperatures above 600°C, sufficient for Nb_3Sn formation. In trilayers with Al_2O_3, the counterelectrode deposited at $\geq 700°C$ was

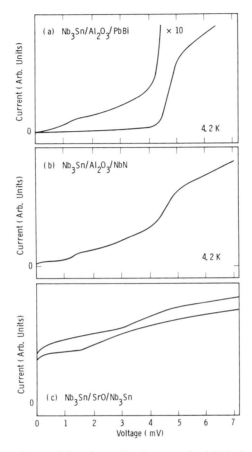

Fig. 15. (a) The $I-V$ characteristics of tunneling into an epitaxial $Nb_3Sn(100)/Al_2O_3$ bilayer using a Pb-Bi counterelectrode. For comparison, characteristics of (b) $Nb_3Sn/Al_2O_3/NbN$ and (c) $Nb_3Sn/SrO/Nb_3Sn$ are also plotted.

epitaxial, according to RHEED, but the reactions at pinholes and Al_2O_3 interfaces were severe enough to result in complete shorts. In one case of a textured SrO barrier, the top Nb_3Sn layer was co-evaporated at ambient temperature and then flash annealed at 650°C. This counterelectrode was polycrystalline by RHEED with no evidence of texturing. The resulting junction contained microshorts and was extremely leaky but a low sumgap voltage of, approximately, 4 meV could be estimated at 4.2 K (Fig. 15c). This suggested that all-Nb_3Sn junctions could be feasible with further process refinements.

In conclusion, the "MBE" approach permitted the authors to grow epitaxial

Nb$_3$Sn of relatively high quality. It appears that the key unresolved issue in all-Nb$_3$Sn epitaxial tunnel junction fabrication is the presence of pinholes and/or thin areas in the barrier. High gap voltage Nb$_3$Sn trilayers would be viable if a very smooth single-crystal base electrode and a completely continuous Al$_2$O$_3$ barrier could withstand processing temperatures high enough to obtain epitaxial top layers of Nb$_3$Sn.

3.4.4. Nb-Ge Films

In the Nb-Ge binary system, the A15 structure is stable between 18.5 and 20 at.% Ge at 900°C [78]. The critical temperature of the stable phase is only 6.5 K, approximately, and the large lattice parameter, $a_0 \geq 0.517$ nm, is typical of anti-site disorder. Even at the peritectic solidus temperature of 1865°C the phase boundary was determined to be substoichiometric, at 23 ± 1 at.% Ge [78]. The nearly stoichiometric, high-T_c A15 phase has been obtained only in the form of films where the presumably metastable material was stabilized. High-T_c film deposition was first achieved by sputtering [79] and later also by co-evaporation and CVD. The highest-T_c films consisted of nearly single A15 phase having $a_0 = 0.512$ to 0.514 nm but often contained small admixtures of the tetragonal or hexagonal Nb$_5$Ge$_3$ phase. The mechanism of the metastable Nb$_3$Ge stabilization, a subject of numerous studies, has not been elucidated satisfactorily. It was, however, observed by Gavaler et al. [80] and confirmed by many authors that either incorporation of some oxygen or the presence of a low partial pressure of oxygen (or another impurity) during the deposition was necessary to stabilize the stoichiometric phase. The typical oxygen partial pressure required was found to be in the range of 10^{-6} to 10^{-7} torr.

Dayem et al. [81] were the first to observe polycrystalline epitaxial growth of high-T_c Nb-Ge on isomorphic A15 Nb$_3$Ir polycrystalline films having a matching, adjustable lattice parameter. Figure 16 shows their data of a_0 and T_c vs. at.% Ge. The A15 phase boundary appeared to have shifted to 26.3 at.% Ge and T_c's were much higher than in the same deposition conditions but without Nb$_3$Ir substrate. This was an indication that epitaxy can contribute to the stabilization of the metastable Nb-Ge phase, at least at background impurity pressures of high 10^{-7} to 10^{-8} torr, which were typically obtained without intentional additions of oxygen. Gavaler et al. [82] observed $T_c = 20$ K even in the thinnest (<20 nm) films co-sputtered on Nb-Ir. On sapphire, however, deposition of a layer at least 100 nm thick was necessary to attain this critical temperature. Evidently, the high-T_c phase was directly nucleated on the Nb-Ir

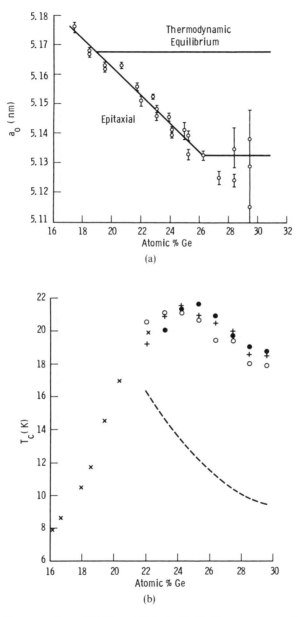

Fig. 16. (a) Lattice parameter and (b) T_c vs. at.% Ge in Nb-Ge films co-evaporated epitaxially on Nb_3Ir [81]. The dotted line represents T_c onset in the absence of epitaxy.

substrate, rather than formed as a result of a gradual, homoepitaxial overgrowth.

Kuwasa and Nakano [83] have shown that in UHV of 10^{-8} to 10^{-9} torr (during co-evaporation) the homoepitaxial, polycrystalline nucleation of Nb_3Ge was assisted by the presence of oxygen in the Nb-Ge substrate layer. Subsequent UHV growth produced a phase having T_c's higher than on oxygen-free Nb-Ge but lower than typically obtained in HV deposition.

Asano et al. [42] reported heteroepitaxial growth of textured Nb_3Ge films on cubic, yttria stabilized, zirconia (YSZ) single crystals with (100) orientation and $a_0 = 0.512$ to 0.516 nm. Films were magnetron-sputtered simultaneously on both YSZ and (1$\bar{1}$02) sapphire substrates, presumably in a HV system. The x-ray diffractometer traces indicated that polycrystalline epitaxy occurred with the orientation A15 $Nb_3Ge(100) \| YSZ(100)$. Diffraction data for films on sapphire were not given. Figure 17 shows T_c vs. the film thickness for YSZ and sapphire substrates. Indeed, in very thin films, higher T_c (and lower a_0) obtained on YSZ confirmed that these films were stabilized by epitaxy. However, with increasing film thickness, this beneficial YSZ substrate effect gradually disappeared, presumably due to the homoepitaxial over-growth of the higher-T_c phase.

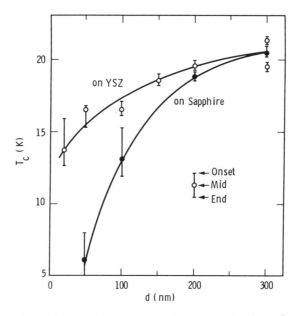

Fig. 17. T_c vs. thickness of Nb-Ge films on YSZ(100) and α-Al_2O_3(1$\bar{1}$02) [42].

8. "MBE" GROWTH OF SUPERCONDUCTING MATERIALS

In our "MBE" experiments, Ge was evaporated from an effusion cell and Nb from an e-gun, as in the case of Nb_3Sn [84]. The substrate temperatures were in the range between 820 and 920°C, optimally 880°C. The background pressure was extremely low, 5×10^{-11} torr, and it was possible to deposit films at various low partial pressures of oxygen, up to 2×10^{-9} torr, bled deliberately into the chamber in the proximity of the substrate holder. The in-situ electron diffraction (LEED and RHEED) of films 50 to 200 nm thick confirmed that in UHV the A15 Nb-Ge phase can grow epitaxially on a variety of single-crystal substrates, including sapphire, as shown in Table 2.

The x-ray diffraction data were obtained by texture camera only and limited to epitaxial films on single crystal (100) Nb_3Ir [50]. The diffraction spots for the film and substrate could not be resolved well enough to obtain the film lattice parameter as a function of oxygen partial pressure during deposition. However, the T_c versus $p(O_2)$ dependence (Fig. 18), obtained by resistive measurements, showed that epitaxy alone did not stabilize the high-T_c phase. The stable, low-T_c A15 phase dominated in films deposited without oxygen injection. The T_c was increasing with $p(O_2)$ but the onsets did not attain 20 K up to $\geq 10^{-9}$ torr. Evidently, much higher oxygen partial pressures are required for stoichiometric Nb_3Ge stabilization. Only at these higher pressures, a close epitaxial lattice parameter match, obtained with Nb-Ir and

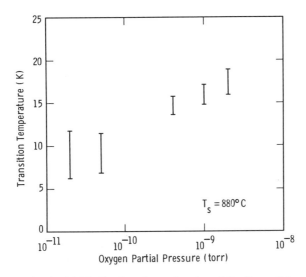

Fig. 18. The T_c of epitaxial Nb_3Ge plotted as a function of the O_2 partial pressure present during deposition.

YSZ substrates but not the sapphire, is contributing to the nucleation and growth of the metastable phase, as shown by the results of Dayem et al., Gavaler et al., and Asano et al., which were summarized above.

In conclusion, we found the UHV "MBE" method to be rather unsuitable for growing epitaxial high-T_c Nb_3Ge films and tunneling structures.

3.5. High-T_c Oxide Superconductors

All high-T_c oxide superconductors (HTS) that have $T_c > 35$ K are cuprates, i.e., oxide compounds of copper, an alkaline earth metal(s), and other elements [85,86]. The number of materials and their confirmed T_c's are growing, due to a very high level of effort worldwide. The confirmed record T_c is presently 125 K, in a thallium-based cuprate, $Tl_2Ca_2Ba_2Cu_3O_y$ [87,89]. To date, the most researched material is $YBa_2Cu_3O_7$ (YBCO) and its derivatives in which Y is replaced by another lanthanide (Ln) rare-earth element. The T_c common to $LnBa_2Cu_3O_7$ is 90–95 K.

All HTS cuprates are ceramic materials, very brittle, and environmentally rather unstable. All have tetragonal or orthorhombic crystal structures which are perovskite-related. Each of the Bi- and Tl-based cuprates (BSCCO and TBCCO) crystallizes in several structures in which T_c increases with the number of Cu-O planes in the elementary cell n that are stacked along the c-axis normal to the basal plane [88,89]. The stability of these structures appears to decrease with increasing n, so that a mixture of phases with different n values is being obtained in the highest-T_c compound synthesis. In the case of TBCCO, the number of Tl planes in the unit cell can also vary [89]. At present, the LnBCO family is the most reproducible object of application studies, although the interest in applications of BSCCO and TBCCO is increasing rapidly.

At a first glance, co-evaporation in a "MBE"-type system does not appear to be a method suitable for deposition of oxide films several hundreds of nanometers thick. Sputtering and ion-beam sputtering are more typical methods of oxide deposition. However, the "MBE" hardware has been used by Chaudhari et al. [44] to, for the first time, co-evaporate YBCO films on $SrTiO_3$ single-crystal substrates. After their report, the "MBE" of YBCO has been pursued successfully by many experimenters. Thereafter, sputtering and laser evaporation became equally effective, as documented in a multitude of publications in the 1987–88 period. However, as it did not involve "MBE" techniques, most of that work will not be reviewed here.

8. "MBE" GROWTH OF SUPERCONDUCTING MATERIALS

The particular mode of "MBE" use has been determined by the following structural, chemical, and electronic properties, which are, at least qualitatively, common to all high-T_c cuprates.

(1) The cuprate crystalline phase(s) can, obviously, form only in the presence of oxygen. In addition, the crystal structure and electrical characteristics of the film depend strongly upon deviations of the oxygen concentration from the stoichiometric value. This effect is most pronounced in LnBCO in which oxygen atoms intercalate reversibly into the crystal lattice at relatively low temperatures. The tetragonal $LnBa_2Cu_3O_6$ crystal is an insulator. Upon oxygenation, it transforms into an orthorhombic modification at 6.3 to 6.6 oxygen atoms per unit cell, depending upon the conditions of the process and, at slightly higher oxygen contents, the c-axis lattice parameter exhibits a nonlinear reduction. At this point the crystal becomes superconducting [90]. The highest T_c and optimum superconducting properties of $LnBa_2Cu_3O_x$ are attained when approaching $x = 7.0$ oxygens per unit cell. By controlling x locally, patterned regions of a YBCO film can be easily transformed from superconductor, to semiconductor, to insulator. This may offer a unique flexibility in tailoring devices.

(2) Some constituent elements, especially Tl, have high vapor pressures and are also weakly bound in the crystal lattice. Consequently, synthesis of a stoichiometric compound is only possible at high ambient pressure of the element or with a large excess of Tl in the source. The best near-stoichiometric TBCCO films were obtained by post-annealing in sealed ampoules containing an excess amount of bulk TBCCO [91].

(3) The constituent elements, spurious phases and, to a lesser degree, the compound itself react at high temperatures with most substrate materials and with CO_2 and water vapor in the ambient atmosphere.

(4) The coherence length is extremely short and anisotropic. In these extreme type II superconductors, it decreases with increasing T_c and, along the c-axis, is comparable to the interatomic distance (Table 1). Due to this short range of superconducting interactions, atomic-scale defects disrupt superconductivity. For example, the transfer of current between crystallites in polycrystalline materials is severely limited by imperfect grain boundaries. Tunneling along the c-axis would require crystalline perfection up to the first atomic monolayer at the surface.

(5) The critical parameters, especially J_c and H_c are strongly anisotropic. While the intrinsic (depairing) J_c limit is high (in the 10^8 A/cm^2 range) and

self-field values $J_c \leq 5 \times 10^6$ A/cm² have been indeed obtained at 77 K in the a–b plane of nearly single-crystalline films of YBCO on $SrTiO_3$ or MgO crystal substrates, the experimental J_c's along the c-axis are lower, typically by one to two orders of magnitude.

Relatively high partial pressures of oxygen are required to deposit epitaxial cuprate oxide(s) directly from vapor phase or to oxygenate an oxygen-deficient deposit by *in-situ* post-annealing. This is a significant deviation from standard MBE practice which involves UHV. However, the rationale for "MBE" is still to maintain undegraded interfaces in multilayer deposition and *in-situ* surface/interface characterization. A low background pressure of gas-phase impurities is also a requirement well satisfied by "MBE".

Following Chaudhari et al. [44], some experimenters used "MBE"-type evaporation chambers to deposit epitaxial films of LnBCO from e-beam or effusion cell sources and MBE was specifically identified as the preparation method. The various film fabrication procedures, however, did not follow the "MBE" methodology, except for *in-situ* substrate characterization [92]. The range of deposition temperatures and pressures was broad. On one extreme was the co-deposition of constituent metals in UHV at temperatures between ambient and 400°C, followed by *ex-situ* oxidation, annealing in oxygen between 850 and 900°C (to crystallize the compound) and by either post-annealing at a lower temperature or cooling slowly in O_2 to approach the desired concentration of 7 oxygen atoms per unit cell [93]. On another extreme was the reactive e-beam co-evaporation on substrates hot enough to obtain a crystalline deposit: $T_s = 560$ to $650°C$ at an oxygen partial pressure so high (up to 10^{-3} torr) that the evaporation paths were no longer line-of-sight, making rate control difficult, and the lifetime of e-gun filaments limited to a few hours [94]. In the latter case, Lathrop et al. [94] backfilled the chamber with 1 to 20 torr of O_2 at the end of each deposition so that the deposit was slowly cooled in oxygen. Even so, an additional *ex-situ* post-anneal in O_2 was usually necessary.

Generally, in experiments involving co-evaporation from e-gun sources in the presence of oxygen, even at low partial pressures, the precision of the evaporation rate control is impaired due to the oxidation of the molten pool and a possible effect on flux sensor calibration. Specific data are scarce but Terashima et al. [109] reported reproducibility within $\pm 5\%$ from the YBCO stoichiometry when operating two e-guns to evaporate Y and Ba in a background pressure of 10^{-5} torr of oxygen. The common characteristic of all films processed *ex-situ* has been a resistive, nonsuperconducting surface

8. "MBE" GROWTH OF SUPERCONDUCTING MATERIALS

attributed to the interaction with CO_2 and H_2O in the ambient atmosphere. The first, to our knowledge, experiments to *in-situ* fabricate and characterize YBCO film surfaces prior to the deposition of an overlayer were performed in the system of Figs. 2 and 3. Films deposited as an amorphous phase were annealed *in-situ* in the introduction chamber at 1 atm of O_2 pressure and their surfaces characterized by XPS before and after annealing. The two main results were

(1) Oxidation of an oxygen-deficient amorphous deposit caused a segregation of Ba and Y atoms to or toward the surface, thus resulting in a non-superconducting surface layer even in the absence of reactions with H_2O or CO_2 [95]. Rapid thermal ramping to the crystallization temperature was found to minimize the segregation.

(2) Films annealed in this manner were epitaxial, with c-axis in the film plane, and their surfaces were at least in part superconducting [96]. Thin Au-overlayers deposited *in-situ* at room temperature after the complete annealing cycle conducted, at 4.2 K, a supercurrent between the YBCO and Nb counterelectrodes. With tunnel barriers of MgO *in-situ* evaporated on Au, unambiguous evidence of tunneling into the YBCO film surface was obtained in YBCO/MgO/Nb junctions with a YBCO gap voltage of 19 mV at 4.2 K, although device-quality junctions have not yet been obtained [97].

These experiments demonstrated again the advantages of an "MBE" approach, although the method of film synthesis by solid state epitaxy was not a desirable one. The post-deposition crystallization annealing of amorphous deposits, at temperatures $>800°C$, was promoting interdiffusion and reactions of substrate and film components. This was producing interface degradation in excess of that found in crystalline deposits (by vapor phase epitaxy), even if post-annealing at the same temperature and time was employed [32,98]. Solid state epitaxy is also unlikely to result in films having single-crystalline quality comparable to that of films deposited from vapor phase, monolayer by monolayer. Finally, the use of high oxygen pressure at high temperatures is hardly compatible with standard "MBE" equipment and requires modifying many components (Section 2) or even the deposition chamber itself. For example, Silver *et al.* [99] constructed a differentially pumped evaporation chamber shown schematically in Fig. 19.

At present, a very promising approach for "MBE", permitting the use of standard equipment, is reactive epitaxial growth from the vapor phase. It does not require high oxygen pressures due to the use of activated or atomic

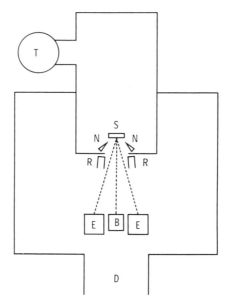

Fig. 19. Schematic of a differentially pumped evaporator [100]. The substrates (S) are in a subchamber pumped by a turbopump (T). Activated oxygen is sprayed on the substrate surface through two nozzles (N). The main part of the chamber, pumped by a diffusion pump (D), contains two e-beam sources (E), one evaporation boat (B), and rate monitors (R).

oxygen. Chamber pressures of $p(O_2) = 10^{-6}$ to 10^{-5} torr appear to be sufficient in this case. Several methods of activated or atomic oxygen generation have been reported as successful in producing YBCO films that did not require any post-annealing at a temperature higher than the deposition temperature. These are

- rf-plasma activation and atomic oxygen generation [99–103],
- atomic oxygen generation by an ozone source [104],
- atomic oxygen generation by NO cracking [105],
- low-energy ion beam generation by a Kaufman source [103] or electron cyclotron resonance (ECR) [106].

Each of the methods permitted the respective authors to deposit superconducting ($T_c \geq 80$ K, $R = 0$), highly epitaxial YBCO films at low substrate temperatures, between 450 and 650°C, either without any post-annealing or with oxygen annealing below the deposition temperature only. It is not clear yet which of these methods will turn out to be the most advantageous and, therefore, generally accepted.

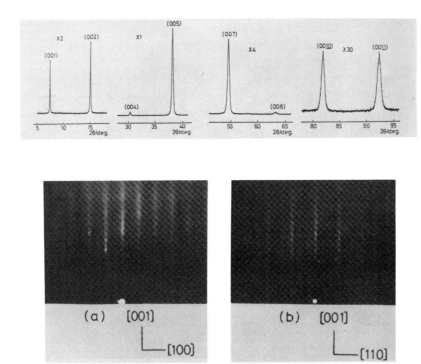

Fig. 20. RHEED and x-ray diffraction patterns for as-grown YBCO(100) 100-nm thick [100,101].

According to Bando et al. [100, 101] the oxygen activation by rf-plasma resulted in YBCO films that are probably the best to date, independent of the fabrication method. These films were co-evaporated from two e-guns and one thermal source on single-crystalline $SrTiO_3$ with (100) and (110) orientation and on MgO(100). The deposition temperature was $T_s = 500$ to $700°C$, and the chamber pressure was 10^{-5} torr of O_2, while the local pressure of rf-plasma-activated O_2 sprayed at the substrate was 10^{-2} torr. *In-situ* post-annealing in O_2 was limited to temperatures lower than T_s. Figure 20 shows x-ray and electron diffraction patterns of 100-nm thick films deposited on (100) $SrTiO_3$ that suggest a very high degree of texturing and a relatively high surface quality. LEED patterns indicate that these films are, indeed, single crystals [107]. These films had a residual resistivity of only 63 $\mu\Omega$-cm (resistivity ratio of 3.4), were superconducting at 90 K ($R = 0$), and at 77 K exhibited a $J_c = 4 \times 10^{-6}$ A/cm^2. The *in-situ* RHEED pattern of a film growing on MgO was characteristic of a perovskite at thicknesses exceeding 0.6 nm. On $SrTiO_3$, a film 10-nm thick was not only crystalline by x-ray diffraction but had a $T_c = 82$ K ($R = 0$) after low-temperature oxidation. Data by Bando et al. and by other authors referenced above offer a strong argument for the use of activated oxygen in vapor-phase epitaxy of YBCO.

Progress in HTS epitaxial film growth and characterization attained by summer–fall of 1988 suggests that "MBE" employing activated oxygen sources may become a leading technique in YBCO layered film device fabrication. It is not clear yet, whether or not a similar approach will be successful with other high-T_c cuprates, and especially with TBCCO. A severe loss of Tl occurs at high temperatures while, thus far, crystallization of high-T_c phases has only been observed close to the compound's melting point.

Section 3.5 was prefaced with a brief comparison of "MBE" growth versus higher-oxygen-pressure techniques for growth of high-T_c films. The successful use of activated oxygen sources to grow epitaxial YBCO films in high vacuum has differentiated "MBE" growth from higher-pressure methods. The benefits of operation in high vacuum are those traditionally associated with MBE: high surface mobility for adatoms, long mean-free paths, and access to the growth surface by various beam techniques. Examples of the last benefit are real-time monitoring by RHEED [102,108] and ion-assisted deposition.

Perhaps the work of Fujita et al. [108] is an example that indicates the future direction for superconductor "MBE", although they used dual ion-beam sputtering rather than evaporation. They created artificial superlattices in the Bi-Sr-Ca-Cu-O system on an atomic scale and attempted to grow metastable structures with c-axis unit cell lengths halfway between those of $Bi_2Sr_2Ca_nCu_{n+1}O_x$ ($n = 0, 1, 2$). They failed to stabilize any new phases, but succeeded in performing growth experiments on an atomic scale. A similar approach may be the best method for growing a tunnel barrier in high-T_c Josephson junctions, creating new device structures, stabilizing single-phase films, or searching for new superconductors.

A possibly intrinsic drawback of "MBE" fabrication is the loss of oxygen from the surface of HTS films exposed to UHV and the resulting degradation of the order parameter. Such loss was observed for YBCO by List et al. [112]. Use of cold substrate holders may be required to minimize the degradation prior to low-temperature deposition of a protective layer.

In closing, we should note that inherently non-"MBE," high-growth-pressure methods can be incorporated into separate chambers of a "MBE" multi-chamber system, as long as these chambers can be pumped down, after the deposition, to the requisite background pressure of the system. This was demonstrated for sputtering (Fig. 2) but can be equally well done for high-oxygen-pressure laser evaporation or even CVD. The surface oxygen loss problem will persist in these cases.

4. Conclusions

This chapter would have been very brief if we had confined the discussion to a description of technologically useful superconductors grown by MBE techniques defined in a narrow, traditional sense. Instead, we defined "MBE growth" broadly so that it encompassed all growth of highly oriented superconductors that were both deposited by PVD and maintained in an environment suitable for surface analysis and subsequent deposition of oriented film layers. Within the broader scope, MBE techniques have been used successfully to address scientific issues related to electronic applications of conventional superconductors.

However, it is the short coherence lengths and anisotropic properties of HTS materials that will move the use of MBE techniques to the forefront of superconductor technology. Passive electronic devices using HTS will require highly oriented films for low dissipation, low noise, and high current densities. Active devices will require, in addition, attention to problems with surfaces and interfaces that could be ignored in cases where long-coherence-length superconductors such as Nb or Pb were used. The development of active HTS devices will employ the experimental techniques and reopen materials issues already explored for short-coherence-length conventional superconductors—NbN and the A15 compounds.

References

1. Beasley, M.R., and Kircher, C.J. Josephson junction electronics: Material issues and fabrication techniques. *In* "Superconductor Materials Science" (S. Foner and B.B. Schwartz, eds.), pp. 605–684. Plenum Press, New York, (1981).
2. Raider, S.I. Josephson tunnel junctions with refractory electrodes. *IEEE Trans. Magn.* **21**, 110–117 (1985).
3. Braginski, A.I., Gavaler, J.R., Janocko, M.A., and Talvacchio, J. New materials for refractory tunnel junctions: Fundamental aspects. *In* "SQUID '85: Superconducting Quantum Interference Devices and their Applications" (H.D. Hahlbohm and H. Luebbig, eds.), pp. 591–630. Walter de Gruyter, Berlin, 1985.
4. Gallagher, W.J. Theory of Josephson tunneling into proximity-effect sandwiches. *Physica B* **108**, 825–826 (1981).
5. Cuomo, J.J., Laibowitz, R.B., Mayadas, A.F., and Rosenberg, R. Single crystal tunnel devices. United States Patent No. 3,816,845; 1974.
6. Luscher, P.E., and Collins, D.M. Design considerations for molecular beam epitaxy systems. *In* "Molecular Beam Epitaxy" (B.R. Pamplin, ed.), pp. 15–30. Pergamon Press, Oxford, 1980.
7. Talvacchio, J., Janocko, M.A., Gavaler, J.R., and Braginski, A.I. UHV deposition and analysis of thin-film superconductors. *In* "Advances in Cryogenic

Engineering—Materials, Vol. 32" (A.F. Clark and R.P. Reed, eds.), pp. 527–541. Plenum, New York, 1986.

8. Hammond, R.H. Electron beam evaporation synthesis of A15 superconducting compounds: Accomplishments and prospects. *IEEE Trans. Magn.* **11**, 201–207 (1975).
9. Bean, J.C., and Sadowski, E.A. Silicon MBE apparatus for uniform high-rate deposition on standard format wafers. *J. Vac. Sci. Technol.* **20**, 137–142 (1982).
10. Schellingerhout, J., Janocko, M.A., Klapwijk, T.M., and Mooij, J.E. Rate control for electron gun evaporation. *Rev. Sci. Instrum.* **60**, 1177–1183 (1988).
11. Becker, C.H., and Gillen, K.T. (1984). Surface analysis by nonresonant multiphoton ionization of desorbed or sputtered species. *Anal. Chem.* **56**, 1671–1674 (1984).
12. Dushman, S. "Scientific Foundations of Vacuum Technique." Wiley, New York, 1962.
13. Houdy, Ph., Sirat, J.A., Theeten, J.B., Landesman, J.P., Baudry, H., Monneraye, M., Schiller, C., and Patillon, J.N. Amorphization evidence from kinetic ellipsometry in monolayer-controlled deposition of RF sputtered YBaCuO compounds. *In* "Thin Film Processing and Characterization of High-Temperature Superconductors" (J.M.E. Harper, R.J. Colton, and L.C. Feldman, eds.), pp. 122–129. Am. Inst. of Phys., New York, 1988.
14. Cohen, P.I., Pukite, P.R., van Hove, J.M., and Lent, C.S. Reflection high energy electron diffraction studies of epitaxial growth on semiconductor surfaces. *J. Vac. Sci. Tech. A* **4**, 1251–1258 (1986).
15. Park, S.I., Marshall, A.F., Hammond, R.H., Geballe, T.H., and Talvacchio, J. The role of ion-beam cleaning in the growth of strained-layer epitaxial thin transition metal films. *J. Mater. Res.* **2**, 446–455 (1987).
16. Adachi, H., Setsune, K., and Wasa, K. Superconductivity of La-Sr-Cu-O single-crystal thin films. *Japan. J. Appl. Phys.* supp. **26-3**, 1139–1141 (1987).
17. Enomoto, Y., Murakami, T., Suzuki, M., and Moriwaki. K. Largely anisotropic superconducting critical current in epitaxially grown $Ba_2YCu_3O_{7-y}$ thin film. *Japan. J. Appl. Phys.* **26**, L1248–1250 (1987).
18. Talvacchio, J., Sinharoy, S., and Braginski, A.I. Surface stability of NbN single-crystal films. *J. Appl. Phys.* **62**, 611–614 (1987).
19. Lagally, M.G. Diffraction Techniques. *In* "Methods of Experimental Physics, Vol. 22" (R.L. Park and M.G. Lagally, eds.), p. 237. Academic Press, New York, 1985.
20. Talvacchio, J., Gavaler, J.R., and Braginski, A.I. Epitaxial niobium nitride/insulator layered structures. *In* "Proceedings TMS–AIME: Metallic Multilayers and Epitaxy" (M. Hong, D.U. Gubser, and S.A. Wolf, eds.), pp. 109–134. The Metallurgical Society, Pittsburgh, 1987.
21. Jona, F. LEED crystallography. *J. Phys. C* **111**, 4271–4306, (1987).
22. Talvacchio, J., Sinharoy, S., and Takei, K. Low-energy electron diffraction study of epitaxial NbN tunnel junctions. In preparation.
23. Talvacchio, J., and Braginski, A.I. Lattice-matched oxide barriers for NbN tunnel junctions. *In* "Proc. Intl. Superconductivity Electronics Conference" (S. Hasuo, ed.), pp. 309–312. Tokyo, 1987.

24. Laiho, R., Heikkila, L., and Snellman, H. Microstructural investigation of the high-T_c superconductor $YBa_2Cu_3O_x$ with a scanning tunneling microscope. *J. Appl. Phys.* **63**, 225–227 (1988).
25. Kellogg, G.L., and Brenner, S.S. Investigations of superconducting and non-superconducting $YBa_2Cu_3O_{7-x}$ by field ion microscopy, atom-probe mass spectroscopy and field electron emission. *J. de Phys.* **49**, C6, 465–475 (1988).
26. Siegbahn, K. "ESCA Applied to Free Molecules." North Holland, Amsterdam, 1969.
27. Kwo, J., Wertheim, G.K., Gurvitch, M., and Buchanan, D.N.E. X-ray photoelectron study of surface oxidation of Nb/Al overlayer structures. *Appl. Phys. Lett.* **40**, 675–678 (1982).
28. Briggs, D., and Seah, M.P., eds. "Practical Surface Analysis." Wiley, New York, 1983.
29. Laibowitz, R.B., and Cuomo, J.J. Tunneling sandwich structures using single-crystal niobium films. *J. Appl. Phys.* **41**, 2748–2750 (1970).
30. O'Neal, J.E., and Wyatt, R.L. Hetero-epitaxial films of niobium on sapphire. *Thin Films* **2**, 71–81 (1971).
31. Durbin, S.M., Cunningham, J.E., and Flynn, C.P. Growth of single-crystal metal superlattices in chosen orientations. *J. Phys. F: Met. Phys.* **12**, L75–L78 (1982).
32. Talvacchio, J., Janocko, M.A., Braginski, A.I., and Gavaler, J.R. Comparison of $YBa_2Cu_3O_7$ films grown by solid state and vapor-phase epitaxy. *IEEE Trans. Magn.* **25**, 2538–2541 (1989).
33. Sosniak, J. The deposition of niobium thin films by dc diode and substrate bias sputtering. *J. Appl. Phys.* **39**, 4157–4163 (1968).
34. Eizenberg, M., Smith, D.A., Heiblum, M., and Segmuller, A. Electron beam evaporation of oriented Nb films onto GaAs crystals in ultrahigh vacuum. *Appl. Phys. Lett.* **49**, 422–424 (1986).
35. Talvacchio, J., Janocko, M.A., and Greggi, J. Properties of evaporated Mo-Re thin-film superconductors. *J. Low Temp. Phys.* **64**, 395–408 (1986).
36. Oya, G., and Onodera, Y. Transition temperatures and crystal structures of single-crystal and polycrystalline NbN films. *J. Appl. Phys.* **45**, 1389 (1974).
37. Talvacchio, J., and Braginski, A.I. Tunnel junctions fabricated from coherent $NbN/MgO/NbN$ and $NbN/Al_2O_3/NbN$ structures. *IEEE Trans. Magn.* **23**, 859–862 (1987).
38. Noskov, V.L., Titenko, Y.V., Korzhinskii, F.I., Zelenkevich, R.L., and Komashko, V.A. Heteroepitaxial layers of niobium nitride on sapphire. *Sov. Phys. Crystallogr.* **25**, 504–508 (1980).
39. Marshall, A.F., Hellman, F., and Oh, B. Epitaxy of Nb_3Sn films on sapphire. In "Layered Structures, Epitaxy, and Interfaces; MRS Vol. 37" (J.M. Gibson and L.R. Dawson, eds.), p. 517. Materials Research Society, Pittsburgh, 1985.
40. Janocko, M.A., Braginski, A.I., Gavaler, J.R., Talvacchio, J., and Walker, E. Properties of epitaxial Nb_3Ge films on Nb_3Ir and Nb_3Sn single crystals. *Bull. Am. Phys. Soc.* **31**, 238 (1986).
41. Gavaler, J.R., Braginski, A.I., Janocko, M.A., and Talvacchio, J. Epitaxial growth of high-T_c superconducting films. *Physica* **135B**, 148–153 (1985).
42. Asano, H., Tanabe, K., Katoh, Y., and Michikami, O. Epitaxial growth of

superconducting Nb_3Ge Films on YSZ single-crystal substrates. *Japan. J. Appl. Phys.* **27**, 35–39 (1988).

43. Suzuki, M., Enomoto, Y., Moriwaki, K., and Murakami, T. Anisotropic properties of superconducting $(La_{1-x}Sr_x)_2CuO_4$ single-crystal thin films. *Japan. J. Appl. Phys.* **26**, L1921–L1924 (1987).
44. Chaudhari, P., Koch, R.H., Laibowitz, R.B., McGuire, T.R., and R.J. Gambino Critical current measurements in epitaxial films of $YBa_2Cu_3O_{7-x}$. *Phys. Rev. Lett.* **58**, 2684–2687 (1987).
45. Simon, R.W., Platt, C.E., Lee, A.E., Lee, C.S., Daly, K.P., Wire, M.S., Luine, J.A., and Urbanik, M. Low-loss substrate for epitaxial growth of high-temperature superconductor thin films. *Appl. Phys. Lett.* **53**, 2677–2679 (1988).
46. Sandstrom, R.L., Giess, E.A., Gallagher, W.J., Segmuller, A., Cooper, E.I., Chisholm, M.F., Gupta, A. Shinde, S., and Laibowitz, R.B. Lanthanum gallate substrates for epitaxial high-T_c superconducting thin films. *Appl. Phys. Lett.* **53**, 1874–1876 (1988).
47. Feenstra, R., Boatner, L.A., Budai, J.D., Christne, D.K., Galloway, M.D., and Poker, D.B. Epitaxial superconducting thin films of $YBa_2Cu_3O_{7-x}$ on $KTaO_3$ single crystals. *Appl. Phys. Lett.* **54**, 1063–1065 (1989).
48. Höhler, A., Guggi, D., Neeb, H., and Heiden, C. Fully textured growth of $YBa_2Cu_3O_{7-\delta}$ films by sputtering on $LiNbO_3$ substrates. *Appl. Phys. Lett.* **54**, 1066–1067 (1989).
49. Royer, L. Experimental research on parallel growth or mutual orientation of crystals of different species. *Bull. Soc. Fr. Mineral. Crystallogr.* **51**, 7 (1928).
50. Sinharoy, S., Braginski, A.I., Talvacchio, J., and Walker, E. A LEED, AES and XPS study of single crystal Nb_3Ir substrates. *Surf. Science* **167**, 401–416 (1986).
51. Braginski, A.I., Gavaler, J.R., and Schultze, K. Formation of A15 phase in expitaxial and polycrystalline Nb-Sn and Nb-Al diffusion couples. *In* "Advances in Cryogenic Engineering—Materials, Vol. 32" (A.F. Clark and R.P. Reed, eds.), pp. 585–592. Plenum, New York, 1986.
52. Claassen, J.H., Wolf, S.A., Qadri, S.B., and Jones, L.D. Epitaxial growth of niobium thin films. *J. Cryst. Growth* **81**, 557–561 (1987).
53. McWhan, D.B. Structure and coherence of metallic superlattices. *In* "Layered Structures, Epitaxy, and Interfaces; MRS Vol. 37" (J.M. Gibson and L.R. Dawson, eds.), p. 483. Materials Research Society, Pittsburgh, 1985.
54. O'Neal, J.E., and Rath, B.B. Crystallography of epitaxially grown molybdenum on sapphire. *Thin Solid Films* **23**, 363–380 (1974).
55. Durbin, S.M., Buchanan, D.S., Cunningham, J.E., and Ginsberg, D.M. Observation of tunneling anisotropy in superconducting niobium crystals. *Phys. Rev. B.* **28**, 6277–6280 (1983).
56. Darlinski, A., and Halbritter, J. Angle-resolved XPS studies of oxides at NbN, NbC, and Nb surfaces. *Surf. Inter. Anal.* **10**, 223–237 (1987).
57. Pollak, R.A., Stolz, H.J., Raider, S.I., and Marks, R.F. Chemical composition and interface chemistry of very thin Nb_2O_5 films prepared by rf plasma oxidation. *Oxidation of Metals* **20**, 185–192 (1983).
58. Raider, S.I., Johnson, R.W., Kuan, T.S., Drake, R.E., and Pollak, R.A.

Characterization of Nb/Nb oxide structures in Josephson tunnel junctions. *IEEE Trans. Magn.* **19**, 803–806 (1983).
59. Celaschi, S., Geballe, T.H., and Lowe, W.P. Tunneling properties of single crystal Nb/Nb$_2$O$_5$/Pb Josephson junctions. *Appl. Phys. Lett.* **43**, 794–796 (1983).
60. Henkels, W.H. and Kircher, C.J. Penetration depth measurements on type II superconducting films. *IEEE Trans. Magn.* **13**, 63–66 (1977).
61. Braginski, A.I., Talvacchio, J., Janocko, M.A., and Gavaler, J.R. Crystalline oxide tunnel barriers formed by thermal oxidation of aluminum overlayers on superconductor surfaces. *J. Appl. Phys.* **60**, 2058–2064 (1986).
62. Kodama, J., Itoh, M., and Hirai, H. Superconducting transition temperature versus thickness of Nb films on various substrates. *J. Appl. Phys.* **54**, 4050–4052 (1983).
63. Reible, S.A. Wideband analog signal processing with superconductive circuits. *In* "1982 Ultrasonics Symposium Proceedings," pp. 190–201. IEEE, New York, 1982.
64. McAvoy, B.R., Wagner, G.R., Adam, J.D., and Talvacchio, J. Superconducting stripline resonator performance. *IEEE Trans. Magn.* **25**, 1104–1106 (1989).
65. Face, D.W., Prober, D.E., McGrath, W.R., and Richards, P.L. High quality tantalum superconducting tunnel junctions for microwave mixing in the quantum limit. *Appl. Phys. Lett.* **48**, 1098–110 (1986).
66. Kosaka, S. and Onodera, Y. Epitaxial deposition of niobium nitride by sputtering. *Japan. J. Appl. Phys. Suppl.* **2-1**, 613–616 (1974).
67. Yamashita, T., Hamasaki, K., and Komata, T. Epitaxial growth of NbN on MgO film. *In* "Advances in Cryogenic Engineering—Materials, Vol. 30" (A.F. Clark and R.P. Reed, eds.), pp. 616–626. Plenum, New York, 1986.
68. Shoji, A., Aoyagi, M., Kosaka, S., Shinoki, F., and Hayakawa, H. NbN Josephson tunnel junctions with MgO barriers. *Appl. Phys. Lett.* **46**, 1098–1100 (1985).
69. Knapton, A.G. The Mo-Re system. *J. Inst. Metals* **87**, 62–64 (1958).
70. Gavaler, J.R., Janocko, M.A., and Jones, C.K. A-15 structure Mo-Re superconductor. *Appl. Phys. Lett.* **21**, 179–180 (1972).
71. Postnikov, V.S., Postnikov V.V., and Zheleznyi, V.S. Superconductivity in Mo-Re system alloy films produced by electron beam evaporation in high vacuum. *Phys. Stat. Solidi* **A39**, K21–23 (1977).
72. Testardi, I.R., Hauser, J.J., and Read, M.H. Enhanced superconducting T_c and structural transformation in Mo-Re alloys. *Solid State Commun.* **9**, 1829–1831 (1971).
73. Brophy, J.H., Rose, R.M., and Wulff, J. On the solubility of interstitial elements in binary transition metal alloys. *J. Less-Common Metals* **5**, 90–91 (1963).
74. Rudman, D.A., Hellman, F., Hammond, R.H., and Beasley, M.R. A15 tunnel junction fabrication and properties. *J. Appl. Phys.* **55**, 3544–3553 (1984).
75. Hellman, F., Talvacchio, J., and Geballe, T.H. A new look at the growth of thin films of Nb-Sn. *In* "Advances in Cryogenic Engineering—Materials, Vol. 32" (A.F. Clark and R.P. Reed, eds.), pp. 593–602. Plenum, New York, 1986.
76. Talvacchio, J., Braginski, A.I., Janocko, M.A., and Bending, S.J. Tunneling and

interface structure of oxidized metal barriers on A15 superconductors. *IEEE Trans. Magn.* **21**, 521–524 (1985).
77. Schellingerhout, J. Unpublished analytical data. Technological University of Delft, Netherlands, 1986.
78. Jorda, J.L., Fluekiger, R., and Muller, J. The phase diagram of the Nb-Ge system. *J. Less-Common Metals* **62**, 25–37 (1978).
79. Gavaler, J.R. Superconductivity in Nb-Ge films above 22 K. *Appl. Phys. Lett.* **23**, 480–482 (1973).
80. Gavaler, J.R., Miller, J.W., and Appleton, B.R. Oxygen distribution in sputtered Nb-Ge films. *Appl. Phys. Lett.* **28**, 237–239 (1976).
81. Dayem, A.H., Geballe, T.H., Zubeck, R.B., Hallak, A.B., and Hull, Jr. G.W. Epitaxial growth of high T_c superconducting Nb_3Ge on Nb_3Ir. *Appl. Phys. Lett.* **30**, 541–543 (1977).
82. Gavaler, J.R., Braginski, A.I., Ashkin, A., and Santhanam, A.T. Thin films and metastable phases. In "Superconductivity in d- and f-Band Metals" (H. Suhl and M.B. Maple, eds.), pp. 25–36. Academic Press, New York, 1980.
83. Kuwasa, Y., and Nakano, S. Correlation of homoepitaxial growth of high-T_c A15 Nb_3Ge with characteristics of substrate surface. *J. Low Temp. Phys.* **61**, 45–53 (1985).
84. Braginski, A.I., Janocko, M.A., Gavaler, J.R., and Talvacchio, J. Unpublished results, presented at the 1987 CEC–ICMC, 1987.
85. Bednorz, J.G., and Müller, K.A. Possible high T_c superconductivity in the Ba-La-Cu-O system. *Z. Phys. B* **64**, 189–193 (1986).
86. Wu, M., Ashburn, J.R., Torng, C.J., Hor, P.H., Meng, R.L., Gao, L., Huang, Z.J., Wang, Y.Q., and Chu, C.W. Superconductivity at 93 K in a new mixed phase Y-Ba-Cu-O compound system at ambient pressure. *Phys. Rev. Lett.* **58**, 908–910 (1987).
87. Parkin, S.S.P., Lee, V.Y., Engler, E.M., Nazzal, A.I., Huang, T.C., Gorman, G., Savoy, R., and Beyers, R. Bulk superconductivity at 125 K in $Tl_2Ca_2Ba_2Cu_3O_x$. *Phys. Rev. Lett.* **60**, 2539–2542 (1988).
88. Tarascon, J.M., McKinnon, W.R., Barboux, P., Hwang, D.M., Bagley, B.G., Greene, L.H., Hull, G.W., LePage, Y., Stoffel, N., and Giroud, M. Preparation, structure and properties of the superconducting compound series $Bi_2Sr_2Ca_{n-1}Cu_nO_y$ with $n = 1,2$ and 3. *Phys. Rev. B.* **38**, 8885–8892 (1988).
89. Parkin, S.S.P., Lee, V.Y., Nazzal, A.I., Savoy, R., Beyers, R., and La Placa, S.J. $Tl_1Ca_{n-1}Ba_2Cu_nO_{2n+3}$ ($n = 1,2,3$): A new class of crystal structures exhibiting superconductivity at up to 110 K. *Phys. Rev. Lett.* **61**, 750–753 (1988).
90. Cava, R.J., Battlog, B., Sunshine, S.A., Siegrist, T., Fleming, R.M., Rabe, K., Schneemeyer, L.F., Murphy, D.W., van Dover, R.B., Gallagher, P.K., Glarum, S.H., Nakahara, S., Farrow, R.C., Krajewski, J.J., Zahurak, S.M., Waszczak, J.V., Marshall, J.H., Marsch, P., Rupp, Jr., L.W., Peck, W.F., and Rietman, E.A. Studies of oxygen-deficient $YBa_2Cu_3O_{7-\delta}$ and superconducting Bi(Pb)-Sr-Ca-Cu-O. *Physica C* **153–155**, 560–565 (1988).
91. Lee, W.Y., Lee, V.Y., Salem, J., Huang, T.C., Savoy, R., Bullock, D.C., and Parkin, S.S.P. Superconducting Tl-Ca-Ba-Cu-O thin films with zero resistance at temperatures of up to 120 K. *Appl. Phys. Lett.* **53**, 329–331 (1988).

92. Kwo, J., Hsieh, T.C., Hong, M., Liou, S.H., Davidson, B.A., and Feldman, L.C. Structural and superconducting properties of orientation-ordered $YBa_2Cu_3O_{7-x}$ films prepared by molecular-beam epitaxy. *Phys. Rev. B* **36**, 4039–4042 (1987).
93. Gavaler, J.R., and Janocko, M.A. Data quoted by Braginski A.I., Study of superconducting oxides at Westinghouse. In "Novel Superconductivity" (S.A. Wolf and V.Z. Kresin, eds.), pp. 935–950. Plenum Press, New York, 1987.
94. Lathrop, D.K., Russek, S.E., and Buhrman, R.A. Production of $YBa_2Cu_3O_{7-y}$ superconducting thin films *in situ* by high-pressure reactive evaporation and rapid thermal annealing. *Appl. Phys. Lett.* **51**, 1554–1556 (1987).
95. Gavaler, J.R., and Braginski, A.I. Near surface atomic segregation in YBCO thin films. *Physica C* **153–155**, 1435–1436 (1988).
96. Braginski, A.I., Talvacchio, J., Gavaler, J.R., Forrester, M.G., and Janocko, M.A. *In-situ* fabrication, processing and characterization of superconducting oxide films. In "SPIE Proceedings Vol. 948, High-T_c Superconductivity: Thin Films and Devices" (R.B. van Dover and C.C. Chi, eds.), pp. 89–98. SPIE, Bellingham, Washington, 1988.
97. Gavaler, J.R., Forrester, M.G., and Talvacchio, J. Properties of YBCO-based tunnel junctions. *Physica C* in press (1989).
98. Wu, X.D., Venkatesan, T., Inam, A., Chase, E.W., Chang, C.C., Yeon, Y., Croft, M., Magee, C., Odom, R.W., and Radicati, F. Versatility of pulsed laser deposition technique for preparation of high-T_c superconducting thin films. In "SPIE Proceedings Vol. 948, High-T_c Superconductivity: Thin Films and Devices (R.B. van Dover and C.C. Chi, eds.), pp. 50–65. SPIE, Bellingham, Washington, 1988.
99. Silver, R.M., Berezin, A.B., Wendman, M., and de Lozanne, A.L. As-deposited superconducting Y-Ba-Cu-O thin films on Si, Al_2O_3, and $SrTiO_3$ substrates. *Appl. Phys. Lett.* **52**, 2174–2176 (1988).
100. Bando, Y., Terashima, T., Iijima, K., Yamamoto, K., Hirata, K., and Mazaki, H. Single crystal $YBa_2Cu_3O_{7-x}$ thin film by activated reactive evaporation. *Physica C* **153–155**, 810–811 1988).
101. Bando, Y., Terashima, T., Iijima, K., Yamamoto, K., Hirata, K. and Mazaki, H. Single crystal $YBa_2Cu_3O_{7-x}$ thin film by activated reactive evaporation. *In* "FED-65: Extended Abstracts of Future Electron Devices Workshop," pp. 11–16. Miyagi-Zao, 1988.
102. Kwo, J., Hong, M., Trevor, D.J., Fleming, R.M., White, A.E., Farrow, R.C., Kortan, A.R., and Short, K.T. *In-situ* epitaxial growth of $YBa_2Cu_3O_{7-x}$ films by molecular beam epitaxy with an activated oxygen source. *Appl. Phys. Lett.* **53**, 2683–2685 (1988).
103. Missert, N., Hammond, R.H., Mooij, J.E., Matijasevic, V., Rosenthal, P., Geballe, T.H., Kapitulnik, A., Beasley, M.R., Laderman, S.S., Lu, C., Garwin, E., and Barton, R. *In-situ* growth of superconducting YBaCuO using reactive electron-beam coevaporation. *IEEE Trans. Magn.* **25**, 2418–2421 (1989).
104. Berkley, D.D., Johnson, B.R., Anand, N., Beauchamp, K.M., Conroy, L.E., Goldman, A.M., Maps, J., Mauersberger, K., Mecartney, M.L., Morton, J., Tuominen, M., and Zhang, Y.-J. *In situ* formation of superconducting

$YBa_2Cu_3O_{7-x}$ thin films using pure ozone vapor oxidation. *Appl. Phys. Lett.* **53**, 1973–1975 (1988).
105. Evetts, J. Unpublished data (Cambridge University), 1988.
106. Moriwaki, K., Enomoto, Y., Kubo, S., and Murakami, T. As-deposited superconducting $Ba_2YCu_3O_{7-y}$ films using ECR ion beam oxidation. *Japan. J. Appl. Phys.* **27**, L2075–L2077 (1988).
107. Sakisaka, Y., Komeda, T., Maruyama, T., Onchi, M., Kato, H., Aiura, Y., Yanashima, H., Torashima, T., Bando, Y., Iijima, K., Yamamoto, K., and Hiata, K. Angle-resolved photoemission investigation of the electronic band properties of $YBa_2Cu_3O_{7-x}$ (001). *Phys. Rev.* B **39**, 9080–9090 (1989).
108. Fujita, J., Tatsumi, T., Yoshitake, T., and Igarashi, H. Film fabrication of artificial (BiO)/(SrCaCuO) layered structure. *In* "Science and Technology of Thin-Film Superconductors" (R. McConnell and S.A. Wolf, eds.), Plenum, New York, 175–184, 1989.
109. Terashima, T., Iijima, K., Yamamoto, K., Takada, J., Hirata, K., Mazaki, H., and Bando, Y. Formation and properties of $YBa_2Cu_3O_{7-x}$ single-crystal thin films by activated reactive evaporation. *J. Crystal Growth* **95**, 617–620 (1989).
110. Park, S.I., and Geballe, T.H. Superconductive tunneling in ultrathin Nb films. *Phys. Rev. Lett.* **57**, 901–904 (1986).
111. Kwo, J., Hong, M., and Nakahara, S. Growth of rare-earth single crystals by molecular beam epitaxy: The epitaxial relationship between hcp rare earth and bcc niobium. *Appl. Phys. Lett.* **49**, 319–321 (1986).
112. List, R.S., Arko, A.J., Fisk, Z., Cheong, S.-W., Conradson, S.D., Thompson, J.D., Pierce, C.B., Peterson, D.E., Bartlett, R.J., Shinn, N.D., Schirber, J.E., Veal, B.W., Paulikas, A.P., and Campuzano, J.C. Photoemission from single crystals of $EuBa_2Cu_3O_{7-x}$ cleaved below 20 K: Temperature-dependent oxygen loss. *Phys. Rev. B* **38**, 11966–11969 (1988).
113. Gavaler, J.R., Talvacchio, J., and Braginski A.I. Epitaxial growth of NbN films. *In* Advances in Cryogenic Engineering—Materials, Vol. 32 (A.F. Clark and R.P. Reed, eds.), pp. 627–633. Plenum Press, New York, (1986).
114. Poppe, U., Prieto, R., Schubert, J., Soltner, H., Urban, K., and Buchal C. Epitaxial multilayers of $YBa_2Cu_3O_7$ and $PrBa_2Cu_3O_7$ as a possible basis for superconducting electronic devices. *Solid State Commun.* **71**, 569–572 (1989).
115. Rogers, C.T., Inam, A., Hegde, M.S., Dutta, B., Wu, X.D., and Venkatesan, T. Fabrication of heteroepitaxial $YBa_2Cu_3O_{7-x}$-$PrBa_2Cu_3O_{7-x}$-$YBa_2Cu_3O_{7-x}$ Josephson devices grown by laser deposition. *Appl. Phys. Lett* **55**, 2031–2034 (1989).
116. Shoji, A., Aoyagi, M., Kosaka, S., and Shinoki, S. Temperature-dependent properties of niobium nitride Josephson tunnel junctions. *IEEE Trans. Magn.* **23**, 1464–1467 (1987).
117. Song, S.N., Jin, B.Y., Yang, H.Q., Ketterson, J.B., and Schuller, I.K. Preparation of large-area NbN/AlN/NbN Josephson junctions. *Jpn. J. Appl. Phys. Suppl.* **26-3**, 1615–1616 (1987).

CHAPTER 9

Three-Terminal Devices

A.W. KLEINSASSER
and
W.J. GALLAGHER

IBM Research Division
T.J. Watson Research Center
Yorktown Heights, New York

1.	Introduction	325
2.	Hybrid Superconductor–Semiconductor Devices	328
	2.1. Introduction	328
	2.2. Junction Transistors	329
	2.3. Field Effect Transistors	338
	2.4 Summary	352
3.	Nonequilibrium Superconducting Devices	353
	3.1. Introduction	353
	3.2. Stacked-Junction Devices	355
	3.3. Injection-Controlled Links	360
	3.4. Conclusions on Nonequilibrium Devices	363
4.	Magnetic and Other Devices	364
5.	Conclusions	366
	Acknowledgements	367
	References	367

1. Introduction

Over the last three decades, superconductive electronics has developed along two lines: digital applications based on high switching speed and low power dissipation, and analog applications based on extremely high sensitivity and/or frequency response extending into the far infrared. Virtually all of superconductive electronics depends on the use of tunneling and/or the Josephson effect in two-terminal devices. Without a three-terminal transistor-like device, superconductive electronics will undoubtedly remain relegated to specialized niches. There has been no lack of proposals for three-terminal devices. Some of these devices are not truly transistor-like, and none have yet been developed to a practical stage. Progress has been made, however, in a number of areas since the last general review of this subject [1]. This chapter will discuss the subject of three-terminal superconducting devices with an emphasis on recent progress.

The discovery of superconductivity at temperatures exceeding 100 K in copper oxide-based materials broadens the scope of potential superconducting device applications considerably. There is a convergence of semiconductor technology, with its growing focus on low temperatures (particularly 77 K) for advanced computer applications, and superconductor technology [2,3]. Although the most obvious application might be to use superconducting interconnects in advanced semiconductor circuits, there is considerable interest in using the unique aspects of superconducting devices at higher temperatures, or in looking for new devices that take advantage of, for example, a larger superconducting energy gap. To date, there has been little work on high temperature three-terminal superconducting devices, with most thinking aimed at extending old ideas to a new temperature range. This article will therefore focus on low (helium) temperature three-terminal devices, extrapolating to higher temperatures where possible. Of course, it is possible that truly novel high temperature applications will emerge in the future.

Research on three-terminal superconducting devices has been aimed at demonstrating basic operating principles and is not focused on specific applications, although digital devices appear to be the most common aim. The discussion in this chapter is thus intended to be fairly general, with a slant towards digital applications. The basic function of a transistor-like device can be understood through use of Landauer's simple fluid-actuated valve analogy [4], illustrated in Fig. 1. A piston, actuated by a "control" fluid in one pipe, is used to control the flow of fluid in another pipe. The two fluids must be separate, with little mixing. Three-terminal devices such as vacuum tubes, junction transistors, and FETs can be described by this model, making it a useful guide for evaluating new device ideas. For example, the deficiencies of some proposed non-equilibrium superconducting devices can be clearly illustrated using this fluid analogy [1]. For actual applications, of course, there is an additional list of desirable properties for three-terminal devices. In digital applications, for example, one needs current and voltage gain, high speed, compatibility with line impedances, and manufacturability, among other properties [5].

For applications involving high levels of circuit integration, removal of the heat dissipated by device operation is a particular concern, especially for low operating temperatures. The product of power dissipated per switching cycle

9. THREE-TERMINAL DEVICES

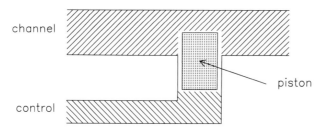

Fig. 1. Schematic of a simple fluid model for a three-terminal device. The fluid in one (gate) pipe controls a piston that varies the flow of fluid in a second pipe. Separation of the two fluids is important. Three-terminal devices such as vacuum tubes, junction transistors, and FETs can be described by this model. Landauer's model [4]

and switching delay is thus an important figure of merit for digital devices. This power-delay product is orders of magnitude smaller in Josephson devices than in competing semiconductor devices, a much-touted advantage of Josephson technology. This is often illustrated with plots of power versus gate delay [2], in which Josephson devices are far removed from semiconductor devices. However, the ability to remove heat from a wafer scales linearly with temperature [6], and device power dissipation must be reduced along with the operating temperature. The standard plots of power versus delay should really be replaced by plots in which power is scaled by kT, and the comfortable margin enjoyed by Josephson technology is actually much smaller than is generally appreciated. Clearly, low temperature digital devices must operate at low voltages. Conventional semiconductor devices dissipate too much power to be of interest for dense circuits at low (~ 10 K) temperatures. The superconducting energy gap (<5 meV for conventional superconductors) sets a suitably small voltage scale for dense low temperature circuits. However, many potential applications for low temperature devices, both analog and digital, do not require high circuit density, so the requirement of low operating voltage is not a general one.

In this chapter, we discuss devices that are intended to act like transistors. The major subject will be hybrid superconductor–semiconductor devices, which are discussed in Section 2. This has been the most active research area in recent years. Section 3 deals with devices based on nonequilibrium superconductivity. Such devices have received considerable attention, but the prospects for useful devices appear to be narrowing, with the remaining interest focused on low carrier density materials. A brief discussion of magnetic and other devices is included in Section 4, and Section 5 contains concluding remarks.

2. Hybrid Superconductors–Semiconductor Devices

2.1. Introduction

In stark contrast to the situation with superconducting devices, there are several entrenched classes of three-terminal semiconductor devices, the most significant being bipolar and field effect transistors. It is reasonable to take working device principles from semiconductor technology and apply them in hybrid superconductor–semiconductor devices; indeed, several such devices have been proposed. In this section, we discuss these devices in the context of the familiar types of semiconductor devices, addressing the questions of what advantages the incorporation of superconductivity into a device can offer and whether any new device principles result.

Semiconductor transistors are by themselves interesting for low temperature applications. This is becoming particularly evident at 77 K, but is also true down to helium temperatures. A comprehensive review of low temperature semiconductor electronics is clearly beyond the scope of this article. The interested reader is referred to the literature, which includes compilations of recent articles on low temperature electronics [7,8], a review of high speed semiconductor electronics [9], a comparison of semiconductor and Josephson technologies [10], and two short reviews of hybrid superconductor–semiconductor device work [11,12].

For our purposes, existing semiconductor devices can be placed into two categories, junction transistors and field effect transistors (FETs). Junction transistors include both bipolar (BJT) and unipolar (UJT) devices, the latter being hot electron (majority carrier) devices which are similar in concept and structure to bipolar devices. FETs are also unipolar (current is carried by one type of carrier, in contrast to bipolar transistors, in which both electrons and holes are important). BJTs and FETs are familiar in semiconductor technology, while UJTs remain in an exploratory stage. Most hybrid superconductor–semiconductor device work has been in these categories. Numerous other novel devices such as tunneling transistors [13,14], resonant tunneling transistors [15,16], and devices based on real space charge transfer [17,18] have been proposed. Such devices are of current research interest, but no attempts have been made to consider superconducting versions.

The scale of operating voltages in superconducting devices is usually of the order of the superconducting energy gap, which is a few meV or less for conventional superconductors and, being proportional to T_c, is expected to be a few tens of meV for the new oxide materials (i.e., in a material with $T_c \approx$ 90–120 K). Typical voltages required for semiconductor device operation

9. THREE-TERMINAL DEVICES

are of order 1 V, but scaling down of device dimensions in order to allow denser circuits requires a reduction of device voltages. This scaling is most successful with MOS devices, Si CMOS [19] for example, leading to voltages of the order of a few hundred mV in 0.25 μm circuits, and tending to force a reduction of the operating temperature for CMOS circuits. This may lead to some convergence of the voltage scales for superconducting and semiconducting technologies, although they are at present still an order of magnitude or so apart. As will emerge in the following discussion, the operating voltages of hybrid devices may or may not be determined by the same physics that applies to the related semiconductor devices.

2.2. Junction Transistors

Junction transistors have three active regions: emitter, collector, and base. Device operation consists of the injection of charge carriers from emitter to base, transit across the base, and collection [20]. The injected carriers are energetically separate from the charge carriers in the base, conforming to the fluid analogy described earlier. In an npn (pnp) bipolar transistor, electrons (holes), which are majority carriers in the emitter, are injected into the base, where they are minority carriers, the majority carriers in the base being the control fluid. In order for the device to have gain, the transport of the injected carriers must be controlled with only a small base current.

Typical Si bipolar transistors do not work at low temperatures [9], although this is not a fundamental limitation. However, at low voltages no current can flow until the emitter–base bias exceeds a voltage that is essentially the band gap of the base. Bipolar devices are thus not competitive for applications that require low power dissipation (even in small gap materials, this voltage is several hundred mV). For the purposes of the present discussion, we need only note that the bipolar transistor is a purely semiconductor device, and there is no obvious way to incorporate superconductivity into a hybrid device. However other devices of interest were inspired by the bipolar transistor concept.

2.2.1. Hot Electron Transistors

The speed of bipolar devices is limited by the time it takes minority carriers to diffuse across the base. The use of a thinner base reduces this time, but this is of limited use because of the resulting increase in base resistance. In 1960, Mead [21] proposed the metal base transistor (MBT), a unipolar analog of the

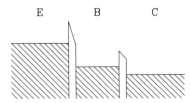

Fig. 2. Schematic band diagram of a MIMIM hot electron transistor. Carriers from the emitter (E) are injected by tunneling into the base (B) (alternatively, thermionic emission over the top of a barrier can be used). Under certain conditions, these "hot" carriers must be collected over the top of the collector (C) barrier before relaxing down to the base Fermi energy and contributing to the base current.

bipolar transistor. Such a device is shown schematically in Fig. 2. The device is based on the injection of majority carriers (electrons) into the base film; these carriers have a large kinetic energy in the base (i.e., they are "hot") and are thus energetically separate from the equilibrium carriers in the base. The control and controlled fluids in Landauer's model are the majority equilibrium and minority hot carriers in the base (in analogy to majority holes and minority electrons in the base of an npn bipolar transistor). These hot carriers must be collected before they lose so much energy or normal component of momentum that they cannot surmount the base–collector barrier, in which case they scatter and become mixed with the controlling equilibrium carriers. A number of versions of this basic device have been introduced [20]. For example, it is possible to use a semiconductor as the collector, with the base–collector barrier formed by the Schottky barrier existing at the interface. Thus, NININ, NINSm, and SmNSm are among the possible structures.* Hot carrier injection into the base can be accomplished by tunneling through, or thermionic emission over, a barrier. All of these devices are "hot electron transistors" (HETs). The ultimate goal in an HET device is ballistic transport, in which electrons traverse the base region without scattering, and emerge from the base with their full injection energy, since this promises the ultimate in device speed (or at least base transit time) and current gain. For a thorough

---

* A comment on notation: In describing the various device structures of interest here, N denotes a (normal) metal, M a metal which can be either normal or superconducting, Su a superconductor, Sm a semiconductor, and I an insulator. Thus, normal and superconducting oxide tunnel junction structures are NIN and SuISu, while the general case is MIM. A Schottky contact is NSm. In Sections 3 and 4, S will be used to denote a superconductor since it is unambiguous in absence of semiconductors in the device structure.

discussion of HETs, the interested reader is referred to the literature, accessible through various recent reviews [22–25].

The early interest in hot electron transistors waned, due to fundamental problems with achieving current gain. The common-base current transfer ratio α (the ratio of collector current to emitter current) must approach unity if the device is to have a useful current gain $\beta = \alpha/(1 - \alpha)$, the ratio of collector to base current in the common-emitter configuration. Losses in the base and quantum mechanical reflections at the base–collector barrier tend to limit α to rather small values in MBTs. Large α values were occasionally reported, but such results were usually attributed to pinholes; measurement of α alone is of limited value. The permeable base transistor (PBT) [26], a device of substantial current interest for high frequency applications, uses intentional pinholes in the base to achieve a vertical FET with many parallel channels.

Interest in hot electron transistors has revived in recent years, with impressive progress due largely to technological advances that make it possible to form an entire multilayer device structure from epitaxial, lattice-matched heterojunctions, allowing a reduction of interfacial scattering, electron–electron scattering in the base, and quantum mechanical reflections. A version of Mead's tunnel injection device can be made using doped GaAs for the metal layers and undoped AlGaAs (which has a larger bandgap) for the insulating barriers [22]. Injection via thermionic emission over a barrier is used in planar doped barrier (PDB) device [27]. Experimental demonstrations of these devices have been made, as described in recent reviews [23,24,25]. In particular, ballistic operation in devices with significant current gain has been demonstrated. In vertical pseudomorphic devices [28], β values of 27 and 41 have been reported at 77 and 4.2 K, while β's in excess of 100 have been obtained in lateral 2DEG devices [29] at 4.2 K.

Much of the interest in ballistic devices has centered on the extremely short (<1 ps) base transit time that should be possible. The response of a real device is limited by the time required to charge the device capacitance, which depends upon base resistance. Estimates of the ultimate switching time of these devices range from less than a picosecond [24] to several picoseconds [23].

Large base resistance, due to the limitations imposed by the trade-off between base doping and base thickness, is a significant potential problem in these devices. Two possible solutions are the use of materials that have long ballistic mean free paths [30], and the use of a high mobility two-dimensional electron gas (2DEG) in the base to improve conductivity [31] and reduce impurity scattering. The recent work of Levy and Chiu [32] on a device structure aimed at room temperature operation has a number of features that make it interesting for this discussion. The device consists of a AlSb$_{0.92}$As$_{0.08}$ emitter, a thin InAs (2DEG) base, and a GaSb collector on a GaSb substrate. The use of a wide gap emitter allows room temperature operation. Non-alloyed barrier-free ohmic contacts are made to the InAs base. In principle, scattering rates for hot electrons are low in InAs [30], although scattering for the case of transport normal to a 2DEG is not presently understood.

Quantum mechanical reflections at the base–collector interface represent a fundamental problem for hot electron devices [31,33]. The problem is much larger with a metal than with a semiconductor base, due to the large value of the Fermi energy in a metal. The ratio of barrier height to electron energy is the important parameter in a simple analysis [31]; this ratio must be small to minimize reflections. The point is that these energies are measured from the bottom of the conduction band, and the large value of the Fermi level makes the ratio large. The use of a collector material having a small effective mass (compared with that of the base material) can greatly reduce the reflection problem [34], as originally indicated by Crowell and Sze [35]. In fact, a simple analysis indicates that reflections can be virtually eliminated by matching the hot electron velocities in the base and collector layers [32] by a proper choice of materials (effective masses). Grading of the base–collector barrier, a straightforward matter with heterojunction devices, can also help in minimizing reflections. If, however, carriers are largely unscattered in traversing the base, the effect of reflections on current gain should be drastically reduced anyway [25], because the reflected electrons will also be largely unscattered, either re-entering the emitter or making another pass through the base (in neither case do they contribute to losses in the base). In many cases, this amounts to allowing reflections to be virtually ignored. If this prediction is correct, it is very important in that it removes a major objection to hot electron transistors in general, and metal base transistors in particular. This may be key to the operation of a superconducting version of the metal base transistor.

2.2.2. Superconducting-Metal Base Transistor

The fact that base resistance has always been an issue with junction transistors leads to consideration of a superconducting base, which would eliminate base resistance althogether (making base inductance an important parameter). Aside from considerations of operating temperature, this is an obvious step forward for metal base transistors. However, it is backward step along the path that hot electron device work has followed in recent years, in that the tendency has been to use low carrier densities, or even a 2DEG, in the base to minimize both electron–electron and impurity scattering. The mean free path for ballistic motion should be very short in a metal due to the large carrier density, and the presence of superconductivity does nothing to alter this fact. In fact, mean free paths in high-T_c oxide materials are extremely short, a significant drawback for higher temperature devices. Of course, it is the ratio of base thickness to effective mean free path that matters, and zero base resistance should allow the base layer to be extremely thin, down to tens of angstroms.

Tonouchi et al. have studied a metal base device, which they named Super-HET, both theoretically [34] and experimentally [36]. The device has a SmSuSm structure. Referring to Fig. 2, the emitter is a degenerate semiconductor (GaAs), the emitter–base insulator is a depletion region (GaAs–Nb Schottky barrier), the base is a superconductor (Nb or NbN), the base–collector barrier is a second depletion region (Nb–InSb Schottky barrier), and the collector is a second degenerate semiconductor (InSb). The device has an inverted structure: InSb, which forms low Schottky barriers to metals, is deposited in polycrystalline form on the base film; GaAs is the substrate.

Electrical measurements of the common-base characteristics of Nb (20–40-nm thick) and NbN (60-nm thick) based devices at 4.2 K were consistent with α values of ≈ 0.6–0.8 and ≈ 0.6, respectively. Assuming that $\alpha \propto \exp(-d/\lambda)$, where d is the base thickness, the effective hot electron mean free path λ was inferred to be 70 nm or greater in the Nb case (and even larger for NbN). This hot electron mean free path was measured to be 110 nm in Nb at 4.2 K in a separate set of related experiments [37]. Of course, a simple measurement of common-base characteristic does not differentiate between ballistic electrons and electrons that have undergone scattering but which still have sufficient forward momentum to surmount the collector barrier (thus, the claim of a measurement of a ballistic mean free path, implied by the title of reference [36], was later withdrawn [34]. For Nb, the same group calculated a ballistic mean free path [34] of ≈ 14 nm at low temperatures, limited by phonon emission (the component due to electron–electron scattering was estimated to be exceed of 200 nm for all energies of interest).

It is difficult to establish the absence of pinholes that may allow PBT action. A similar situation exists in the case of epitaxial Si–CoSi$_2$–Si devices [38], and analysis of electrical measurements in conjunction with careful electron microscopy studies allows some light to be shed on the pinhole question. It is clear that this sort of analysis or, even better, hot electron spectroscopic experiments [24], are needed on the Super-HET structure to eliminate the possibility of pinholes, and more experimental work is needed to establish reliable values for both ballistic and non-ballistic mean free paths.

The behavior of Nb-polycrystalline InSb–Au structures used to study the base–collector interface was non-ideal, although this was blamed on alloying at the interface with the Au contact and not the Nb. The emphasis of recent experimental work [39] has been on epitaxial growth of Nb on the GaAs emitter. A decrease of Nb film quality with decreasing film thickness for non-epitaxial Nb causes a degradation of performance in thin base devices [37]. However, much earlier hot electron transistor work emphasized the base–collector interface as being most important.

The work described above used an inverted SmSuSm structure with the emitter on the bottom. The device can also be made with the collector as the bottom layer. It is also possible to use a NISu emitter–base structure [40]. With this structure, the emitter, emitter–base barrier, and base can be polycrystalline or amorphous layers deposited on the single-crystal semiconductor collector (alternatively, the base film could be grown epitaxially on the collector). Techniques for making well-controlled barriers and thin superconducting films are readily available from Josephson technology. The emitter can even be superconducting. The collector can then be any suitable semiconductor, either the substrate itself or a lattice-matched or strained (pseudomorphic) layer [41]. Heterojunction collector structures are then possible, allowing low voltage devices and hot electron spectroscopy. The possibility of studying ballistic transport in metals is interesting even if these transistors do not prove to have practical applications.

The operating voltage of heterojunction devices can be quite small, far below the intrinsic voltage scale of bipolar devices, making them interesting from the point of view of low power dissipation. The voltage scale is set by barrier heights, which are controlled by composition and doping. In principle, arbitrarily low barriers are possible. Experimental explorations of low barriers have been made [14,42,43]; however, the question of fundamental limits is still open. Control of barrier heights, doping uniformity (including random density variations), quantum mechanical reflections, and base resistance represent significant problems.

The voltage scale for a semiconductor–collector metal base devices is determined by the Schottky barrier height of the base–collector contact, typically hundreds of meV. This is far larger than the characteristic voltage associated with superconductivity, the energy gap. It is a limitation for low temperature applications, where a meV (or few tens of meV) scale is more appropriate. The base–collector barrier height is a more or less intrinsic property of the particular metal–semiconductor system used. With most III–V semiconductors, Schottky barrier heights do not vary much for different metals. They are fixed by the pinning position of the Fermi level at surfaces or interfaces with metals. This pinning usually occurs at about midgap; however, it varies widely with composition. In some materials, such as InAs, pinning occurs in the conduction band, resulting in a negative Schottky barrier (accumulation layer) to n-type material. In ternary system, such as $In_xGa_{1-x}As$, the Schottky barrier height varies with composition x, from roughly 0.7 to -0.2 eV. Thus, ternary materials are of interest for structures in which low barrier heights are desirable.

2.2.3. SUBSIT

Even prior to the work discussed above, Frank et al. [44] proposed SUBSIT (SUperconducting-Base Semiconductor-Isolated Transistor), a superconducting metal base transistor that avoids the difficulties of ballistic mean free path and quantum mechanical reflections and actually resembles a bipolar transistor. The device structure is illustrated in Fig. 3a. The device has a MISuSm structure, with the MISu being the emitter–base section (the emitter can be superconducting). Quasiparticles (essentially unpaired electrons) are injected into the base film by tunneling. Unlike the usual metal base transistor, these quasiparticles do not need to be "hot". They can be injected with energies just above the superconducting energy gap in the base, or they can relax down to this level by scattering. The injected quasiparticles diffuse across the base, and gain in the device depends upon their being collected before they recombine into pairs. Unlike hot electron transistors, however, they do not have to enter the collector on the first attempt. This improves the collection

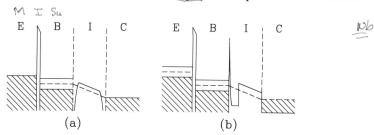

Fig. 3. (a) Schematic diagram of SUBSIT [44]. Single electron tunneling is used to inject electrons into the superconducting base. The injected quasiparticles diffuse across the base and are collected before they can recombine into pairs that contribute a base current. The isolator (I) allows a collector bias without imposing a barrier between base and collector that would tend to trap carriers in the base. (b) Schematic of recent realization of SUBSIT [47]. The emitter and base are Nb, with an Al oxide barrier. The isolator and collector are an n^+InGaAs–InAlGaAs–n^+InGaAs heterostructure. The barrier between base and isolator is very thin in order to minimize its role in current transport.

efficiency, but increases the base transit time (it is not a ballistic device). However, as mentioned above in connection with hot electron transistors, this time is not the dominant one in determining device speed in the picosecond regime, and simulations that include transport and charging time indicate that SUBSIT should be a fast device [44], with response in the 10 ps range. In contrast to the Super-HET, the operating voltage scale for SUBSIT is set by the base energy gap voltage. Current gain depends on the base transit time, the effective base quasiparticle recombination time, and a transmission factor. Rough estimates indicate that this gain can be large [44].

The two "fluids" in the base are the pairs (majority carriers) and quasi-particles (minority carriers) in the superconductor, in close analogy to the bipolar transistor. This device is unique in that it is consistent with an old suggestion by Landauer [4] that that analogy might somehow be exploited, and it is probably as close as one can get to having a superconducting bioplar transistor, that is, one in which the base "fluid" has no electrical resistance. Complementary devices (i.e., both hole and electron based) are possible.

As in other metal base transistors, the key to SUBSIT operation is the base–collector contact, the realization of which will be key to making the proposed device work. Frank et al. [44] refer to the semiconductor layer in contact with the base as the "isolator", the collector being the metal (or heavily doped semiconductor) contact to this layer. Contacts to the isolator must not only be ohmic, but barrierless (on the scale of the superconducting energy gap), since the presence of a barrier would tend to trap quasiparticles in the base, decreasing the current gain and slowing down the device. The isolator also serves the function of preventing direct pair-breaking tunneling between base and collector; this process represents a potential leakage current. The isolator must be frozen out since thermally generated carriers also represent a source of leakage. Of course, the operating temperature must be low enough that thermally generated quasiparticles in the base do not contribute a significant leakage current or source of quasiparticles for recombination.

A metal (superconducting base)–semiconductor (isolator) contact having a barrier height smaller than the superconducting energy gap has yet to be realized. A material in which the Fermi level pinning position is in the conduction band (or valence band for p-type material) forms contacts with negative barriers, so that an accumulation layer forms between base and collector (this type of contact was assumed by Frank et al. [44]). However, the effect of such a layer, which may act as a two-dimensional proximity effect superconductor, has not been explored. A contact having no barrier (either positive or negative) is possible in principle using ternary semiconductors [45], such as $In_xGa_{1-x}As$ (see above discussion). However, the barrier height change with composition x is of order 10 meV/at.%, so that requirements on the ability to control composition (spatial distribution on a wafer and between fabrication runs) are stringent. Also, given the discrete nature of dopants and the relatively large local variations of potential on the meV scale of the superconducting gap, it is not really permissible to use the usual continuum picture for the spatial variation in semiconductor bands. High-T_c superconductors allow some relaxation of requirements, due to the presumed order of magnitude or so increase in the gap, but the problem is still present.

Tamura et al. [46] proposed a similar structure and made several experimental attempts at realizing an operating device. Their original structure consisted of Nb–Al_2O_3–Nb–nInSb (note the use of a superconducting emitter) and had a common base current transfer factor α of order 10^{-4}. There was no reference to an isolator in their original proposal. (The energy band diagram which appears in [46] should include a substantial Schottky barrier, much larger than the superconducting energy gap, between base and collector.) A later design [47] that included an isolator replaced the InSb collector with a heterostructure based on nInGaAs and InAlGaAs, lattice-matched to an InP substrate. The structure is shown schematically in Fig. 3b. An α of ≈ 0.3 was reported for this device in a preliminary experiment.

The philosophy followed by this group has been to accept a small, and therefore very transmissive, barrier at the base–collector interface, rather than to aim for a barrierless contact. (Note that there are no ternary compounds that give barrierless contacts and also lattice-match to reasonable substrates, such as GaAs and InP, so that pseudomorphic structures would have to be used [41].) As mentioned above, the barrier should affect speed and current gain, which will presumably be studied in future experiments. This group has also done modeling of the device [48] and has predicted large achievable gain and ≈ 10 ps response.

In summary, the present situation for superconducting junction transistors is as follows: There are two devices, one (Super-HET) simply substitutes a superconductor for the normal layer in a well-known device (MBT) and the other (SUBSIT) involves a novel control principle. For the SUBSIT, it is the case that the operating voltage is set by the superconducting energy gap, but this is also the more difficult device to realize, the principal problem being the very low barrier metal–semiconductor contact for the base–isolator junction. In fact, no convincing demonstration has been made to date.

2.3. Field Effect Transistors

2.3.1. Josephson Field Effect Transistors

Field effect transistors are pervasive in low temperature semiconductor electronics because their power dissipation is small (relative to bipolar devices), making them attractive for high circuit density applications. Lowering the operating temperature (at least down to 77 K) results in improved performance [8], particularly in such devices as high electron mobility transistors. A number of years ago, a field effect transistor with superconducting source and drain electrodes was independently proposed by Silver et al. [49] and Clark et al. [50]. The device is a gate-controlled Josephson weak link (JOFET), illustrated schematically in Fig. 4. The weak link is SuNSu-like (often denoted SNS [51]), with a semiconductor as the normal material [52] (i.e., the structure is SuSmSu). The link (channel) is superconducting due to the proximity effect [53]. Cooper pairs leak from the superconducting source and drain electrodes into the semiconductor (the semiconductor must contain mobile carriers at the operating temperature, which can be achieved by using heavy doping or an inversion layer). Overlap of the proximity effect regions of the electrodes results in Josephson coupling in a suitably short link, with a coupling strength (i.e., critical supercurrent) that depends on the carrier density in the link region. Use of the field effect to control carrier concentration is one means of controlling both super and normal currents in the device.

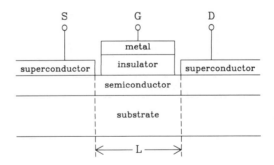

Fig. 4. Schematic diagram of a Josephson FET (JOFET). Superconductivity is induced in the semiconductor by the superconducting source (S) and drain (D) electrodes. A supercurrent is possible if the separation L of these electrodes is no more than a few coherence lengths in the semiconductor. The gate (G) controls both super and normal currents in the device by modulating the properties of the semiconductor channel. Although a metal–insulator–semiconductor (MIS or MOS) FET is illustrated, other types of gates are suitable as well.

9. THREE-TERMINAL DEVICES

Although a MISm gate structure (MISFET) is illustrated in Fig. 4, pn junction (JFET), metal–semiconductor (MESFET), high electron mobility (HEMT), and other devices are also possible. The JOFET represents a limiting case of conventional (nonsuperconducting) FETs. Any FET with superconducting source and drain metallizations should, in principle, exhibit a supercurrent (at low enough temperature), although it will be trivially small unless the channel is quite short and the contacts carefully made. The basic physics and expected device properties from JOFETs have been discussed in a number of papers [11,50,54,55,56]. We review here the experimental results to date and then discuss the relevant physics and device properties.

The first experimental demonstrations of gate-controlled supercurrents were made on MISFET structures. Nishino et al. [57,58] used an Al gate and thermal oxide on the back of a heavily doped Si membrane that had a Pb-alloy weak link formed on top (the Si membrane acted as the link material). Although the normal conductance was negligibly affected by the gate, the critical current was varied by more than an order of magnitude with a gate voltage swing of ~ 100 mV. The membrane structure was not well enough characterized to model it in detail; however, the large response to a gate voltage that only slightly perturbed the channel carrier density was quite surprising [59] (the device would not be expected to function as an inverted-gate FET with even several volts applied to the gate), and the experiment has evidently not been reproduced.

In another MISFET experiment at about the same time, Takayanagi and Kawakami [60,61] used deposited gate dielectrics and Nb electrodes in both n- and p-type InAs weak links. Unfortunately, it is not presently possible to fabricate high quality gates on InAs, and the devices responded only weakly to rather large gate voltages, a major drawback.

InAs does have two attractive features for JOFET applications. The first is a low effective electron mass, which means that the coherence length can be larger than in a higher mass material, allowing a larger spacing between source and drain electrodes and/or a larger supercurrent, as discussed below (other III–V materials, such as GaAs and InGaAs, also have relatively small electron masses). The second is strong pinning of the Fermi level in the conduction band at interfaces, which makes barrier-free metal–InAs contacts possible, a major advantage. Unfortunately, this same pinning severely limits the effectiveness of NISm gates. Also, a conducting layer is present over the entire surface of an InAs wafer, making it difficult to isolate devices from each other. This difficulty was avoided by Kleinsasser et al. [62], who used an insulating GaAs substrate and obtained excellent characteristics for mesa-isolated weak links in unstrained InAs layers grown by MBE. It was intended to extend this structure to FETs using pseudomorphic heterostructure gates, with InGaAs or InAlAs as the dielectric, allowing the problem of Fermi level pinning at the interface between InAs and the gate insulator to be circumvented. Unfortunately, although initial results were promising, MISm capacitors of sufficiently high quality to be of real interest for gate dielectrics were not produced [63], and it is not known whether or not this approach is really viable. Future use of InAs as an FET material depends on the development of an adequate gate structure. One possibility based on the unique properties of InAs [49], an InAs JFET, has been proposed [64] that would eliminate the need for an insulated gate.

More recently, what might be called the first superconducting FET, in which both normal and super currents exhibited a significant response to a gate, was reported by Ivanov and Claeson [65]. The device was a GaAs–AlGaAs HEMT with Nb electrodes. This materials system, like $Si-SiO_2$, is widely used for semiconductor devices, and the device behaved qualitatively as expected, although only a single device was reported. A serious problem with this type of structure is the Su–Sm contact. The JOFET requires a strong proximity effect coupling between the superconducting electrodes and the 2DEG channel, and the alloy contact scheme that was used appeared to seriously limit the output voltage range. This issue is discussed in more detail below.

Subsequent to this initial work, a self-aligned coplanar Si-coupled weak link structure was introduced [66] that featured a direct contact between the superconductor and the channel and which should be applicable to coplanar FET devices. This promised to be a major improvement on the original Si membrane device; however, the work was never extended to FETs. On the other hand, channel lengths in more normal FET structures have been pushed to sub-100 nm dimensions. The use of a self-aligned structure in which superconducting source and drain metallizations are as close as possible to the gate does indeed yield devices that act as JOFETs, as demonstrated by Nishino et al. [67], who recently reported n-channel Si MOSFETs with 100-nm gate lengths (and source–drain metallization spacings). Nb electrodes contacted implanted source and drain regions, which in turn contacted the 2DEG, which was produced by application of a gate bias in excess of a 0.3 V threshold. Further increases in gate bias increased both the supercurrent and normal conductance in accordance with qualitative expectations.

The materials system that has the lowest Schottky barrier heights and smallest effective mass with which high quality FETs have been produced is

In$_{.53}$Ga$_{.47}$As, grown epitaxially on InP substrates. Recently, superconducting junction FETs with Nb electrodes that exhibit gate-controlled supercurrent and normal conductance have been demonstrated by Kleinsasser et al. [68] with this material. The structure is intended as a model for investigations of physics and device properties. The pn junction gate is below the thin epitaxial channel layer, allowing exploration of the effect of the superconductor–semiconductor contact properties on device performance without affecting gate quality.

2.3.2. Semiconductor-Coupled Weak Links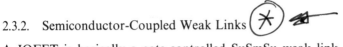

A JOFET is basically a gate-controlled SuSmSu weak link. Much of the relevant physics can be studied using ungated links, for which there is a considerable body of literature. While there has long been work on both SuISu and SuNSu-like sandwich structures [69], which are one-dimensional (1D) in the sense that the superconducting order parameter varies in only one direction, we are principally concerned with SuNSu-like bridge (2D) structures such as that of Fig. 4 (without a gate). Virtually all experimental work on bridges has been on Si [52,58,66,70–73], with two experiments on InAs bridges [62,74] and none on other materials. The focus has been on studying the critical current, which depends on device length, temperature, carrier concentration, magnetic field, and contact properties.

There is no lack of theoretical work in this area. Again, almost all attention has focused on the critical current. The critical current of SuNSu junctions was first worked out by de Gennes [75] for long junctions near T_c. Subsequently, Likharev [51,76] worked out a theory for a class of SuSmSu junctions for all lengths and temperatures. Seto and Van Duzer [70] explicitly discussed SuSmSu structures; their result is consistent with de Gennes, except for an unspecified constant prefactor that arises from tunneling through the Schottky barriers at the SuSm interfaces. The problems stemming from the fact that all of the early theoretical work assumed a one-dimensional (1D) device geometry was discussed by van Dover et al. [77], who also considered arbitrary boundary conditions in SuNSu structures. It has been argued that this result can be extended to SuSmSu devices [78]. A nice feature of this approach is that a hard to calculate prefactor in the expression for the critical current, which is related to the transmission probability for carriers at the SuSm contacts, can be obtained from independent experiments. In the past ten years, there has also been a significant effort in the Soviet Union on SuSmSu theory, which has been virtually ignored by experimentalists in the West. Both

1D [79] and 2D [80] device geometries have been considered. Finally, there has been significant interest recently in devices in which the Sm layer is a 2DEG, although only 1D device geometries have been considered [81]. (Note that there are a number of differences between various theoretical papers which we do not discuss here: the full temperature range may be covered or only that near T_c, Schottky barriers may or may not be included, the treatment may assume the clean or dirty limit, etc.)

The experimental results can be said to agree with theory only in a very limited sense. Virtually nothing in the experiments performed on SuNSu-like SuSmSu junctions to date tests any theory that goes beyond the original work of de Gennes. There is general agreement that the critical current of a long link is exponential in device length scaled by the (temperature-dependent) coherence length in the semiconductor. Little else has been clearly established. For example, the absolute magnitude of the critical current (i.e., the critical current–resistance product) is not understood. This is due in part to a lack of definitive experiments on well-characterized device structures and basic studies of the proximity effect in semiconductors, a situation which has been improving in the past several years. Nevertheless, the size of the critical current (and critical current–resistance product) in SuSmSu weak links [78] and the magnitude of the proximity effect in SuSm structures (as indicated by the reduction of transition temperature [82]) are too large to be consistent with existing theory, which thus appears to be inadequate even though experiments have so far made contact with only the most basic theoretical predictions.

What emerges clearly from the literature on this subject is that the properties of SuSmSu weak links and JOFETs depend critically on two major factors, the SuSm contact and the coherence length in the semiconductor. It is the former factor which has received inadequate attention. We will concentrate on the SuSm contact issue first, followed by a discussion of the length scale.

2.3.3. Superconductor–Semiconductor Contacts

SuSmSu and SuNSu devices are often pictured as being equivalent; however, the presence of (insulating) Schottky barriers at the SuSm contacts invalidates this simple picture, making SuSmSu equivalent to SuINISu, and not SuNSu. The penetration of Cooper pairs into the neutral (conducting) region of the semiconductor is limited by the requirement that they must tunnel through these barriers [70], drastically reducing the proximity effect. This implies that SuSmSu weak links should have very small critical currents (this was recognized in the early FET proposals [49,50], which emphasized the use of materials such as InAs that do not form Schottky barriers at metal contacts). It has been well established that the proximity effect can be observed in SuSm contacts, through the experiments involving SuSmSu weak links. However, the simplest demonstration of the effect is the depression of the Su transition temperature of SuSm bilayers due to leakage of Cooper pairs into into the Sm layer; this is a widely used technique in SuN systems [53]. Leakage of pairs out of the superconductor into a normal material can cause an observable lowering of T_c if the superconductor is only a few coherence lengths thick. For SuSm systems, such measurements have been reported for Nb on p-Si [82], for which a value for the coherence length was extracted, directly verifying the proximity effect in an SuSm system. However, it is the number of pairs that penetrate into the normal layer, rather than how far they penetrate, that is most important for the proximity effect [83], a point which is generally ignored.

Study of the T_c dependence of thin superconductor films on thick semiconductor layers with varying doping can directly study the boundary conditions for the pair amplitude (the variation of barrier thickness or tunnel probability with doping is inherently more important than the variation of normal coherence length). This is qualitatively evident in the experiment of Hatano et al. [82], which showed T_c reductions (from the bulk value) of 7–23% for 40 nm Nb films on 4.5–30 × 10^{18} cm^{-3} pSi. The tunneling probability through a Schottky barrier for a particle of mass m and charge q is (in the WKB approximation) [84] $P \sim \exp(-E_B/E_{00})$, where E_B is the barrier height and $E_{00} = (q\hbar/2)(N_D/\varepsilon_s m)$. Here N_D, ε_s, and m are dopant density, dielectric constant, and effective mass, respectively. Using a value of $E_B = 0.41$ eV for the Schottky barrier height of Nb on p-Si [85], $\mu = 100$ cm^2/V-s, and $m = 0.16 m_e$, one obtains $E_{00} = 0.074$ eV and P ranges from $2 \times 10^{-7} - 2 \times 10^{-3}$ in these samples. This implies a significant and sample-dependent reduction of the pair amplitude in crossing the SuSm boundary, which is inconsistent with the large observed changes in T_c. Thus, Cooper pairs appear to penetrate more easily into the semiconductor than expected. It may well be that what is being tested in all of these experiments is the breakdown of the simple tunneling theory for Schottky contacts. Recent theoretical work [86] suggests that the zero bias conductance of a Schottky contact may be significantly higher at large dopings than is generally believed [87]. One potentially important effect is the breakdown of the continuum band picture for the barrier, due to the discrete nature of the charged dopants that are responsible for the depletion region; this is important for thin barriers. Experimental measurements of contact resistances are difficult in the high doping limit and proximity effect-related measurements are more sensitive, making them rather interesting in this context.

In a second direct demonstration of the proximity effect in SuSm contacts, Nishino et al. [88] used tunneling spectroscopy to study SuSm bilayers, observing both the induced gap in the semiconductor and the dependence of the gap in the superconductor on changes in carrier concentration in the semiconductor (due to changes in dopant density and applied gate voltages). Again, the size of the effects is surprising, given the expectation that the pair potential in the semiconductor should be reduced by tunneling, making these effects negligibly small. This type of bilayer T_c experiment can also be used to study contact processes, for example, chemical or sputter etching to remove oxides, since the T_c reduction rapidly disappears if an oxide (or other insulating) layer exists at the SuSm contact. Also, the sensitivity to Schottky barrier thickness makes this a potentially interesting probe of ohmic contacts (limited, of course, to superconducting metals).

The critical current I_c and the critical current–resistance product $I_c R_n$ should be reduced by a factor of order P^2 from the values expected from a perfectly transmissive barrier [78]. Yet the observed $I_c R_n$ products [52,58,68] are not much smaller than the value expected for ideal Josephson junctions [89], consistent with the above discussion. Experiments on Super-Schottky diodes [90] demonstrate that SuSm interfaces act as SuIN tunnel junctions. For example, both SuSmSu Josephson junctions and Super-Schottky diodes have been made with Pb (or Nb) on p-Si on the same device structure [72,85]. In this light, the observation of a crossover between SuINISu and SuNSu behavior with increasing channel doping (increasing barrier transmittance), observed in nInGaAs JOFETs [68] with Nb electrodes, is significant. What is needed in the future is more quantitative experiments that relate a measured or inferred barrier transmittance to the JOFET (or weak link) critical currents, or to the strength of the proximity effect (e.g., T_c depression). Such work should lead to understanding of the role of Schottky barrier tunneling in the SuSm proximity effect.

To conclude our discussion of SuSm contacts, we mention some recent work on achieving good superconducting contacts to conventional semiconductors. The role of damage produced by sputter "cleaning" in reducing the barrier height in Nb–Si contacts [91], while not explaining why a proximity effect is seen at all, points up the need for further process studies. In the case of GaAs, Au–Ge contacts with an Nb overlayer have been demonstrated [92]. This, along with the above-mentioned GaAs/AlGaAs MODFET experiment

[65], raises the question of how thick a normal layer between the Su and Sm layers can be tolerated before a significant reduction of the critical current results. Finally, we mention the recent work on InGaAs [68], which takes advantage of reduced effective mass and barrier height to increase the expected tunneling probability by orders of magnitude.

2.3.4. Normal Coherence Length

We now discuss the issue of length scale. Near an Su–N (or Su–Sm) interface, the Cooper pair amplitude falls off exponentially with distance into the N (or Sm) layer with a characteristic length [53] which is the distance an electron travels in time $\hbar/kT$. We will refer to this decay length as the normal coherence length ξ_n. In the clean limit ($l \gg \xi_n$), $\xi_{nc} = \hbar v_F/2\pi kT$, while in the dirty limit ($l \ll \xi_n$), $\xi_{nd} = (\hbar D/2\pi kT)^{1/2}$, where l, v_F, and D are the electron elastic mean free path, Fermi velocity, and diffusion constant ($D = v_F l/d$ in a d-dimensional material), respectively. In general [83], $\xi_n \approx (\xi_{nc}^{-2} + \xi_{nd}^{-2})^{-1/2}$. In the case of extreme disorder, localization corrections to ξ_{nd} are predicted [93]. The critical current of a proximity effect Josephson weak link (SuNSu, SuINISu, SuSmSu) of length L is exponential in L/ξ_n (for $L \gg \xi_n$ and temperature not too close to T_c or 0), which provides a means for determining the normal coherence length in weak link experiments.

All experiments performed to date have been in the dirty limit. For a three-dimensional semiconductor with carrier density n, effective mass m^*, and mobility μ is

$$\xi_{nd} = \left(\frac{\hbar^3 \mu}{6\pi m^* ekT}\right)^{1/2} (3\pi^2 n)^{1/3}. \tag{2.1}$$

High mobility and low effective mass are desirable since a given coherence length can be obtained at a low carrier density, which is desirable if the FET is to be sensitive to gate voltage. The dependence of normal coherence length on temperature is illustrated in Fig. 5 for $n = 10^{19}$ cm^{-3} in n-Si ($m^* = 0.26 m_e$) and n-InAs ($m^* = 0.023 m_e$), which were chosen because they are materials of experimental interest with widely differing values of effective mass, and therefore of ξ_n. For each material, the solid (upper) curve represents the clean or large mobility limit and gives the upper limit for ξ_n (this limit is not valid for any experiments reported to date). The dashed (lower) curves are for reasonable lower bounds for μ. The shaded region between the two curves indicates the range of possible coherence lengths for the material. ξ_n is significantly smaller in Si than in InAs because of the order of magnitude

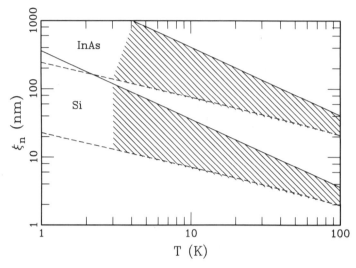

Fig. 5. Temperature dependence of normal coherence length in n-type Si and InAs for $n = 10^{19}$ cm^{-3} (coherence length varies as $n^{1/3}$). The shaded areas indicate ranges of values corresponding to low and high mobility material. Although high mobility is important, the major difference between the materials is the carrier effective mass. This plot is roughly correct for 2DEG materials at mid-10^{12} cm^{-2} densities (the dependence is $N_s^{1/2}$ in the 2D case).

difference in effective mass. Of course, $\xi_n \propto n^{1/3}$. This favors materials that allow the greatest channel carrier density, but increasing n means increasing gate voltage, which, as will be shown below, limits voltage gain, which is already a serious problem.

At 4.2 K, the values for ξ_n in n-Si are in the range 10–100 nm. Useful weak links cannot be longer than a few coherence lengths, so that Si device lengths have an upper limit of a few hundred (mm) for a high-mobility 2DEG. (The dirty limit curves in Fig. 5 are valid for the 2D case with $N_s = 4.6 \times 10^{12}$ cm^{-2}. The clean limit curves are too large by 23%.) Significantly longer devices are possible using low mass materials. For InAs, $\xi_n \approx 0.1$–1 μm (the effective mass used is that corresponding to the conduction band minimum; this is not correct for large carrier densities in such a low mass material, making the InAs ξ_n values optimistic). At 77 K, ξ_n in Si is smaller than ≈ 5 nm, ruling out FET devices for practical purposes at present. In contrast, the value for InAs is still several tens of nm, so that devices up to ≈ 100 nm long are possible. Thus, only III–V or other small effective mass materials appear to be interesting for nitrogen temperature operation in the foreseeable future.

Experimental and theoretical coherence lengths are reasonably consistent [52,58,62,70–74], although careful measurements of μ, m, and n, the parameters needed to calculate the length, are generally lacking. Effects such as the increase of effective mass with carrier concentration due to the breakdown of the parabolic band approximation (particularly significant in low effective mass materials such as InAs) and crossover between dirty and clean limits may account for discrepancies between theoretical and experimental results.

Two-dimensional (2DEG) systems are of interest, and the above coherence length discussion is roughly correct in 2D if $n^{1/3}$ is replaced by $N_s^{1/2}$, N_s being the sheet carrier density. For example, Silver et al. [49] were interested in p-InAs, in which the Schottky barrier height exceeds the band gap due to the pinning of the Fermi level in the conduction band, forming a "complete Schottky barrier" and resulting in an electron-rich inverted surface (2DEG) that is in direct contact with the metal. This was evidently the case in one of the early JOFET experiments [60] and should be relevant to MODFETs [65] and MISFETs [67], which are 2DEG devices. There have been no predictions of qualitatively different behavior between 2DEG and 3DEG weak links, and no clear experimental proof of two-dimensional behavior in weak links.

2.3.5. Gate-Dependent Critical Current

Recent results of Nishino et al. [67] provide, for the first time, data on the gate dependence of JOFET critical currents that can be straightforwardly interpreted. We plot their data, for an n-channel Si MOSFET with Nb electrodes at 4.2 K, in Fig.6 (solid dots). The form

$$I_c \propto \xi_n^{-1} e^{-L/\xi_n} \tag{2.2}$$

for the critical current, derived by Seto and Van Duzer [70], shown by the solid curve, gives an excellent fit. A value of $L/\xi_n = 2.5$ for $V_G = 2$ V was used and the curve was matched to the data at that gate voltage. The gate voltage dependence was assumed to be contained in the coherence length, which, for a 2DEG, is given by

$$\xi_n = \left(\frac{\hbar^3 \mu N_s}{2m^* e k T}\right)^{1/2}, \tag{2.3}$$

where the carrier concentration $N_s = C_G(V_G - V_T)/e$, where C_G is the gate capacitance and V_G and V_T are the gate and threshold voltages. This result is

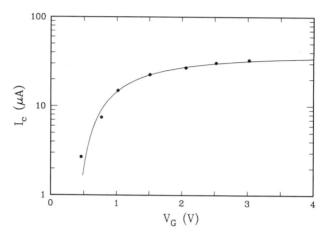

Fig. 6. Dependence of critical current on gate voltage for a n-channel MOSFET with Nb electrodes at 4.2 K [67]. The channel length is less than 100 nm.

important because the data cannot be fit to the form for the critical current derived by Likharev [67]:

$$I_c = R_n^{-1} \frac{4\Delta^2(T)}{\pi e k T_c} \frac{L}{\xi_n} e^{-L/\xi_n}, \qquad (2.4)$$

which applies to SuNSu weak links, but is often used for SuSmSu devices. This expression contains an additional gate voltage-dependent factor $R_n^{-1} = eN_s\mu W/L$, where W and L are the device width and length.

The crucial prefactor in Equation (2.2), which is related to the Schottky barrier tunneling probability, is absent; however, it is not likely to depend strongly on gate voltage. A more complete expression, inspired by earlier SuNSu work [75,77] that dealt with varying boundary conditions was obtained by Kleinsasser [11,78]:

$$I_c = R_n^{-1} \frac{\pi \Delta^2}{2 e k T_c} \frac{\rho_n m_s}{\rho_s m_n} A^2 f^2 \frac{L}{\xi_n} e^{-L/\xi_n}(1 + 2r). \qquad (2.5)$$

In this expression, the resistivities and effective masses in Su and Sm are ρ_s, ρ_n, m_s, and m_n, respectively. $f^2 \sim 1$ is the ratio of the value of the order parameter in Su at the interface to its bulk value (we expect $f \approx 1$), and A is the factor by which the order parameter changes in crossing the SuSm interface. A is related to the Schottky barrier tunneling probability and can be independently determined by experiments such as T_c depression in SuSm bilayers. The last factor in Equation (2.5) is due to the contact resistance, with r the ratio of the resistance of a contact to the channel resistance (assumed to be small in this discussion). The boundary conditions in an SuSmSu bridge and in the SuNSu structure studied by Likharev are different; in the latter case, a cancellation occurs ($A^2 \approx \rho_s m_n / \rho_n m_s$) that causes $I_c R_n$, rather than I_c, to have the dependence given by Equation (2.2). This type of experiment is important because it provides a new means of studying the boundary conditions. Note that the Likharev expression may apply in the case of SuNSu-like SuSmSu structures, i.e., those with barrier-free contacts, as indicated by one experiment on InAs weak links [62]. Equation (2.5) should be valid in SuINISu and SuNSu-like structures, and should be a reasonable basis for future work until a more complete result is obtained.

The effective channel length in this MOSFET is unknown, although it is less than 100 nm. We do know the oxide thickness and the gate bias dependence of I_c and R_n. We can rewrite the expression (2.3) for ξ_n as $\xi_n = (\hbar^3 \sigma / 2m^* e^2 kT)^{1/2}$, where $\sigma = L/R_n W$ is the conductivity. For $V_G = 2$ V, with $m^* = 0.19\, m_e$, $W = 60\,\mu$m, and using $R_n = 8.9\,\Omega$ (obtained from Fig. 3 of Ref. [70]), the above value of 2.5 for L/ξ_n means that $L = 26.7$ nm, and therefore $\xi_n = 10.7$ nm with a 2 V gate bias. Using $C_G = 0.345\,\mu$F/cm^2 and $V_T = 0.3$ V, a mobility value of 85 cm^2/V-s is obtained from $R_n = L/C_G(V_G - V_T)\mu W$. The published resistance and critical current data are all consistent with these parameter values. The channel length inferred is rather short, perhaps due to diffusion of the source and drain implants under the gate.

2.3.6. JOFET Device Properties

Although JOFETs are Josephson junctions, device speed and power dissipation are not the same as for digital Josephson technology. The control mechanism is identical to that in conventional FETs, with the same gate voltage swing required in both cases to produce a given change in channel carrier concentration. Switching the device on and off requires the same CV^2 energy per cycle in both cases. Also, making the channel superconducting does not change carrier transit time, which represents a limit to the speed of FET devices. The role of superconductivity in a superconducting FET is to change the device characteristics, allowing a zero on-state resistance.

Analysis of the prospects for JOFETs as digital devices by Kleinsasser and Jackson [54] concluded that large-signal voltage gain is the major issue. Using a simple model of Si JOFETs, Glasser [56] analyzed the small signal response,

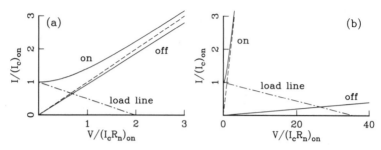

Fig. 7. Schematic IV characteristics of a superconducting FET. (a) If the supercurrent is much more sensitive to the gate field than is the normal conductance, then the resistance change is small between the "on" (large I_c) and "off" (small I_c) states. The dashed line represents the "on" state in the same device in the absence of superconductivity. Output voltage is limited by the device $I_c R_n$ product, which is much smaller than the gate voltage swing. (b) When both normal and supercurrents are greatly altered by the gate field, the resistance change is large between "on" and "off". Then the output voltage can be made arbitrarily large (i.e., the device can have significant large-signal voltage gain).

also concluding that voltage gain is a problem. The natural output voltage scale of these devices is set by the $I_c R_n$ product of the weak link, and thus limited by the superconducting energy gap, while the input (gate) voltage is orders of magnitude larger. Large-signal voltage gain is possible, but it originates from control of channel resistance, as in an ordinary FET [54], rather than from superconducting properties. This is illustrated in Fig. 7. We define "on" and "off" states corresponding to large and negligible critical current. I_c responds exponentially to gate voltage ($I_c \sim \exp(-L/\xi_n)$, $\xi_n \propto n^{1/2}$, $\Delta n \propto \Delta V_{\text{gate}}$), so a large change in I_c can be made with only a small change in R_n, as experimentally observed [57]. The "off" output voltage is of order $I_c R_n$ of the weak link in the "on" state, as shown in Fig. 7a, and the voltage gain is $\ll 1$. However, an optimal link is of order one coherence length long, in which case the critical current response is linear, not exponential. In a well-designed device, the output resistance switches between small and large values, and large voltage gain is possible, as shown in Fig. 7b. Note, however, that the difference between the IV characteristics in the "on" state for superconducting and nonsuperconducting electrodes (solid and dashed curves) is not nearly as dramatic in the latter case (i.e., superconductivity makes little difference in the characteristics).

Experimental IV curves of JOFETs [65,67,68] are similar to those in Fig. 7b. Superconductivity affects the source–drain characteristics only at low (mV) voltages, so that the normal resistance for a given gate voltage is constant. The condition $I_c R_n \ll \Delta V_G$ (the gate voltage in excess of threshold)

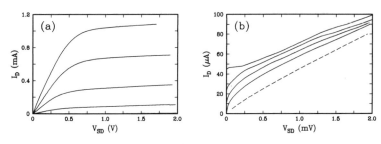

Fig. 8. Current-voltage characteristics of InGaAs JFETs [68]. (a) On a large voltage scale, the devices behave as normal FETs. (b) On a small voltage scale, superconducting effects, such as gate-controlled supercurrent, are evident (the dashed curve was obtained with the electrodes normal).

is equivalent to the condition $I_c \ll I_{sat}$, the saturated source-drain current [54]. Thus, the region of interest for JOFETs is the linear region of the characteristics of a normal FET. Figure 8 shows source-drain current-voltage characteristics for the Nb-InGaAs JFETs of Kleinsasser et al. [68], which illustrate these points. On a coarse voltage scale (Fig. 8a), the JOFET exhibits the familiar saturation of current that is characteristic of semiconductor transistors. On a much finer scale (Fig. 8b), the effects of the superconducting electrodes become evident (e.g., the gate-controlled supercurrent).

The small scale of voltages for which superconductivity is significant in the device characteristics makes it vital that the gate voltage swing be as small as possible; this is limited by the ability to control the inversion threshold (or equivalent threshold voltage) and by the gate capacitance (the number of carriers introduced by a given gate voltage). It is difficult to conceive of mV gate voltages, but tens of mV should be possible. Operation at source-drain and gate voltages corresponding to the energy gap of a conventional superconductor is not feasible, so operation at larger voltages using high-T_c superconductors, which offer the possibility of gap voltages of tens of mV, is potentially very important. Since the size of the $I_c R_n$ product is very important for any potential JOFET applications, the origin of the basic lack of agreement between experiment and theory needs to be understood through experiments on well-defined and characterized device structures. At present, we can say that the ratio of the maximum value of the $I_c R_n$ product to the gate voltage swing required to switch off the device is a figure of merit which should be as large as possible if these devices are to be of significance (values well exceeding ~0.1 would be suitable); this may be possible with thin gates and high T_c materials.

2.3.7. Gate-Controlled Superconductivity

Is it possible to make a FET having a superconductor as the channel (rather than depending on the proximity effect)? An insulated gate FET is basically a capacitor having a current flowing along one of the plates (the channel). Changing the voltage between the capacitor plates (the gate voltage) changes the number of carriers in the plate, thereby changing channel properties. Although the idea of using the field effect to vary the surface properties of a superconductor is not new [94], its possibilities are still being explored [95,96]. The most obvious device would have a metal channel and use the gate to change (either increase or decrease) T_c and/or resistivity. Alternatively, one might use an insulating material and use the gate to induce a superconducting channel [96,97].

The large carrier density in a metal makes it difficult to produce a significant change in carrier density with an attainable gate electric field. Using a ferroelectric gate insulator (high dielectric constant), it may be possible to produce a few percent change in a 1-nm thick metal film with 10^{22} cm^{-3} carriers [96]. For potential applications, low carrier density metals would be important [95]. For low voltage (low power) applications, however, the limitations of devices of this sort are obvious. Using rather aggressive numbers, a relative dielectric constant of 10 and a gate electric field of 10^7 V/cm, a change in sheet carrier density of 5×10^{13} cm^{-2} is possible, representing a change in carrier density of 10^{20} cm^{-3} in a 5 nm metal film. This may be large enough for some interesting physics experiments, but hardly represents an effect that is likely to find practical applications. Considering the possibility of inducing a superconducting surface layer, 5×10^{13} carriers/cm^{-2} implies a resistivity of order 10 k$\Omega/\square$ (using a mobility of 10 cm^2/V-s), making the possibility of inducing even a metallic surface rather marginal. The fact that 5 V is required to produce a field of 10^7 V/cm, even using a 5-nm thick dielectric, indicates that a metal FET will not be a low voltage device.

2.4. Summary

We have described two distinct types of superconductor–semiconductor hybrid devices that are clearly related to well-known semiconductor devices, namely metal base transistors and FETs. Of the two metal base devices, one (Super-HET) is clearly an extension of existing devices and the other (SUBSIT) embodies new device physics. The field effect device (JOFET) may offer some improvement on conventional FETs, although it may be best viewed as a unique type of controlled Josephson junction with possible applications completely different from those associated with ordinary FETs [56]. The voltage scale for operation of the Super-HET is determined by metal–insulator barrier heights, and not by the characteristic superconducting scale (the gap). The JOFET voltage scale is determined by the required gate voltage swing, which is also unrelated to the energy scale associated with superconductivity. However, the superconducting gap does sets the voltage range over which superconductivity makes a significant difference in device characteristics. In the SUBSIT case, the operating voltage is determined by the superconducting energy gap, independent of other parameters, giving it a low natural operating voltage. The principal difficulty with this device is realizing a very low barrier metal–semiconductor contact for the base–isolator junction.

These devices tend to rely on narrow gap III–V semiconductors and/or heterostructures based on ternary III–V alloys, materials that are significant in exploratory device research. The hybrid devices we have discussed tend to emphasize low resistance non-alloyed contacts, low barrier heights (even at metal interfaces), high mobility, and low effective mass, and thus tend to rely on materials such as InAs and InGaAs, rather than GaAs, AlAs, and InP. There has been much more success with the rather familiar device material InGaAs than with the more exotic InAs. The materials problems associated with these devices are closely related, so that study of one device should shed light on the others.

Finally, the study of devices like SUBSIT or JOFET, along with related experiments such as those we discussed, are of basic interest because of their relevance to the fundamental physics of semiconductor contacts on the meV scale, particularly superconductor–semiconductor contacts, and because they probe the proximity effect in new regimes, i.e., in situations in which there is a large mismatch between the Su and N materials or in which there is a fairly transmissive tunnel barrier present.

3. Nonequilibrium Superconducting Devices

3.1. Introduction

Electronic devices inherently operate out of thermodynamic equilibrium. Some devices, such as the Josephson junction, operate very close to equilibrium, the nonequilibrium aspect in this case being very low energy gradients in the phase of the superconducting order parameter (and negligible deviations of quasiparticle densities from their equilibrium values). Other devices, such as bipolar transistors, involve substantial nonequilibrium carrier

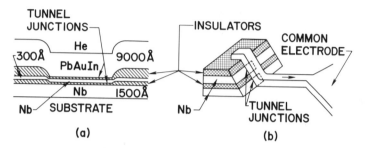

Fig. 9. Schematic representations of nonequilibrium superconducting devices. Planar (a) and edge (b) stacked junction structures. Materials indicated in (a) are those of the quiterons studied in Refs. [1,101], while those indicated for the stacked edge junction structure (b) were used by Buhrman and collaborators [103].

densities. In this section, we focus on attempts to use nonequilibrium superconducting phenomena as the basis for three-terminal devices.

The most commonly investigated nonequilibrium superconducting device structure is stacked back-to-back tunnel junctions, depicted in Fig. 9a. Stacked edge junctions, shown in Fig. 9b, are a variation on this structure that has also been explored. Proposals for stacked-junction superconducting transistor proposals date back to a superconducting gap tunneling device originally proposed by Giaever [98,99], and include a device based on quasiparticle recirculation in the output junction, due to Gray [100], and a gap suppression device called the quiteron, proposed by Faris [101]. Only Giaever's device is based on a microscopic principle that works like the fluid-actuated valve of Fig. 1, which is prototypical of true transistor operation. Consideration of the microscopic mechanism will, however, lead us to propose a second type of stacked junction device that is truly transistor-like.

The other nonequilibrium superconducting device concept that has been investigated is the injection-controlled superconducting link [102]. Planar and edge structures for this device are shown in Fig. 10a,b. These have little microscopic or functional analogy with semiconducting transistors, but have been investigated because devices with substantial gains are easily realized.

We will concentrate here on developments in these classes of nonequilibrium devices that have occurred since the earlier reviews by Gallagher [1] and Buhrman [103]. Thus, we emphasize recent work on quiteron-like devices and injection-controlled links, as well as some new device concepts. The reader is referred to these reviews for more details about earlier work on nonequilibrium superconducting devices. We will have little to say about nonequilibrium devices made with the new high temperature superconduc-

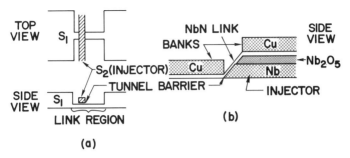

Fig. 10. Planar (a) and edge (b) injection-controlled link structures. The materials for the edge structure are those used by Sprik et al. [119].

tors. There has been no investigation of nonequilibrium superconductivity in these materials yet; neither have there been any convincing realizations of tunnel junctions of reasonable quality, which are elements of all known low-T_c devices.

3.2. Stacked-Junction Devices

3.2.1. Tunneling through the Gap of a Superconductor

The first proposal for a stacked-junction superconducting transistor dates back to Giaever's patent on the superconductor tunnel junction itself [98]. Giaever proposed biasing a normal metal–insulator–superconductor–insulator–normal metal (NISIN) sandwich such that no filled energy states in the outer electrodes line up with empty states in the middle electrode, but filled states in one of the outer electrodes do line up with empty states in the other outer electrode. Ideally in such a situation at low temperatures, no current should flow into the middle electrode, but current can flow between the outer electrodes. The superconducting energy gap forms part of the tunnel barrier between the outer electrodes. By varying the location of the energy gap with a bias on the superconductor, the tunnel current between the outer electrodes should be adjustable . Among the device proposals that we consider in this section, this is the one that could operate closest to equilibrium.

There appears to be no published experimental work directed at observing tunneling through the superconducting gap, which is the phenomenon at the heart of this device. In principle, though, there is no reason to believe the effect will not occur. The technology available at the time of Giaever's original proposal, however, would not have been adequate for observing the effect. The main technological challenge is making the probability of tunneling through the combination of tunnel barriers sufficiently high to pass practical currents while keeping subgap leakage currents through the individual junctions negligibly small. Detailed calculations [99] for ideal quality Nb and NbN junctions indicate that it should be possible to observe the gap tunneling effect at low temperatures (possibly at 1–2 K with NbN), which would be interesting in and of itself.

3.2.2. Gray's Superconducting Transistor

The first demonstration of a stacked-junction superconducting device with some transistor-like properties was made by Gray in 1978 [100]. Gray studied a superconductor–insulator–superconductor–insulator–superconductor (SISIS) stack in which the gain mechanism was based on a novel nonequilibrium superconducting effect involving the admixing of electrons and holes in superconductors and their recirculation in the output, or collector, junction. His device consisted of three stacked 300-Å thick superconducting Al layers, with tunnel junctions separating the layers. One junction, called the injector, had high resistance and was operated with bias voltages above its sum gap. It served to inject excess quasiparticles into the middle film. The other junction, the collector, was biased below its sum gap so as to pass only current due to quasiparticles between its two electrodes. It had a low resistance so that the excess quasiparticle density injected into the middle film was readily shared with the collector film. Due to the dual part hole-like and part electron-like nature of superconducting excitations, it is possible for long-lived excitations to tunnel many times back and forth in the collector junction. Loosely speaking, one can think of electron-like tunneling in one direction and of hole-like tunneling in the other direction. The currents due to these processes add, and the result is that each quasiparticle injected through the injector results in several recirculations through the collector, and thus there is current gain. Gray observed a temperature-dependent current multiplication, which was as large as 4 at the lowest temperature he operated the device, ~ 0.6 K. There was a large signal voltage loss accompanying this, as the injection was above the ~ 0.35 mV sum gap voltage while the collection voltage was below. (A lower gap outer electrode material in the injector could improve the large-signal voltage gain situation.)

Although the electrical characteristics of this device were reminiscent of transistor characteristics, this device, which relies on the quasiparticle recirculation in the output electrodes for gain, has no analog in Landauer's fluid-actuated valve model (Fig. 1). It is not possible to identify separate control and moving charge. As Gray himself pointed out, the device has a common base gain $\alpha > 1$, whereas transistors have $\alpha \leq 1$. Common base gains $\alpha > 1$ result in the possibility of negative current gain in the common emitter and common collector configurations (negative resistance) and these in turn lead to circuit design difficulties, as pointed out by Gallagher [1]. Gain-bandwidth product estimates of only 10^9 Hz by Gray [104] and Frank et al. [105] for devices with Nb electrodes also lead to the conclusion that this device could not be competitive for high performance digital applications. There has been little new work on this sort of device over the last few years, save for a comment that the recirculation effect could enhance the gain of a quasiparticle trapping and multiplying device proposed by Booth [106].

3.2.3. The Quiteron

Faris [101] proposed a SISIS stacked junction device with an operating principle based on nonequilibrium gap suppression in the middle electrode. The gap suppression is caused by injection above the sum gap voltage in one junction, the injector. This modulates the conductivity of the output junction, the acceptor, and its characteristics are expected, at least ideally, to evolve from SIS-like to SIN-like with increasing injection. According to Faris's proposal, there is current gain (over the voltage range at which the output junction modulates) if the output junction can be made to carry substantially more current than the input junction.

Three groups have fabricated and studied quiteron devices, including dc and switching speed measurements [1,101,107,108]. Earlier work [1] had identified isolation as a fundamental problem of this device, symptomatic of the fact that it does not truly function like the fluid actuated valve that is the basis of the transistor. Referring back to the fluid analogy, the lack of isolation of the quiteron device can be seen to stem from the fact that the controlling fluid and the controlled fluid are really the same: the device makes use of only one type of fluid. Whether the source of the fluid is from the injector or the acceptor junction, the gap suppression effects are the same once the fluid is in the middle electrode. The isolation, or nonreciprocity, of a device can be quantified in terms of the ratio of forward to reverse transfer functions. In the only explicit experimental report on this issue, Gallagher [1] measured a 3:1 transresistance ratio for the Nb/I/Nb/I/PbInAu quiterons. This asymmetry was higher than expected and was thought to be due to asymmetric cooling of the outer electrodes of the devices. Unfortunately, this outer electrode response, as we shall discuss in more detail below, is a slow and therefore undesirable part of the device output response.

Recent experimental work has not affected the conclusion that the quiteron lacks isolation, but dynamical studies have confirmed that there is indeed a long time scale associated with a substantial part of the output response. In fact, a fairly consistent picture of the switching performance of quiteron devices emerged from studies of quiteron structures made with three sets of materials: Nb/I/Nb/I/PbInAu studied at IBM [1,101], Sn/I/Sn/I/Sn studied at the University of Tsukuba [107], and NbN/I/PbInAu/I/PbAu studied at Hitachi [108]. The Nb/I/Nb/I/PbInAu quiterons made by IBM, when injected with a ~ 2 ns pulse with a ~ 300 ps rise time, showed a response with a 1 ns rise time and a 1.7 ns fall time [101]. In other measurements [1], these devices when subjected to a step input signal with a 40 ps rise time displayed an output response on two time scales, with about 50% of the response occurring in less than 100 ps and the other 50% of the response occurring over 10 ns [1]. Measurements of the Sn/I/Sn/I/Sn quiterons were made with 2–3 ns time resolution, but showed rise times of ~ 15 ns and decay times of several tens of nanoseconds [107]. The NbN/I/PbInAu/I/PbAu quiterons measured with a Josephson sampler showed a 300 ps turn-on delay, a 300 ps rise time for $\sim 50-70\%$ of the output response, and a slow saturation over several nanoseconds [108]. The response times have been discussed in terms of models for coupled nonequilibrium superconductors. Generally the faster response is due to nonequilibrium effects in the middle electrode, while the slower response reflects heating of the entire structure [109,110,111]. In particular, Iguchi's calculations [111] showed that the first 50% of the response could be quite fast.

One way to eliminate the slow response is by efficiently cooling the two outer electrodes by quasiparticle diffusion, as demonstrated in Hunt and Buhrman's double edge structure [112,113]. On the other hand, the improved cooling in their structure eliminated the negative resistance and precluded the large power gains. Another common suggestion is to make the middle electrode out of a very low carrier density material, such as certain oxide superconductors, which could be driven out of equilibrium with far fewer excess quasiparticles than are necessary to heavily perturb the other electrodes. In either case however, one is still left with a device with a bidirectional response, i.e., a device that lacks isolation, a conclusion that Hunt et al. [113] have independently reached. Buhrman [103] proposed using injection from one NIS junction to switch from the supercurrent state of a second junction to the normal state to get a latching device, but one that at least avoided some of the effects the output has on the input of devices operated in the quiteron mode. This particular proposal does not appear to have been pursued experimentally, but it bears some similarities to injection-controlled links, which are discussed below.

3.2.4. Superconducting Metal Base Tunneling Transistor

Neither Gray's device nor the quiteron truly functions like a transistor, but one can imagine a purely superconducting, nonballistic, SISIS analog of SUBSIT (which was discussed earlier) that does. The base and collector are separated in this case by a highly transmissive, but low leakage SIS tunnel junction, instead of a special superconductor–semiconductor contact. Gain is possible if quasiparticles injected into the base tunnel into the collector before they have a chance to recombine in the base. Transport through the base need not be ballistic, but the injected quasiparticles must not recombine before they tunnel into the collector if the device is to have gain. The principal difficulty is that as injection is made heavier and heavier, recombination, which is a binary process, becomes more likely, and thus gain becomes more difficult to achieve. It is for this reason that a highly transmissive, yet nonleaky, barrier into the collector is needed. Unfortunately, it is more or less universally observed [114] that as tunnel junction specific resistances go below about 10^{-7} Ω-cm^2 (current densities above $\sim 10^4$ A/cm^2 at 1 mV), junction leakage starts to increase rapidly. Whether the degradation mechanism is universal (for example a higher order tunneling process) or is a materials-related problem is not clear.

One variation of this device [115] that could ease the high current density requirements would use spin-polarized quasiparticle injection [116]. In superconductors with low spin-orbit scattering, the injected quasiparticles should stay spin polarized and be long lived. Even under heavy injection, there would be no opposite spin quasiparticles with which they could recombine. Indeed, the nonequilibrium superconducting properties of spin-polarized quasiparticles have never been explored experimentally, and the observation of an enhanced lifetime would be very interesting and could lead to a new way of measuring spin-orbit scattering rates. Recently demonstrated EuS tunnel barriers with substantially different tunneling rates for spin-up and spin-down electrons [117] might lead to a convenient, low field means of inducing spin polarization.

3.3. Injection-Controlled Links

By now it should be clear that there is considerable difficulty in realizing a nonequilibrium superconducting device that truly works like a transistor. Considering the fact that, even including the Josephson device, there is no available high gain digital superconducting device, it is worthwhile asking whether or not there are nonequilibrium devices that, while not possessing all the attributes of transistors, possess high gain or some other property that makes their use attractive. There is indeed one such superconducting device that has demonstrated large power gain, the quasiparticle injection-controlled link operated in a latching mode in which quasiparticle injection through a tunnel junction triggers a superconducting-to-normal transition in a current carrying superconducting line. Wong et al. [102] first demonstrated such a device in which 1.2 mA of quasiparticle current injected at a few millivolts was able to trigger the switching of a 20-fold greater supercurrent in a long line. The line eventually developed up to a 200 mV voltage drop. The voltage gain can be made arbitrarily large by increasing device length, but the response time governed by "hot spot" propagation increases proportionally. The main early interest in this device was to use injection to produce a controlled Josephson element. That is not our concern here. Rather, we will focus on operating these devices in the regime in which large gains are possible. One conclusion of the reviews of both Gallagher [1] and Buhrman [103] was that this class of devices merited more research because of its high gain potential and structural simplicity. To obtain high speed and low power, a very small microbridge was required [1,103], and, even so, the bridge should best be a low carrier density superconductor [103]. In recent years, Takeuchi and Okabe [118] and Sprik et al. [119] have reported on injection-controlled link devices operated in this large signal mode. We will review the results of Sprik et al., which were on submicron devices and included switching speed measurements.

Sprik et al. studied small injection-controlled NbN-links made with an edge junction structure (see Fig. 10b). This structure yielded links with very small dimensions using only optical photolithography. NbN was chosen as the link material because it possesses both high critical supercurrent densities and high normal state resistivities. The combination gives the link both favorable on-state and off-state properties and should lead to considerable Joule heating during switching, which, in turn, it was hoped would enhance the speed of the switching process. The NbN link reported on was 0.3 μm in length, 6 μm in width, and ~350 Å thick and carried a critical supercurrent current density of 3×10^5 A/cm^2 at 4.2 K. The quasiparticles were injected from a Nb-injector

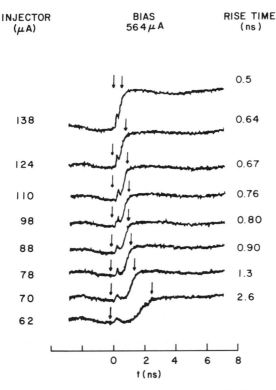

Fig. 11. Observed transient response of the edge injection-controlled link of Sprik et al. [119] shown in Fig. 10b. The indicated injector levels are the amplitude of a step pulse applied with a rise time of ~ 70 ps. The rise times are the delay between the initial spike and the time at which the voltage reaches 80% of the normal state value.

into the NbN-link through a Nb_2O_5 tunnel barrier. The dc-characteristics of the link displayed a critical current of 600 μA, which was reduced to zero by an injection current of 55 μA, for a dc current gain of about 10.

Sprik et al. tested the switching speed of devices incorporated in on-chip $\sim 50\Omega$ coplanar transmission-line structures using a cryogenic probe and standard sampling techniques with a time resolution of ~ 70 ps. Figure 11 shows their data for the development of output voltage across links biased in the superconducting state and triggered by a injection current pulse stepped on with a fast (40–70 ps) rise-time pulse. For low injection amplitudes, the link remained superconducting. When the injector amplitude increased, the link switched into the normal state and a voltage developed. Since the leading edge of the injector pulse was much faster than the response of the output, the latter was a direct measure of the transient response of the device. The leading edge displayed three features: a fast spike followed by a somewhat slower rise over 0.5 to ~2.5 ns followed, in turn, by the final 10–20% of the turn-on that developed rather slowly over several nanoseconds. The overall rise time depended strongly on the injector current and increased when the sum of the injector current and the bias current approached the critical current of the link. The short spike was not due to electrical cross-talk.

Sprik et al. interpreted their results using an extension of a phenomenological model introduced by Tinkham [120] for such supercritical microbridges. Microbridges driven by step-wise supercritical currents display superconducting-to-resistive transitions that are similar in both the overall rise time and in the presence of the initial spike, which was due to a kinetic inductance effect. Tinkham's model combined a time-dependent Ginzburg–Landau equation with an effective temperature description of the quasiparticles, and Sprik et al. added a direct injection term. The predictions for the injection-driven microbridge at least qualitatively described the switching data. The overall time dependence was similar to that of supercurrent-driven microbridges, but the magnitude of the predicted spike was somewhat smaller for the case of quasiparticle injection. Extensive overdrive of the injection enhances speed, but at the expense of current gain. Sprik et al. concluded that the observed switching behavior was governed by condensate dynamics, which in turn was driven by the direct injection of quasiparticles. There was no evidence for a hoped-for enhancement of switching speed due to the large amount of Joule heating that eventually developed in these (short) links. (Hotspot propagation driven by Joule heating would be the dominant factor in the time development of a normal region in a long line.)

This edge injection-controlled link experiment was on about as small a dimensional scale as is practical, but still did not result in sub-100 ps response. The only further means of achieving high speed in such a heavily driven nonequilibrium device is to substantially lower the carrier density. To estimate the requirement, we note that for substantial gap suppression the nonequilibrium quasiparticle density must be of order the critical density $n_c \approx 0.4 \times 2N(0)\Delta(0)$, where $N(0)$ and $\Delta(0)$ are the density of states at the Fermi level and the zero-temperature energy gap of the film. The required injection (input) current is $J_{in} = en_c d/\tau$, where d is the thickness of the perturbed film and τ is the effective relaxation time for the excess quasiparticle population, which should be of order 10 ps. The film thickness has some practical lower limit, say $d \sim 10$ nm. For a reasonably high quality tunnel injector, J_{in} is limited to $\sim 10^5$ A/cm^2, so that the product $N(0)\Delta(0)$ must be less than $\approx 8 \times 10^{18}$ cm^{-3}. For a low-T_c material, $\Delta(0) \sim 2$ meV so that $N(0)$ would have to be about 4×10^{21} cm^{-3}eV^{-1}, which is relatively low, but well within the range of BaPb$_{1-x}$Bi$_x$O$_3$ with $N(0) \approx 10^{21}$ cm^{-3}eV^{-1} [121]. For a high-T_c material, $\Delta(0)$ should exceed 20 mV, so that $N(0)$ must be less than 4×10^{20} cm^{-3}eV^{-1}, which is somewhat below the measured mobile carrier densities of 2×10^{21} cm^{-3}eV^{-1} that are representative of La$_{1.85}$Sr$_{.15}$CuO$_4$, YBa$_2$Cu$_3$O$_{7-x}$, and Bi$_2$Sr$_2$CaCu$_2$O$_8$ (the 80 K phase) [122]. Nevertheless, superconductors with carrier densities close to what would make these devices interesting certainly exist, but it is less clear that device structures with sufficiently high quality elements, e.g., high current density tunnel junctions, can be made with such materials.

With regard to using the high temperature superconductors in nonequilibrium superconducting devices, the lack of any experimental work makes it worthwhile to make only a few additional comments. In general, the times involved for nonequilibrium effects get faster as temperature increases. This implies improved device performance, but it also means that it becomes harder to decouple the electronic excitations from the many additional lattice and other modes that exist on larger energy scales and have considerably more heat capacity at high temperature. This may mean that only heating effects will be possible. For devices such as injection-controlled superconducting links that rely on the resistivity of the superconducting material when it is in the normal state, the high resistivity of the oxide superconductors might be an additional advantage.

3.4. *Conclusions on Nonequilibrium Devices*

The most vigorously pursued nonequilibrium superconducting devices, the quiteron and injection-controlled links, do not function like the fluid-actuated valve that characterizes true transistor-like action. Both of these also involve superconductors driven far from equilibrium. The reciprocity evident in the quiteron rules it out as being of much interest; injection-controlled links might be viable with low carrier density oxide superconductors (low-T_c and high-T_c), but there are serious technological obstacles to incorporating films of these materials in useful device configurations. It may be that basing device operation on superconductors being driven so far from equilibrium will always lead to the slow time constants associated with substantial heating.

Two ways of operating stacked junction devices—by controlled tunneling through the gap of a superconductor or by a tunneling analog of the metal base transistor and SUBSIT—give true transistor action and do not involve driving superconductors so far from equilibrium. Both of these require high current density junctions on very thin electrodes that are of a higher quality than has yet been demonstrated. Further research is required to assess whether or not such high current density, high quality junctions, and the devices they would enable are really viable, and this research appears to be best carried out in a low-T_c context.

4. Magnetic and Other Devices

Superconducting logic gates, beginning with the cryotron and continuing through the use of single Josephson junctions to interferometers (SQUIDs), have relied heavily on magnetic control of supercurrents to provide a pseudo third terminal. However Josephson-based elements are still two-terminal devices, relying on the addition of currents to exceed a threshold value (the use of a magnetic field to reduce the threshold is formally equivalent to simply adding currents). An interesting variation of this type of device is the Super-CIT (Superconducting Current Injection Transistor), proposed and studied by van Zeghbroeck [123], and currently under active investigation by Nordman et al. [124]. The device, discussed in Ref. [1], is electrically a dual of the FET, with a reversal of the roles of current and voltage, magnetic and electric fields, resistance and conductance, etc. Thus, the input voltage, rather than the input current is zero (infinite voltage, rather than current, gain). These devices do not fit the fluid analogy; an appropriate analogy would use flow to control pressure, and the resulting device would have to be used in an entirely new way to do logic. Also, although magnetic devices of this type tend to be very long, restricting their potential application in advanced digital cricuits, analog applications are more appropriate [124]. Finally, this device is related to the vortex flow transistor [125], which is also currently under investigation [126].

Given the difficulties in using magnetic fields to control fast microelectronic devices, the question arises of whether control by magnetic fields can be ruled out on fundamental grounds. For example, the relatively large spatial extent of the quantum of magnetic flux compared to that of an electron has significance for microelectronic devices. Nevertheless, a general dismissal does not come easily, as recognized years ago by Landauer [4]. So it is not surprising that device schemes based upon magnetic control continue to be advanced. Some of these devices may prove useful in applications in which high circuit density is not required.

One novel type of device, proposed by Mannhart et al. [127], uses a magnetic field to control filamentary current flow during avalanche breakdown in a semiconductor (making this a type of superconductor–semiconductor hybrid). In a lightly doped semiconductor, avalanche breakdown occurs at low electric fields, and current flows in well-defined filamentary channels. A perpendicular magnetic field is used to control these currents via the Lorentz force. The device is another magnetic analog of the FET, with nearly zero input impedance. Although operation in the GHz range may be possible, there has been no discussion of the use of these devices in truly high speed circuits, where they presumably suffer from the limitations similar to other magnetic FET analogs [1]. Also, in experiments to date, these devices operate on a scale of volts, and it is not clear that the operating voltage and current can be brought into suitable ranges with voltage and current gain.

Three-terminal devices based on the magnetoelectric effect have also been proposed [128,129]. Goldman [128] suggested a two-junction thin film sandwich, one junction of which contained a magnetoelectric barrier. Switching that junction into the voltage state would produce a magnetic field that is coupled into the other junction. In principle, such a device could be quite fast (~ 10 ps switching time), although the relevant time constant has not been experimentally measured. Unfortunately, the small size of the magnetoelectric coefficient in known magnetoelectric materials makes the required device size rather large (of order mm) and the device margin requirement rather stringent, since the desired junction voltage of order mV limits the available electric field. Prospects for this device would be markedly improved if better magnetoelectric materials were available (the magnetoelectric coefficients of known materials are far from the theoretical upper limit). Sobolewski and Hsiang [129] demonstrated this device experimentally, using a thick magnetoelectric barrier, which had a high input impedance but required several volts on the input to produce switching. Given the known problems with this device, the prospects for a low voltage version appear to be rather bleak.

Finally, there has been considerable recent interest in an area that appears to touch on some of the ultimate limits for such electronic devices, tunnel junctions in which an external electrode controls the transfer of single electrons or Cooper pairs. Three-terminal devices based on these concepts have been proposed in which single electrons control device properties. Fundamental questions remain to be answered prior to any serious discussion of practical applications of such devices. We limit ourselves to referring the reader to the brief discussion of this subject in the chapter in this book by K.K. Likharev, V.K. Semenov, and A.B. Zorin and to the related work of Sugahara and co-workers [130].

5. Conclusions

We have described a number of approaches that have been used in attempts to develop three-terminal superconducting devices, none of which has produced a truly successful device to date. However, the total amount of experimental work in this area is still limited, and the motivation behind attempts to develop these devices remains strong. In some cases, the prospects are very limited, while in others future success depends upon answering several fundamental questions.

Further work towards potential high performance devices based on conventional superconductors driven far from equilibrium does not appear justified. There may be prospects for stacked-junction devices that operate close to equilibrium but require very high quality, high current density junctions. Also, more device research should be directed at exploiting low carrier density superconductors. Some oxide superconductors may have carrier densities low enough to fit into this category.

Hybrid superconductor–semiconductor devices appear promising, but they are in a very early stage of development. It is in this area that progress has been the most significant in the last five years. Based on recent advances in exploratory semiconductor technologies, including FETs and HETs, this progress is likely to continue. SUBSIT is based on an interesting principle, but relies on a special superconductor–semiconductor contact that has yet to be demonstrated. With JOFETs, the major issue is the disparity between input voltage scale and output voltage range over which superconductivity is important, and to what extent this gap can be narrowed.

Other proposed devices are in very rudimentary stages of understanding. Magnetic devices continue to appear, flux flow transistors in particular, that may be useful for specialized analog applications. Explorations of devices based on ultrasmall tunnel junctions are in a very early stage.

Finally, experimental and theoretical explorations related to devices are of interest beyond the potential device applications, as they address fundamentally important issues such as nonequilibrium superconductivity, the proximity effect and the physics of metal–semiconductor and superconductor–semiconductor interfaces.

Acknowledgements

The authors have collaborated with B. Bumble, A. Davidson, C. Jessen, D.P. Kern, T.N. Jackson, D. McInturff, D. Pettit, S.I. Raider, F. Rammo, H. Schmid, and J.M. Woodall in work relating to various aspects of this paper, and wish to acknowledge valuable conversations with R.A. Buhrman, D. Frank, and S. Tiwari.

Partial support of this work was provided by the U.S. Office of Naval Research under contract N00014-85-C-0361.

References

1. Gallagher, W.J. *IEEE Trans. Magn.* **21**, 709 (1985).
2. Nisenoff, M. *Cryogenics* **28**, 47 (1988).
3. Van Duzer, T. *Cryogenics* **28**, 527 (1988).
4. Landauer, R. *Physics Today* **23**, 22 (1970).
5. Keyes, R. *Science* **230**, 138 (1985); *Adv. Electronics and Electron Phys.* **70**, 159 (1987).
6. Keyes, R., Harris, E.P., and Konnerth, K. *Proc. IEEE* **58**, 1914 (1970).
7. Kirschman, R.K., ed. "Low-Temperature Electronics." IEEE Press, New York, 1986.
8. Raider, S.I., Kirschman, R.K., Hayakawa, H., and Ohta, H., eds. "Low Temperature Electronics and High Temperature Superconductors." Electrochem. Soc., Pennington, New Jersey, 1988.
9. Solomon, P.M. *Proc. IEEE* **70**, 489 (1982).
10. Wolf, P. *In* "SQUID '85: Superconducting Quantum Interference Devices and their Applications" (H.D. Halbohm and H. Lübbig, eds.), p. 1127. W. de Gruyter, Berlin, 1985.
11. Kleinsasser, A.W. Superconductor–silicon heterostructures. *In* "Heterostructures on Si: One Step Further with Silicon" (Y.T. Nissim, ed.). Kluwer, The Netherlands, 1988.
12. Frank, D.J., and Davidson, A. *In* "Proc. 5th Intl. Workshop on Future Electron Devices," Miyagi-Zao, Japan, 1988.
13. Chang, L.L., and Esaki, L. *Appl. Phys. Lett.* **31**, 687 (1977).
14. Gueret, P., Kaufmann, U., and Marclay, E. *Electron. Lett.* **8**, 344 (1985).
15. Capasso, F., and Kiehl, R. *J. Appl. Phys.* **58**, 1366 (1985).
16. Yokoyama, N., Imamura, K., Muto, S., Hiyamizu, S., and Nishi, H. *Japan. J. Appl. Phys. Lett.* **24**, L583 (1985); *Sol. State Electron.* **31**, 577 (1988).
17. Hess, K., Morkoc, H., Shichijo, H., and Streetman, B.G. *Appl. Phys. Lett.* **35**, 469 (1979).
18. Kastalsky, A., and Luryi, S. *IEEE Electron Dev. Lett.* **31**, 832 (1983).
19. Dennard, R.H., and Wordeman, M.W. *Physica B* **129**, 3 (1985).
20. Sze, S.M. "Physics of Semiconductor Devices," p. 248. Wiley, New York, 1981.

21. Mead, C.A. *Proc. IRE* **48**, 359 (1960).
22. Heiblum, M. *Sol. State Electron.* **24**, 343 (1981).
23. Luryi, S., and Kastalsky, A. *Physica B* **134**, 453 (1985).
24. Heiblum, M., and Fischetti, M. *In* "Physics of Quantum Electron Devices" (F. Capasso, ed.). Springer-Verlag, Berlin, 1987.
25. Dumke, W.P. *In* "Low Temperature Electronics and High Temperature Superconductors" (S.I. Raider, R. Kirschman, H. Hayakawa, and H. Ohta, eds.), p. 449. Electrochem. Soc., Pennington, New Jersey, 1988.
26. Bozler, C.O., and Alley, G.D. *IEEE Trans. Electron Dev.* **26**, 619 (1980).
27. Malik, R.J., AuCoin, T.R., Ross, R.L., Board, K., Wood, C.E.C., and Eastman, L.F. *Electron. Lett.* **16**, 836 (1980).
28. Seo, K., Heiblum, M., Knoedler, C.M., Oh, J.E., Pamulapati, J., Bhattacharya, P., *IEEE Electron Dev. Lett.* **10**, 73 (1989).
29. Palevski, A., Heiblum, M., Umbach, C.P., Knoedler, C.M., Broers, A.N., and Koch, R.H. *Phys. Rev. Lett.* **62**, 1776 (1989). Palevski, A., Umbach, C.P., and Heiblum, M. *Appl. Phys. Lett.*, **55**, 1421 (1989).
30. Levy, A.F.J. *Appl. Phys. Lett.* **48**, 1609 (1985).
31. Luryi, S. *IEEE Electron Dev. Lett.* **6**, 178 (1985).
32. Levy, A.F.J., and Chiu, T.H. *Appl. Phys. Lett.* **51**, 984 (1987).
33. Sze, S.M., and Gummel, H.K. *Solid State Electron.* **9**, 751 (1966).
34. Tonouchi, M., Sakai, H., and Kobayashi, T. *Japan J. Appl. Phys.* **25**, 705 (1986).
35. Crowell, C.R., and Sze, S. *J. Appl. Phys.* **37**, 2683 (1966).
36. Sakai, H., Kurita, Y., Tonouchi, M., and Kobayashi, T., *Japan J. Appl. Phys.* **25**, 835 (1986).
37. Kobayashi, T., Sakai, H., Kurita, Y., Tonouchi, M., and Okada, M. *Japan J. Appl. Phys.* **25**, 402 (1986).
38. Rosencher, E., Delage, S., Campidelli, Y., and Arnaud d'Avitaya, F. *Electron Lett.* **20**, 762 (1984). Rosencher, E., Badoz, P.A., Pfister, J.C., Arnaud d'Avitaya, F., Vincent, G., and Delage, S. *Appl. Phys. Lett.* **49**, 271 (1986). Hensel, J.C., Levi, A.F.J., Tung, R.T., and Gibson, J.M. *Appl. Phys. Lett.* **47**, 151 (1985). Hensel, J.C. *Appl. Phys. Lett.* **49**, 522 (1986).
39. Tonouchi, M., Hashimoto, K., Sakaguchi, Y., Kita, S., and Kobayashi, T. *Japan J. Appl. Phys.*, to be published (1989).
40. Brosious, P.R., Heiblum, M., and Kleinsasser, A.W. Unpublished work (1984).
41. Heiblum, M., Kleinsasser, A.W., and Woodall, J.M. Unpublished work (1985).
42. Gueret, P., Baratoff, A., Bending, S., Meier, H., Marclay, E., and Py, M. *In* "Proc. Intl. Conf. on High Speed Electronics" (B. Kallback and H. Beneking, eds.), Springer Series on Electronics and Photonics, Vol. 22, p. 24, 1986.
43. Kleinsasser, A.W., Woodall, J.M., Pettit, G.D., Jackson, T.N., Tang, J.Y.-F., and Kirchner, P.D. *Appl. Phys. Lett.* **46**, 1168 (1985).
44. Frank, D.J., Brady, M.J., and Davidson, A. *IEEE Trans. Magn.* **21**, 721 (1985).
45. Davidson, A., Brady, M.J., Frank, D.J., Woodall, J.M., and Kleinsasser, A.W. *IEEE Trans. Magn.* **23**, 727 (1987).
46. Tamura, H., Hasuo, S., and Yamaoka, T. *Japan J. Appl. Phys.* **24**, L709 (1985).
47. Yoshida, A., Tamura, H., Fujii, T., and Hasuo, S. *In* "Proc. 1987 Intl. Superconductivity Electronics Conf.," Tokyo, Japan, 1987.

48. Tamura, H., Fujimaki, N., and Hasuo, S. *J. Appl. Phys.* **60**, 711 (1986).
49. Silver, A.H., Chase, A.B., McColl, M., and Millea, M.F. In "Future Trends in Superconductive Electronics," (B.S. Deaver, C.M. Falco, J.H. Harris, and S.A. Wolf, eds.), p. 364. American Institute of Physics, New York, 1978.
50. Clark, T.D., Prance, R.J., and Grassie, A.D.C. *J. Appl. Phys.* **51**, 2736 (1980). Clark, T.D. PhD Thesis, University of London, 1971.
51. Likharev, K.K. *Rev. Mod. Phys.* **51**, 101 (1979).
52. Van Duzer, T., and Turner, C.W. "Principles of Superconductive Devices and Circuits," p. 302. Elsevier, New York, 1981.
53. Deutscher, G., and de Gennes, P.G. In "Superconductivity" (R.D. Parks, ed.), Vol. 2, p. 1005. Marcel Dekker, New York, 1969.
54. Kleinsasser, A.W., and Jackson, T.N. *IEEE Trans. Magn.* **25**, 1274 (1989).
55. Klapwijk, T., Heslinga, D.R., and van Huffelen, W.M. In "Proc. NATO Adv. Study Institute on Superconducting Electronics," II Ciocco, Italy, June 26–July 6, 1988.
56. Glasser, L. *IEEE J. Sol. State Ckts.*, to be published (1989).
57. Nishino, T., Miyake, M., Harada, Y., and Kawabe, U. *IEEE Electron Dev. Lett.* **6**, 297 (1985).
58. Nishino, T., Yamada, E., and Kawabe, U. *Phys. Rev. B* **33**, 2042 (1986).
59. Kleinsasser, A.W. *Phys. Rev. B* **35**, 8753 (1987).
60. Takayanagi, H., and Kawakami, T. *Phys. Rev. Lett.* **54**, 2449 (1985).
61. Takayanagi, H., and Kawakami, T. In "Intl. Electron Dev. Mtg. Digest," p. 98. IEEE, Piscataway, New Jersey, 1985.
62. Kleinsasser, A.W., Jackson, T.N., Pettit, G.D., Schmid, H., Woodall, J.M., and Kern, D.P. *Appl. Phys. Lett.* **49**, 1741 (1986).
63. Kleinsasser, A.W., Jackson, T.N., McInturff, D., Pettit, G.D., and Woodall, J.M. Unpublished work (1987).
64. Kawakami, T., Akazaki, T., and Takayanagi, H. In "Proc. 1987 Intl. Superconductivity Electronics Conf.," p. 174. Tokyo, Japan, 1987.
65. Ivanov, Z., and Claeson, T. *Japan J. Appl. Phys.* **26**, suppl. 26-3, 1617 (1987).
66. Hiraki, M., and Sugano, T. In "Proc. 1987 Intl. Superconductivity Electronics Conf.," p. 218. Tokyo, Japan, 1987.
67. Nishino, T., Hatano, M., Hasegawa, H., Murai, F., Kure, T., Hirawa, A., Yagi, K., and Kawabe, U. *IEEE Electron Dev. Lett.* **10**, 61 (1989).
68. Kleinsasser, A.W., Jackson, T.N., McInturff, D., Pettit, G.D., Rammo, F., and Woodall, J.M. *Appl. Phys. Lett.* **55**, 1909 (1989).
69. Kroger, H. *IEEE Trans. Electron Dev.* **27**, 2016 (1980).
70. Seto, J., and Van Duzer, T. In "Proc. 13th Intl. Conf. on Low Temp. Phys.," Vol. 3, p. 328. Plenum, New York, 1972.
71. Hatano, M., Nishino, T., Murai, F., and Kawabe, U. *Appl. Phys. Lett.* **53**, 409 (1988).
72. Huang, C.L., and Van Duzer, T. *IEEE Trans. Electron Dev.* **23**, 579 (1976).
73. Gudkov, A.L., Likharev, K.K., and Makhov, V.I. *Sov. Tech. Phys. Lett.* **11**, 587 (1985). Gudkov, A.L., Zhuravlev, Yu.E., Makhov, V.I., and Tyablikov, A.V. *Sov. Tech. Phys. Lett.* **9**, 457 (1983). Huang, C.L., and Van Duzer, T. *Appl. Phys. Lett.*

25, 753 (1974); *IEEE Trans. Magn.* **11**, 766 (1975). Kandyba, P.E., Kolesnikov, D.P., Kolyasnikov, V.A., Koretskaya, S.T., Lavrishchev, V.P., Ryzhkov, V.A., Samus, A.N., and Semenov, V.K. *Sov. Microelectronics* **7**, 140 (1978). Okabe, Y., and Takatsu, M. *Japan J. Appl. Phys.* **24**, 1312 (1985). Ruby, R.C., and Van Duzer, T. *IEEE Trans. Electron Dev.* **28**, 1394 (1981). Schyfter, M., Maah-Sango, J., Raley, N., Ruby, R., Ulrich, B.T., and Van Duzer, T. *IEEE Trans. Magn.* **13**, 862 (1977). Serfaty, A., Aponte, J., and Octavio, M. *J. Low Temp. Phys.* **63**, 23 (1986); **67**, 319 (1987). Seto, J., and Van Duzer, T. *Appl. Phys. Lett.* **19**, 488 (1971).
74. Kawakami, T., and Takayanagi, H. *Appl. Phys. Lett.* **46**, 92 (1985).
75. de Gennes, P.G. *Rev. Mod. Phys.* **36**, 225 (1964).
76. Likharev, K.K. *Sov. Tech. Phys. Lett.* **2**, 12 (1976).
77. van Dover, R.B., De Lozanne, A., and Beasley, M.R., *J. Appl. Phys.* **52**, 7327 (1981).
78. Kleinsasser, A.W., and Jackson, T.N. *Japan J. Appl. Phys.* **26**, suppl. 26-3, 1545 (1987).
79. Aslamazov, L.G., and Fistul, M.V. *JETP Lett.* **30**, 213 (1979); *Sov. Phys. JETP* **54**, 206 (1981). Alfeev, V.N., and Gritsenko, N.I. *Sov. Phys. Sol. State* **22**, 1951 (1980). Itskovitch, I.F., and Shekhter, R.I. *Sov. J. Low Temp. Phys.* **7**, 418 (1981).
80. Alfeev, V.N., Verbilo, A.V., Kolesnikov, D.P., and Ryzhkov, V.A. *Sov. Phys. Semicond.* **13**, 93 (1979). Alfeev, V.N., and Gritsenko, N.I. *Sov. Phys. Sol. State* **24**, 1258 (1982); *Sov. J. Low Temp. Phys.* **9**, 290 (1983); **10**, 218 (1984).
81. Kresin, V.Z. *Phys. Rev. B* **34**, 7587 (1986). Tanaka, Y., and Tsukada, M., *Sol. St. Commun.* **61**, 445 (1987); *Phys. Rev. B* **37**, 5087 (1988); **37**, 5095 (1988).
82. Hatano, M., Nishino, T., and Kawabe, U. *Appl. Phys. Lett.* **50**, 52 (1987).
83. Silvert, W. *J. Low Temp. Phys.* **20**, 439 (1975).
84. Ref. [20], page 264.
85. Roth, L.B., Roth, J.A., and Schwartz, P.M. *In* "Future Trends in Superconductive Electronics" (B.S. Deaver, C.M. Falco, J.H. Harris, and S.A. Wolf, eds.), p. 384. American Institute of Physics, New York, 1978.
86. Boudville, W.J., and McGill, T.C. *J. Vac. Sci. Technol. B* **3**, 1192 (1985).
87. Chang, C.Y., Fang, Y.K., and Sze, S.M. *Sol. St. Electron.* **14**, 541 (1971).
88. Nishino, T., Hatano, M., and Kawabe, U. *Japan J. Appl. Phys.* **26**, suppl. 26-3, 1543 (1987).
89. For a review of Josephson phenomena see, for example, Barone, A., and Paterno, G. "Physics and Applications of the Josephson Effect." Wiley, New York, 1982.
90. McColl, M., Millea, M.F., and Silver, A.H. *Appl. Phys. Lett.* **23**, 263 (1973).
91. Heslinga, D.R., and Klapwijk, T. *Appl. Phys. Lett.* **54**, 1048 (1989).
92. Gurvitch, M., Kastalsky, A., Schwartz, S., Hwang, D.M., Butherus, D., Pearton, S., and Gardner, C.R. *J. Appl. Phys.* **60**, 3204 (1986).
93. Fukuyama, H., and Maekawa, S. *J. Phys. Soc. Japan* **55**, 1360 (1986).
94. Glover, R.E., and Sherrill, M.D. *Phys. Rev. Lett.* **5**, 248 (1960); and references contained in Ref. [1].
95. Hebard, A.F., Fiory, A.T., and Eick, R.H. *IEEE Trans. Magn.* **23**, 1279 (1987).
96. Gurvitch, M., Stormer, H.L., Dynes, R.C., Graybeal, J.M., and Jacobsen, D.C.

9. THREE-TERMINAL DEVICES

In "Superconducting Materials Extended Abstracts." Mat. Res. Soc., Pittsburg, Pennsylvania, 1986.
97. Brazovskii, S.A., and Yakovenko, V.M. *Phys. Lett. A* **132**, 290 (1988).
98. Giaever, I. U.S. Patent No. 3,116,427 (1963).
99. Gallagher, W.J., Kleinsasser, A.W., and Raider, S.I. *IBM Technical Disclosure Bulletin* **27**, 6721 (1985).
100. Gray, K.E. *Appl. Phys. Lett.* **32**, 392 (1978).
101. Faris, S.M. U.S. Patent 4,334,158; *Physica B* **126**, 165 (1984). Faris, S.M., Raider, S.I., Gallagher, W.J., and Drake, R.E. *IEEE Trans. Magn.* **19**, 807 (1983).
102. Wong, T.-W., Yeh, J.T.C., and Langenberg, D.N. *Phys. Rev. Lett.* **37**, 150 (1976); *IEEE Trans. Magn.* **13**, 743 (1977).
103. Buhrman, R.A. Three-terminal non-equilibrium superconducting devices. *In* "SQUID '85: Superconducting Quantum Interference Devices and their Applications" (H.D. Halbohm and H. Lübbig, eds.), pp. 171–188. Walter de Gruyter, Berlin, 1985.
104. Gray, K. Tunneling: A probe of non-equilibrium superconductivity. *In* "Nonequilibrium Superconductivity, Phonons, and Kapitza Boundaries" (K.E. Gray, ed.), pp. 131–168. Plenum, New York, 1981.
105. Frank, D.J., Davidson, A., and Klapwjk, T.M. *Appl. Phys. Lett.* **46**, 603 (1985).
106. Booth, N.E. *App. Phys. Lett.* **50**, 293 (1987).
107. Iguchi, I., and Kashimura, H. *In* "SQUID '85: Superconducting Quantum Interference Devices and their Applications" (H.D. Halbohm and H. Lübbig, eds.), pp. 191–195. Walter de Gruyter, Berlin, 1985.
108. Hatano, M., Hatano, Y., Nishino, T., Harada, Y., and Kawabe, U. *IEEE Trans. Elect. Dev.* **33**, 1286 (1986).
109. Gallagher, W.J. *In* "Proc. 17th Int. Conf. on Low Temp. Phys." (U. Eckern, A. Schid, W. Weber, and H. Wühl, eds.), p. 441. Elsevier, 1984.
110. Frank, D.J., *J. Appl. Phys.* **56**, 2553 (1984).
111. Iguchi, I. *J. Appl. Phys.* **59**, 533 (1986).
112. Hunt, B.D., and Buhrman, R.A. *IEEE Trans. Magn.* **19**, 1155 (1983).
113. Hunt, B.D., Rabertazzi, R., and Buhrman, R.A. *IEEE Trans. Magn.* **21**, 717 (1985).
114. Raider, S.I. *IEEE Trans. Magn.* **21**, 110 (1985).
115. Gallagher, W.J. Unpublished (1989).
116. Tedrow, P.M., and Meservey, R. *Phys. Rev. B* **7**, 318 (1973).
117. Moodera, J.S., Hao, X., Gibson, G.A., and Meservey, R. *Phys. Rev. Lett.* **61**, 637 (1988).
118. Takeuchi, K., and Okabe, Y. *IEEE Trans. Magn.* **25**, 1282 (1989).
119. Sprik, R., Gallagher, W.J., Raider, S.I., Bumble, B., and Chi, C.-C. *Appl. Phys. Lett.*, **55**, 489 (1989).
120. Tinkham, M. Heating and dynamic enhancement in metallic weak links. *In* "Nonequilibrium Superconductivity, Phonons, and Kapitza Boundaries" (K.E. Gray, ed.), pp. 231–262. Plenum, New York, 1981.
121. Batlogg, B. *Physica B* **126**, 275 (1984).
122. Shafer, M.W., Penney, T., Olson, B.L., Greene, R.L., and Koch, R.H. *Phys. Rev. B* **39**, 2914 (1989).

123. van Zeghbroeck, B.J. *Appl. Phys. Lett.* **42**, 736 (1984); *IEEE Trans. Magn.* **21**, 916 (1984).
124. McGinnis, J.P., Beyer, J.B., and Nordman, J.E. *IEEE Trans. Magn.* **25**, 1262 (1989); and references therein.
125. Rajeevakumar, T.V. *Appl. Phys. Lett.* **39**, 439 (1981).
126. McGinnis, J.P., Hohenwarter, G.K.G., and Nordman, J.E. *IEEE Trans. Magn.* **25**, 1258 (1989). Hashimoto, T., Enkupu, K., and Yoshida, K. *IEEE Trans. Magn.* **25**, 1266 (1989).
127. Mannhart, J., Huebener, R.P., Parisi, J., and Peinke, J. *Sol. State Commun.* **58**, 323 (1986); Mannhart, J., and Huebener, R.P. *J. Appl. Phys.* **60**, 1829 (1986); Mannhart, J., Parisi, J., Mayer, K.M., and Huebener, R.P. *IEEE Trans. Electron. Dev.* **34**, 1802 (1987).
128. Goldman, A.M. *IEEE Trans. Magn.* **21**, 928 (1985).
129. Sobolewski, R., and Hsiang, T.Y. *In* "Proc. 1987 Intl. Superconductivity Electronics Conf.," p. 167. Tokyo, Japan, 1987.
130. Sugahara, M., Yoshikawa, N., and Murakami, T. *IEEE Trans. Magn.* **25**, 1278; 1286 (1989); and references therein.

CHAPTER 10

Artificial Tunnel Barriers

S.T. RUGGIERO

Department of Physics
University of Notre Dame
Notre Dame, Indiana

1. Introduction 373
2. Early Work 374
3. Semiconductor and Surface-Layer Barriers 377
4. Deposited-Oxide Barriers 378
   4.1. Introduction 378
   4.2. Deposition Techniques and Materials Properties 382
   4.3. Electrical Properties 382
5. Conclusions 386
   References 387

1. Introduction

The use of artificial barriers dates back to the original tunneling studies by Giaever [1,2], which took advantage of the natural oxidation of such metals as Al, Cr, Ni, etc. Besides the use of natural oxides, a variety of semiconductors that could be thermally evaporated (such as Ge) and other materials that could be deposited as thin insulating films (such as polymers, formvar, etc.) were tried in place of naturally grown oxide barriers. First noted by these experimenters—and, as we shall see, this will turn out to be a recurrent theme in the study of artificial barriers—was that although intrinsic tunneling properties could be attributed to these materials, they tend to contain microscopic pin holes, the deleterious effects of which would become apparent for sufficiently thin films. The recognition of this effect, and the means of addressing it, is an important element in both the early and later work on artificial barriers.

The second generation of work that emerged was the use of surface-layer junctions in which a base electrode (initially principally Nb) was coated with a thin layer of a second metal (at first aluminum) that itself formed a robust,

---

Support for the author during the preparation of this manuscript was provided by the National Science Foundation through Grant No. DMR-8610375.

well-defined native oxide. The success of this method is now generally recognized and a variety of successful base-electrode/surface-layer combinations have been recorded. An extension of this method, in which high-quality, high-critical-current junctions have been fabricated, involves the active oxidation of the surface-layer metal by which a thinner barrier can be formed.

The final method used to create artificial barriers has involved the direct deposition of a barrier as an oxide. This can be done reactively, by deposition of a metal in the presence of an oxidizing agent, or by deposition directly from an oxide target by electron-beam (e-beam) or RF sputter deposition.

2. Early Work

In the present day, the term "artificial barrier" has come to have a number of meanings in which advanced materials-deposition techniques have made it possible to play many novel "tricks" with tunnel barriers that would have been out of the question just twenty years ago.

"Artificial" now refers to tunnel barriers that are not principally comprised of the native oxide formed on the material to be tunneled into. The need for artificial barriers was recognized early on in the history of tunneling because of the limited number of materials that form native oxides suitable for tunneling.

After Giaever's landmark results with $Al/Al_2O_3/Pb$ junctions [3,4], demonstrating the now well-known manifestation of the superconducting gap in the tunneling density of states, many materials were tried in place of aluminum oxide. Very early on Jaklevic et al. [5] and Taylor and Burstein [6] used the now so-called surface-layer technique of depositing a very thin (roughly 10 Å or more) surface layer of Al on a material of interest and oxidizing away all or most of this layer to leave an Al_2O_3 barrier on the underlying material. In particular, their work involved CdS, which was of interest because of its photoelectric properties.

In the same paper, less successful experiments with other materials that could be put down as ultra-thin insulating (50–150 Å) films were also described. Included in their Edisonian-like list were such things as evaporated SiO_x, collodion layers, and even formvar, which can be "painted" on to materials as very thin layers. What was evident from the beginning was that the latter class of materials tended to break down as barriers for very thin layers (in the vicinity of 10–20 Å) because of pinholes that produced alternate conduction paths in addition to tunneling.

Giaever [7] repeated this approach and included polymerized oil to the list of potential barrier materials with identical results. That is, it was seen that all artificially applied materials tended to have pinholes that interfered with their tunneling properties when prepared as thin layers.

In the same work, using evaporated Ge films, it was noted that in putting Ge over oxidized Pb films, the situation was improved because of the presence of the PbO. That is, even though PbO was noted to be, by itself, a generally problematical tunneling oxide, it can in effect serve to electrically block the pinholes in the Ge overlayer. The important role played by the PbO was clarified by experiments with Au/Ge/Pb systems wherein the absence of base-electrode oxide, however imperfect, was seen to be crucial in producing acceptable tunnel barriers. This technique of base-electrode oxidation before barrier deposition was also used with Sn/CdSe/Sn junctions [8].

In somewhat later work [9,10], it was shown that post oxidation of Ge barriers was also efficacious in "patching" pinholes in the barrier, presumably by oxidizing the small areas of the base electrode left exposed by the pinholes. It was also demonstrated that barriers could either be put down directly or, in the case of CdS, CdO, ZnS, and ZnO, by the post-deposition reaction of Cd and Zn to sulfur and oxygen. It is also interesting to note that early work of Laibowitz and Eldridge [11] showed that even "glass" barriers could be prepared on silicon by exposure to and reaction with P_2O_5 vapor to form $Si_xP_yO_z$ barriers.

In results along a similar vein, a variety of compound semiconductors were tried by MacVicar et al. [12,13] with Cu base electrodes. The limited success of this work, attributed to chemical interactions between the Cu and some of the barrier materials, could perhaps have also been due in part to the absence of insulating base-electrode oxides.

Unique to this work, though, was the use of amorphous C layers as barriers, thus eliminating the possibility of reaction with the base (or counter) electrodes. It was again found, however, that for barrier thicknesses below 90 Å, the recurrent problem of shorts once again became manifest. Although of limited success in and of itself, this and related work [14] did foreshadow and perhaps help inspire that of more than a decade later which took advantage of another class of chemically stable oxide barriers, as will be discussed in Section 4.

With the technique of sputtering brought to bear on the creation of artificial barriers [15,16], Ge and other technologically interesting compound semiconductors like GaAs and InSb could be deposited directly from targets. At the same time, the use of active (plasma) post-oxidation of barriers was discussed

Table 1. Early Work with Artificial Tunnel Barriers

Junction	Preparation technique	Deposited barrier thickness (Å)	Reference
$CdS/SiO_x/Au$	thermal	50–150	[5]
CdS/collodion/Au	evaporation		
CdS/formvar/Au	and painting		
$Pb/Al/Al_2O_3/Pb$	thermal evaporation		[6]
$Bi/Al_2O_3/Al$	thick-film e-beam		[14]
Pb/CdS/Pb	thermal evaporation		[9]
Sn/Ge/Sn	thermal evaporation		[10]
Mg/MgO/Sn	and reaction of		
Sn/CdS/Sn	Cd and Zn with		
Sn/ZnS/Sn	O_2 and S		
Cd/CdO/Pb			
Zn/ZnO/Pb			
Au/Ge/Pb	thermal evaporation		[7]
Pb/Ge/Pb			
$Si/Si_xP_yO_z/Au$	reaction of Si	30–100	[11]
$Si/Si_xP_yO_z/Al$	with P_2O_5 vapor		
Cu/InSb/C.E.	thermal evaporation	90–270	[12]
Cu/GaAs/C.E.			
Cu/PbS/C.E.			
Cu/HgS/C.E.			
$Cu/As_2Se_3/C.E.$			
B.E./Ge/C.E.			
B.E./Te/C.E.			
B.E./CdS/C.E.			
Pb/C/Pb			
Sn/C/Pb			
Sn/C/Pb			[13]
Al/C/Al			
Al/C/Pb			
Pb/C/Pb			
Rh/C/Pb			
Nb/Ge/Pb	sputter		[15]
Nb/Si/Pb			
Nb/GaAs/Pb			
Nb/InSb/Pb			
Pb/Te/Pb		400–800	[19]
Sn/CdSe/Sn			[8]
Nb/CdSe/Pb			[17]
Nb/Ge/Pb			[16]
Nb/InSb/Pb			
Nb/Te/Pb		10–100	[18]
Nb/Te/Nb			

in connection with CdSe on Nb base electrodes [17]. These two ideas of (1) the application of "advanced" deposition techniques (sputter or, yet to come, electron-beam) and (2) some form of active post-oxidation ushered in the modern era of artificial barriers.

Indeed, by the late '70s to early '80s, the general success of the preparation of artificial semiconducting barriers would become routine and Si barriers would become a workhorse for both basic studies of the A-15 superconductors and barriers in Josephson logic devices. Because of the emergence of Nb, the A-15's, and NbN, artificial barriers became increasingly important because of the desire to create junctions comprised entirely of these "high-temperature" superconductors, which do not have satisfactory native oxides. Artificial barriers also offered the opportunity to employ materials with desired electrical properties, such as Te which possesses a low dielectric constant [18] and are thus preferred for microwave applications.

In Table 1, we list the early work on the preparation of artificial barriers. Modern techniques of artificial barrier deposition will be addressed in the sections to follow.

3. Semiconductor and Surface-Layer Barriers

Modern work with semiconducting barriers was motivated in large part by a desire to form a good quality tunnel barrier on the A-15 superconductors.

This work began with that of the Stanford group [20,21,22] who showed that Si could form a barrier far superior to what could generally achieved with native oxides alone. It was also again seen that the addition of base-electrode native-oxide growth (by post-oxidation) indeed played a supporting role in improving barrier characteristics.

On aspect of the Si (and other semiconducting) barriers that is distinctly different from the native-oxide materials is its small barrier height. This was shown in systematic studies of pure Si [23] and in the case where oxidation was shown to increase the effective barrier height [24]. This amplifies the idea that post-oxidation indeed improves overall junction characteristics, again by plugging pinholes in the artificial barrier.

The intrinsic nature of Si barriers also surely differs from that of native oxides in that while good quality native-oxide barriers (with barrier heights in excess of 1 eV) can be approximated by a simple rectangular quantum-mechanical barrier with a relatively small effective width (in the vicinity of 15 Å), Si barriers are probably much more complex. This is reflected in the relatively large effective tunneling lengths for the material with respect to intrinsic oxide, as will be discussed in more detail in Section 4.

For example, studies have shown that in oxidized (amorphous) Si barriers [25,26], states within the barrier contribute to the observed tunneling density of states, and that tunneling through Si barriers on Ni [27] produces attenuated spin-polarized tunneling peaks, implying electron scattering within the barrier. In this regard, Kroeger *et al.* [28,29] have shown that these states can be reduced to some degree—with a concomitant improvement in tunneling characteristics—by hydrogenation of the Si barrier, which presumably ties up dangling bonds within the barrier that act as trapping sites.

Another approach to forming artificial barriers is the so-called surface-layer technique. The idea is to cap a metal base electrode that does not form a good native oxide, like Nb, with a thin layer (5–80 Å) of metal that does, like Al. This technique was used first with Al for both basic tunneling studies of Nb [30] and with the goal in mind of making practical Nb-based tunnel-junction technology [31,32,33].

Other materials were also successfully employed in this way including Mg [34], Zr [35,36,37], Ta [38,39,40], and a variety of rare-earth metals [41,42].

One problem with this technique is that residual capping material can be left on the surface of the underlying material, complicating the tunneling density of states with undesirable proximity effects. However, surface layers may be the barrier technology of choice when high critical-current density is required, in that dense plasma-grown oxide barriers can be formed with these systems [43].

4. Deposited-Oxide Barriers

4.1. Introduction

In this section, we discuss work involving barriers directly deposited from compound oxide (or fluoride) sources by sputtering or electron-beam deposition. This work is related to the artificial semiconductor work in the sense that the barrier material itself is directly deposited on the base electrode and related to the surface-layer work in that the new barrier formed is an oxide (or fluoride) and hence generally more chemically stable.

This work began with the use of aluminum oxide by the MIT group [44] who electronbeam deposited Al_2O_3 on both Au and Ni liquid-nitrogen-cooled base electrodes. With $Ni/Al_2O_3/Al$ junctions, low leakage ($<1\%$) below the superconducting energy gap was observed. In addition, spin

Table 2. Artificial Metal-Oxide, -Fluoride, and -Nitride Tunnel Barriers

Junction	Preparation technique	Deposited barrier thickness (Å)	Barrier decay length (Å)	Reference
$Ni/Al_2O_3/Al$ $Au/Al_2O_3/Al$	e-beam	14–24		[44]
$NbN/MgO/NbO$	sputter	10		[45]
$Nb/AlF/Pb$ $Nb/ZrF/Pb$	thermal evaporation and reactive sputter	15–80	5 28	[47]
$Au/MgO/Pb$	ion-beam	40–400		[49]
$NbN/MgO/NbN$	sputter	10		[46]
$NbN/CaF_2/PbBi$ $NbN/CaF_2/NbN$	e-beam	80		[48]
$NbN/MgO/NbN$	sputter			[52]
$NbN/AlN/CE$	ion-beam	5–48	15	[50]
$Cu/Al_2O_3/Pb$ $Cu/Al_2O_3/PbBi$	sputter	12–50	1	[61]
$NbN/MgO/NbN$ $NbN/Al_2O_3/NbN$	sputter and e-beam	70		[56]
$NbN/MgO/NbN$	e-beam	15–35		[60]
$NbN/MgO/NbN$	sputter	5–10	1	[53]
$NbN/MgO/NbN$	sputter			[54]
$NbN/MgO/NbN$	sputter	8–60	7	[59]
$NbN/MgO/NbN$ $NbN/Al_2O_3/NbN$ $NbN/MgO-CaO/NbN$ $NbN/MgO/Pb$ $NbN/Al_2O_3/Pb$ $NbN/MgO-CaO/Pb$	sputter/ e-beam of MgO and Al_2O_3	12–50	7 (MgO-CaO)	[70]

splitting due to the presence of the Ni was clearly observed in dI/dV traces, indicating that true tunneling was occurring with the absence of significant scattering within the barrier. Studies of $Au/Al_2O_3/Al$ systems were also performed to eliminate the possibility that base-electrode oxide was a component of the barrier and showed similar low-leakage characteristics.

One strong motivating factor for the re-emergence of the use of directly deposited oxides was the desire to make all-high-temperature superconductor tunnel junctions; in particular, those with NbN base and counter electrodes.

Shoji et al. [45] were the first to report the successful application of this procedure. They prepared large-area samples of NbN/MgO/NbN trilayers and used reactive etching to define individual junctions. X-ray diffraction and RHEED studies showed these MgO films to be amorphous. The utility of this technique was further demonstrated by Niemeyer et al. [46] who produced a series array of over 1000 NbN/MgO/NbN junctions to create a Josephson voltage standard with an output of a full volt.

The inherent chemical stability and low dielectric constants of the metal fluorides, important in limiting junction capacitance for high-speed operation, was recognized by Asano et al. [47] who experimented with barriers of Al and Zr fluoride. Their barriers were deposited by both simple heating of tungsten boats containing AlF_3 and ZrF_4 powders and, in the case of Al fluoride, reactive sputtering. Thick films of the materials analyzed by X-ray diffraction were found to be amorphous.

A second group also looked at fluoridized materials. This was the work of Talvacchio et al. [48] who prepared polycrystalline CaF_2 barriers on preoxidized NbN base electrodes. As in the work of Asano et al. discussed above, they recognized the need for a blocking layer of some type to bolster the overlying barrier layer. In any case, however, junction quality was degraded when NbN counter-electrodes were used.

In an approach that deviates somewhat from the stricture of direct-deposition, Hebard et al. [49] used ion-beam deposition of Mg in an oxygen background to prepare MgO_x barriers on Au substrates at room temperature. The films were found to be polycrystalline and to have a thickness-dependent oxygen content. Nonetheless, the authors indicate that $Au/MgO_x/Pb$ junctions show excellent characteristics, comparable to those for Al_2O_3 deposited on Au [44].

In other experiments with ion-beam techniques, Track et al. [50] created base-electrodes of NbN using two ion sources running in tandem, the first to deposit Nb and the second to form NbN by directing an N_2^+ beam on the growing Nb film. Noting that AlN is also a wide bandgap (≈ 6.3 eV) material and of potential interest as an artificial barrier, they extended their dual-ion-beam deposition technique to create AlN barriers on the NbN base-electrodes by using Al as a starting material.

As reflected in part by the number of papers presented at the 1986 Applied Superconductivity Conference [51], interest in artificial barriers gained momentum during the 1985–1987 time period. Most of this work involved the preparation of NbN/MgO/NbN junctions because of the then recognized compatibility of MgO with NbN to serve as both a substrate (MgO, like NbN, is a B1 material and has a lattice mismatch of roughly 4% with respect to NbN) and a barrier, as judged by the ability to form high sum-gap junctions and tolerate heating. Therefore, junctions were often comprised of a series of layers as Substrate/MgO/NbN/MgO/NbN [52].

The use of NbN/MgO/NbN systems was extended to whole-wafer deposition by Shoji et al. [53]. In this work, the entire wafer was covered with the trilayer and individual junctions were later defined by chemical etching, reminiscent of previous work with Si barriers. Large-area circuits were also prepared by the whole-wafer process at Hypress by Radparvar et al. [54], again with sputtered NbN/MgO/NbN junctions. Here, good critical-current uniformity was attained for J_c's in the vicinity of 10^3 A/cm^2.

Based on the success of MgO, Talvacchio et al. [55] experimented with the idea of improved lattice match with the NbN counter-electrode beyond that provided by MgO by using compound barriers of MgO and CaO. Although the desired effect of improved lattice match was indeed achieved, it would appear from their work that only a marginal advantage in junction characteristics can be gained in this way. Thus, MgO would seem to retain its "preferred-barrier status" at least with respect to artificial barriers on (or under) NbN.

In related work, the Westinghouse group [56] also demonstrated the growth of epitaxial barrier films of MgO and Al_2O_3 by elevated-temperature sputter- and electron-beam deposition to form crystalographically coherent NbN/barrier/NbN structures. They emphasized that although intrinsically better quality barriers could be grown at high temperatures—a parameter they had under their control—it was the degree of complete barrier *coverage* of the base electrode that was more important in obtaining good junction quality. Indeed, it was recognized throughout the course of their work on deposited-insulator barriers that, reminiscent of the earlier work with Ge, Te, and other semiconductor barriers, the need remained to employ base-electrode oxide or other material to plug barrier pinholes.

Reiterating this idea is the work by Asano et al. [57] with Al-based barriers. Here, the process of cleaning Nb with CF_4 apparently leaves a thin insulating layer on the Nb surface. It is believed that this layer, itself a Nb-F compound, is also responsible for suppressing subsequent suboxide formation on the Nb surface, hence forming a stable, insulating blocking layer beneath the barrier layer.

In the NbN/barrier/NbN work, one approach used (Talvacchio and Braginski [58]) was to sputter barriers in a Ar–1% CH_4 gas mixture to form niobium carboxide, which apparently formed in the pinholes during and grew with the barrier layer. This theme of systematic improvement of junction characteristics by pinhole filling was repeated in other NbN/MgO/NbN work by Thakoor et al. [59] with sputter-deposited and by LeDuc et al. [60] with electron-beam-deposited MgO. To plug pinholes, they used plasma oxidation after barrier deposition.

As we see, much of the work on MgO was related to interest in NbN junctions. An additional system, Al_2O_3 was also explored by the Notre Dame group [61]. One difference in these studies was the use of Cu base electrodes, chosen because Cu does not form an insulating oxide. It was found that, as with previous work, high quality junctions could be prepared on such base electrodes by sputtering—in the usual way—directly from a pressed-powder Al_2O_3 target. These studies revealed a clear threshold for barrier thickness in the vicinity of 10 Å below which the failure rate of junctions became high. Indeed, the failure rate is reportedly in accord with a simple statistical model that indicates that as soon as the junction surface is covered by at least one layer of randomly placed atoms, a good junction is produced.

4.2. Deposition Techniques and Materials Properties

As for the deposition of the MgO or Al_2O_3 barriers themselves, a general consensus has emerged for sputter-deposition parameters. Normally, sputtering takes place in 3–20 μm of Ar with RF power in the range of 200–500 W. The deposition rate at the substrate depends on whether the substrate remains fixed over the sputter gun, oscillates over it, or simply passes over it periodically. Reported rates thus cover the range of 5–100 Å/min.

4.3. Electrical Properties

As discussed by Hebard et al. [63], the overall quality of deposited-oxide materials, from the standpoint of their dielectric properties, can be excellent. The absolute magnitude of the dielectric constant of a given material is also of importance, smaller values of which are preferred since this in turn implies smaller values of junction capacitance for a given barrier thickness—an advantage *vis a vis* speed of operation.

This is a distinct advantage for artificial barriers in that materials can be selected for their dielectric properties. Defining $\varepsilon = \kappa\varepsilon_0$, it was found for Al_2O_3 that $\kappa = 9.03$ and that the ratio of thickness to dielectric constant was $t/\varepsilon = 6$ Å. Specifically emphasizing the desirability of low-dielectric materials, Asano et al. [64] reported values of 2–4 Å for t/ε for Zr- and Al-fluoride.

Thus values of this ratio for artificial barriers compare favorably with, for example, Nb oxide which has poor dielectric properties based on its ratio $t/\varepsilon \approx 1$ Å.

A second important point that was raised by the latter paper was the interpretation of barrier heights. For native oxides, barrier height is normally determined with a modified version of the Simmon's model [65,66], which treats a barrier as a perfect quantum-mechanical potential of finite height and width. Therefore, by measuring the rate of increase of the tunnel conductance near zero bias and the junction area and resistance, self-consistent values of the barrier height and width can be obtained from this model.

However, the range of applicability of this approach is thought to be limited. In the case of Nb, for example—as with other natural-oxide barriers—there is a systematic decrease in barrier height as derived by this procedure with increasing barrier width [67]. Thus, one cannot eliminate the possibility that the derived barrier width is exaggerated in magnitude because of a degradation of the barrier as its physical thickness increases. Thus, especially for lower height barriers (say < 0.5 eV), these numbers may perhaps be most usefully viewed as a parameterization of a barrier that is probably quite complicated, perhaps including traps, etc., which decrease the effective barrier height.

One advantage in the interpretation of barrier properties that the artificial barriers have over native-oxides is that the (deposited) barrier thickness is a directly accessible parameter. Since in the case of artificial barriers a sequence of samples can be prepared with varying thickness, in principle the barrier height can be obtained from a plot of the product of Resistance × Area (RA) versus barrier thickness, as discussed below.

A note of caution must again be expressed, however, since once again the situation is more complicated than the simple paradigm would suggest—because again the barriers themselves are not intrinsically simple. Clearly, as noted by both modern and early artificial-barrier investigations, the "plugging" of pinholes is normally required to achieve good quality barriers—either by generating native oxide or a surface passivation layer, prior to barrier deposition or plasma or other active oxidation process after barrier deposition, to plug the pinholes by oxidizing the exposed base electrode beneath the barrier. Therefore, the barrier is at best comprised of two components: small areas of native oxide in parallel with deposited barrier material. One possible exception is the case of barrier material deposited on non-oxidizing base-electrodes in which this process does not occur and which should exhibit unique characteristics.

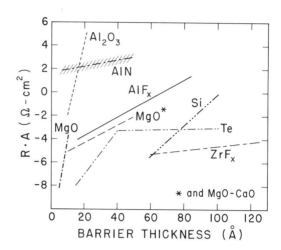

Fig. 1. Resistance × Area as a function of deposited barrier thickness for a variety of deposited-oxide barrier tunnel junctions. Results for Al_2O_3 from Refs. [38,39,40]. Results for AlN from Ref. [50]. Results for MgO from Ref. [53]. Results for MgO* from Refs. [69,70]. Results for AlF_x and ZrF_x from Ref. [47]. Results for Si from Ref. [23]. Results for Te from Ref. [18].

With these considerations in mind, we examine studies of RA as a function of deposited barrier thickness for a variety of barrier materials. These data have been collected in Fig. 1 where we have indicated the trends only of reported data.

If we make the assumption that the barriers are simple rectangular quantum-mechanical potential barriers with a width given by the deposited barrier thickness t and a barrier height ϕ, then in the limit where bias voltage $V \to 0$ it can be shown that [68]

$$RA = 3.17 \times 10^{-11} \left(\frac{t}{\phi^{1/2}}\right) \exp(1.025\, t\phi^{1/2}),$$

where ϕ is expressed in units of eV and t in Å.

Considering only the dominant, exponential behavior of RA with ϕ, we can rewrite the above equation in the form

$$RA = C \exp\left(\frac{t}{\alpha}\right),$$

where α represents an effective tunneling length into the material.

The rapid rise of RA with barrier thickness for the Al_2O_3 and low-thickness MgO films suggests that this is a manifestation of the increase in resistance expected for the respective intrinsic materials. Indeed, values of ϕ

obtained from the above relation are in the range of 1–2 eV, which approach those for native-oxide barriers.

In light of the fact that the Al_2O_3 barriers were deposited on Cu—which presumably does not form a good blocking oxide—this suggests that complete barrier coverage is occurring for $t > 9$–10 Å since the slope of RA versus t for the data appear to give a value for α comparable to that for good quality native-oxide material. As was indeed noted by Barner and Ruggiero [62] for $t < 9$ Å, the barriers tend to fail electrically, suggesting that complete coverage of the base electrode occurs in the vicinity of this thickness value. Furthermore, a study of barrier yield as a function of thickness was shown to be consistent with this idea and with a simple model of barrier formation based on the random buildup of (zero-mobility) Al_2O_3 molecules on the base-electrode surface.

In the case of the MgO barriers, however, the model is that incomplete coverage is accompanied by the filling in of the associated pinholes by oxides of the NbN base electrode. Therefore, for small MgO coverage, the rate of change of RA with t would be large and reflect the large barrier height of the MgO itself, whereas for large thicknesses the tunneling resistance from the MgO will be dominant and the observed tunneling conductance will be associated with the higher conductivity (poorer quality) pinhole material.

In reviewing the MgO data, we see that indeed Shoji et al.'s results [53], which are for 5 Å $< t < 15$ Å, show a relatively large slope and a correspondingly small value of α of 1.3 Å. Alternatively, the results of Thakoor et al. [69] on NbN/MgO/NbN systems and with NbN/MgO-CaO/NbN by the Westinghouse Group [70] show very consistent results in the range $t > 10$ Å, represented by a line with a far lower slope than the low barrier thickness data. In this case, the effective barrier height is far smaller, in the vicinity of 0.4 eV with a correspondingly larger effective tunneling length of 7 Å.

A similar phenomenon appears to be occurring for the Te data which shows a similar crossover behavior from a large slope, in this case for $t < 40$ Å, which implies a tunneling length of roughly 2 Å and a very much smaller, virtually flat, behavior above this thickness.

A baseline for semiconducting material is the Si data, which ranges in thickness from 60–100 Å. These data show a somewhat larger but similar slope as compared with the low thickness Te data and yield an effective tunneling length of 3 Å.

With even larger characteristic lengths are the Al- and Zr-fluoride barriers, showing values α's of 5 and 28 Å, respectively. Therefore, these materials, as prepared, are more in the class of semiconductor materials. This long tunneling length, although it does not appear to degrade junction characteristics, is thought to arise from localized states in the material, tending to decrease the effective barrier height. Such a barrier could have an advantage in controlling junction characteristics since junction parameters are a weaker function of barrier thickness for larger characteristic lengths. In a similar class are the AlN barriers, which have a relatively large characteristic length of 15 Å.

Collectively, then, we see that the deposited-oxide data exhibit two distinct types of behavior. The Al_2O_3 (deposited on Cu) and thin MgO barriers (on NbN) exhibit a rapid rise in RA with thickness and corresponding barrier heights in the vicinity of 1–2 eV and tunneling lengths in the vicinity of 1 Å. For these materials, it appears that this behavior is a reflection of intrinsic oxide material, because of the similarity of these parameters to those for native-oxides. All the other materials, both oxides and semiconductors, for which systematic studies have been performed show distinctly different behavior and generally fall into the class of materials with semiconductor-like behavior. That is, they exhibit barrier heights of less than 1 eV and tunneling lengths in excess of 1 Å.

Certainly for the semiconductor-like barriers, the derived parameters are again not meant to be taken as suggesting that the material represents a clean quantum-mechanical rectangular barrier with a well-defined height and width. They are most likely far more complicated—containing traps and other electronic states and physical inclusions of native base-electrode oxide. Nonetheless, because of the advantages of chemical and structural compatibility with base and counter-electrode materials and the good tunneling characteristics that can be obtained with them, low-barrier-height deposited-barrier materials will continue to be the materials of choice in many applications. Indeed, the very fact that tunneling lengths are large represents a potential advantage in that barrier resistance is a weaker function of deposited barrier thickness; this is important where reproducibility and/or selectivity in resistance is desired.

5. Conclusions

In the end, then, we have seen that artificial tunnel barriers have been with us for two decades. Indeed, since the time the first thin film tunneling experiments were performed to the present day, artificial barriers have been used to create tunnel junctions on a variety of materials.

What we have seen in this brief and selective review is that in the evolution of the subject much has been learned about the preparation and the materials and electrical properties of artificial tunnel barriers. Barriers have been selected to achieve low dielectric loss, large tunneling lengths, high current density, materials compatibility with electrodes, etc.

Ultimately, it appears that the direct deposition of oxides or fluorides of metals has produced the best results for creating barriers on arbitrary materials and will serve future needs in this regard. The problem that remains is to achieve a better understanding of the way in which more dense, uniform, pinhole-free barriers can be produced without resorting to natural oxidation growth as a component to the barrier, or else to minimize this dependency.

Finally, we note that it appears certain that artificial barriers will continue to play a role in tunneling in superconductors, even the now stubborn high-T_c materials—perhaps in a form evolved from those presently available.

References

1. Giaever, I. *Phys. Rev. Lett.* **5**, 147 (1960).
2. Giaever, I., and Megerle, K. *Phys. Rev.* **122**, 1101 (1961).
3. Giaever, I. Energy in superconductors measured by electron tunneling. *Phys. Rev. Lett.* **5**, 147 (1960).
4. Giaever, I. Electron tunneling between two superconductors. *Phys. Rev. Lett.* **5**, 464 (1960).
5. Jaklevic, R.C., Donald, D.K., Lambe, J., and Vassel, W.C. Injection electroluminescence in CdS by tunneling films. *Appl. Phys. Lett.* **2**, 7 (1963).
6. Taylor, B.N., and Burstein, E. Excess currents in electron tunneling between superconductors. *Phys. Rev. Lett.* **10**, 14 (1963).
7. Giaever, I. Metal–insulator–metal tunneling. In "Tunneling Phenomena in Solids" (E. Burstein and S. Lundqvist, eds.). Plenum, New York, 1969.
8. Lubberts, G., and Shapiro, S. Cadmium selenide films at high substrate temperature as tunneling barriers between tin electrodes. *J. Appl. Phys.* **43**, 3958 (1972).
9. Giaever, I. Photosensitive tunneling and superconductivity. *Phys. Rev. Lett.* **20**, 1286 (1968).
10. Giaever, I., and Zeller, H.R. Optical phonons in some very thin II–VI compound films. *Phys. Rev. Lett.* **21**, 1385 (1968).
11. Laibowitz, R.B., and Eldridge, J.M. Tunneling in ultrathin glass films. *J. Vac. Sci. Technol.* **6**, 714 (1969).
12. MacVicar, M.L.A., Freake, S.M., and Adkins, C.J. Thin semiconducting films as tunneling barriers. *J. Vac. Sci. Technol.* **6**, 717 (1969).
13. MacVicar, M.L.A. Amorphous carbon films: Conduction across metal/carbon/metal sandwiches. *J. Appl. Phys.* **41**, 4765 (1970).
14. Esaki, L., and Stiles, P.J. Study of electronic band structures by tunneling spectroscopy: Bismuth. *Phys. Rev. Lett.* **14**, 902 (1965).

15. Keller, W.H., and Norman, J.E. Sputtered thin-film superconductor–semiconductor tunnel junctions. *J. Appl. Phys.* **42**, 137 (1971).
16. Keller, W.H., and Nordman, J.E. Niobium thin-film Josephson junctions using a semiconducting barrier. *J. Appl. Phys.* **44**, 4732 (1973).
17. Rissman, P. Photosensitivity in superconducting tunnel junctions with a cadmium selenide barrier. *J. Appl. Phys.* **44**, 1893 (1973).
18. Nagata, K., Uehara, S., Matsuda, A., and Takayanagi, H. Nb based Te barrier Josephson junctions, *IEEE Trans. Magn.* **17**, 771 (1981).
19. Seto, J., and Van Duzer, T., *Appl. Phys. Lett.* **19**, 488 (1971).
20. Moore, D.F., Zubeck, R.B., Rowell, J.M., and Beasley, M.R. *Phys. Rev. B* **20**, 2721 (1979).
21. Rudman, D.A., and Beasley, M.R. *Appl. Phys. Lett.* **36**, 1010 (1980).
22. Rudman, D.A., Howard, R.E., Moore, D.F., Zubeck, R.B., and Beasley, M.R. *IEEE Trans. Magn.* **15**, 582 (1979).
23. Meservey, R., Tedrow, P.M., and Brooks, J.S. *J. Appl. Phys.* **53**, 1563 (1982).
24. Celaschi, S. *J. Appl. Phys.* **60**, 296 (1986).
25. Bending, S.J., and Beasley, M.R. *Phys. Rev. Lett.* **55**, 324 (1985).
26. Bending, S., Brynsvold, R., and Beasley, M.R. *Adv. in Cryo. Eng. Mater.* **32**, 499 (1986).
27. Meservey, R., Tedrow, P.M., and Brooks, J.S. *J. Appl. Phys.* **53**, 1563 (1982).
28. Kroger, H., Potter, C.N., and Jillie, D.W. *IEEE Trans. Magn.* **15**, 488 (1979).
29. Kroger, H. *IEEE Trans. Electron. Dev.* **27**, 2016 (1980).
30. Wolf, E.L., Zasadinski, J., Osmun, J.W., and Arnold, G.B. *J. Low Temp. Phys.* **40**, 19 (1980).
31. Gurvitch, M., Washington, M.A., and Huggins, H.A. *Appl. Phys. Lett.* **42**, 472 (1983).
32. Tanabe, K., Asano, H., and Michikami, O. *Japan J. Appl. Phys.* **25**, 183 (1986).
33. Yuda, M., Kuroda, K., and Nakano, J. *Japan J. Appl. Phys.* **26**, L166 (1987).
34. Talvacchio, J., Gavaler, J.R., Braginski, A.I., and Janocko, M.A. *J. Appl. Phys.* **58**, 4638 (1985).
35. Celaschi, S., Geballe, T.H., and Lowe, W. *In* "Proc. of the Material Research Society," p. 241, 1982.
36. Celaschi, S., Hammond, R., Geballe, T.H., Lowe, W.P., and Green, A. *Bull. Am. Phys. Soc.* **28**, 423 (1983).
37. Asano, H., Tanabe, K., Katoh, Y., and Michikami, O. *Japan J. Appl. Phys.* **25**, L261 (1986).
38. Ruggiero, S.T., Face, D.W., and Prober, D.E. *IEEE Trans. Magn.* **19**, 960 (1983).
39. Ruggiero, S.T., Arnold, G.B., Track, E., and Prober, D.E. *IEEE Trans. Magn.* **21**, 850 (1985).
40. Ruggiero, S.T., Track, E., Prober, D.E., Arnold, G.B., and DeWeert, M.J. *Phys. Rev. B* **34**, 217 (1986).
41. Umbach, C.P., Goldman, M.A., and Toth, L.E. *Appl. Phys. Lett.* **40**, 81 (1982).
42. Kwo, J., Wertheim, G.K., Gurvitch, M., and Buchanan, D.N.E. *IEEE Trans. Magn.* **19**, 795 (1983).
43. Face, D.F. Thesis, Yale University, 1987.

10. ARTIFICIAL TUNNEL BARRIERS   389

44. Moodera, J.S., Meservey, R., and Tedrow, P.M. Artificial tunnel barriers produced by cryogenically deposited Al_2O_3. *Appl. Phys. Lett.* **41**, 488 (1982).
45. Shoji, A., Aoyagi, M., Kosaka, S., Shinoki, F., and Hayakawa, H. Niobium nitride Josephson tunnel junctions with magnesium oxide barriers. *Appl. Phys. Lett.* **46**, 1098 (1985).
46. Niemeyer, J., Sakamoto, Y., Vollmer, E., and Hinken, J.H. Nb/Al-oxide/Nb and NbN/MgO/NbN tunnel junctions in large series arrays for voltage standards. *Japan J. Appl. Phys.* **25**, L343 (1986).
47. Asano, H., Tanabe, K., Michikami, O., and Beasley, M.R. Fluoride barriers in Nb/Pb Josephson junctions. *Japan J. Appl. Phys.* **24**, 289 (1985).
48. Talvacchio, J., Janocko, M.A., Gavaler, J.R., and Braginski, A.I. UHV deposition and *in-situ* analysis of thin-film superconductors. *In* "Advances in Cryogenic Materials" (R.P. Reed and A.F. Clark, eds.), Vol. 32, p. 527. Plenum, New York, 1986.
49. Hebard, A.F., Fiory, A.T., Nakahara, S., and Eick, R.H. Oxygen-rich polycrystalline magnesium oxide—A high quality thin-film dielectric. *Appl. Phys. Lett.* **48**, 520 (1986).
50. Track, E.K., Lin, L.-J., Cui, G.-J., and Prober, D.E. Dual ion-beam deposition of superconducting NbN films. *In* "Advances in Cryogenic Materials" (R.P. Reed and A.F. Clark, eds.), Vol. 32, p. 635. Plenum, New York, 1986.
51. See *IEEE Transactions on Magnetics* **23**, (1987).
52. Yamashita, T., Hamasaki, K., and Komata, T. Epitaxial growth of NbN on MgO films. *In* "Advances in Cryogenic Materials" (R.P. Reed and A.F. Clark, eds.), Vol. 32, p. 617. Plenum, New York, 1986.
53. Shoji, A., Aoyagi, M., Kosaka, S., and Shibnoki, F. Temperature-dependent properties of niobium nitride Josephson tunnel junctions. *IEEE Trans. Magn.* **23**, 1464 (1987).
54. Radparvar, M., Berry, M.J., Drake, R.E., Faris, S.M., Whitley, S.R., and Yu, L.S. Fabrication and performance of all NbN Josephson tunnel junction circuits. *IEEE Trans. Magn.* **23**, 1480 (1987).
55. Talvacchio, J., Gavaler, J.R., and Braginski, A.I. *In* "Proceedings TMS–AIME: Metallic Multilayers and Epitaxy" (M. Hong and S.A. Wolf, eds.), 1988.
56. Talvacchio, J., and Braginski, A.I. Tunnel junctions fabricated from coherent NbN/MgO/NbN and NbN/Al_2O_3/NbN structures. *IEEE Trans. Magn.* **23**, 859 (1987).
57. Asano, H., Tanabe, K., Michikami, O., and Beasley, M.R. Fluoride Barriers in Nb/Pb Josephson junctions. *Japan J. Appl. Phys.* **24**, 289 (1985).
58. Talvacchio, J., and Braginski, A.I. Tunnel junctions fabricated from coherent NbN/MgO/NbN and NbN/Al_2O_3/NbN structures. *IEEE Trans. Magn.* **23**, 859 (1987).
59. Thakoor, S., Leduc, H.G., Stern, J.A., Thakoor, A.P., and Khanna, K. Insulator interface effects in sputter-deposited NbN/MgO/NbN (superconductor–insulator–superconductor) tunnel junctions. *J. Vac. Sci. Technol. A* **5**, 1721 (1987).
60. LeDuc, H.G., Stern, J.A., Thakoor, S., and Khanna, S.K. All refractory NbN/MgO/NbN tunnel junctions. *IEEE Trans. Magn.* **23**, 863 (1987).

61. Barner, J.B., and Ruggiero, S.T. RF sputter-deposited aluminum oxide films as high quality artificial tunnel barriers. *IEEE Trans. Magn.* **23**, 854 (1987).
62. Barner, J.B., and Ruggiero, S.T. Tunneling in artificial Al_2O_3 tunnel barriers and Al_2O_3-metal multilayers. *Phys. Rev. B* **39**, 2060 (1989).
63. Hebard, A.F., Ajuria, S.A., and Eick, R.H. *Appl. Phys. Lett.* **51**, 1349 (1987).
64. Asano, H., Tanabe, K., Michikami, O., and Beasley, M.R. Fluoride barriers in Nb/Pb Josephson junctions, *Japan J. Appl. Phys.* **24**, 289 (1985).
65. Rowell, J.M. Chapter 27 *in* "Tunneling Phenomena in Solids" (E. Burstein and S. Lundqvist, eds.). Plenum, New York, 1969.
66. Brinkman, W.F., Dynes, R.C., and Rowell, J.M. *J. Appl. Phys.* **41**, 1915 (1970).
67. See discussion in Ref. [61].
68. Simmons, J.G. *In* "Tunneling Phenomena in Solids" (E. Burstein and S. Lundqvist, eds.), p. 135. Plenum, New York, 1969.
69. Thakoor, S., Leduc, H.G., Stern, J.A., Thakoor, A.P., and Khanna, K. Insulator interface effects in sputter-deposited NbO/MgO/NbN (superconductor–insulator–superconductor) tunnel junctions. *J. Vac. Sci. Technol. A* **5**, 1721 (1987).
70. Talvacchio, J., Gavaler, J.R., and Braginski, A.I. *In* "Proceedings TMS–AIME: Metallic Multilayers and Epitaxy" (M. Hong and S.A. Wolf, eds.), 1988.

INDEX

A

Admittance
    imbedding, 179, 180
    matrix, 175
    RF, 174, 180
Amplifier, radio frequency, 81, 82, 93
Analog-to-digital converter, 27, 199
    accuracy, 29
    counting type, 199
    flash-type, 207
        bit-parallel, 208
        fully parallel, 215
        quantum flux parametron, 218
    incremental type, 205
    low frequency, 30
Antenna
    bow-tie, 188, 189
    long-periodic, 188, 189
    planar, 188
    self-complementary, 188
    spiral, 188, 189
Aperture, 204
    time, 207, 211, 215
    TRAP, 205

B

Backward-wave coupler, 244
Bandwidth, 207
    of bridge-type comparator, 216
    of multi-junction SQUID comparator, 210, 213
    relation to sampling frequency, 208
Break junction, 89

C

Cavity, toroidal, 73
Coherence length
    2D electron gas, 347
    in normal metal, 345
    in superconductor, 343
Comparator
    bridge-type, 215
    edge-triggered latch, 210
    multi-junction SQUID, 21, 208, 209
    one-junction SQUID, 21, 213, 217
    sensitivity, 21
    time resolution, 21
Contacts
    ohmic, 332, 340
    Schottky, 339
    superconductor–semiconductor, 336, 342, 344
Convolver, 232, 248, 249
Cooling fin, 64
Correlator, 234, 250
    time-integrating, 261
Counter, 204
    bipolar, 205
Cryotron, 197

D

Delay line
    dispersive, 232, 264
    tapped, 231, 237, 243, 247
        programmable, 248
Detector
    Josephson, 170

photon, 193
saturation power, 171, 191
SIS direct, 171, 189, 190
Dynamic range, 207, 216

E

Electromagnetic shield, 12
Encoder
  for bridge-type comparator, 216
  for one-junction SQUID comparator, 218

F

Filter
  chirp, 246, 261
  Hamming, 246, 255
  transversal, 231
Flip-flop, 203, 204
Flux quantization, 52, 58, 66, 75
Flux quantum, 52, 91

G

Gradiometer, 75, 76, 77, 78
Gravity
  gradiometer, 86
  wave antenna, 83, 84
Gray code, 208, 211

H

High T_c superconductor
  anisotropy, 3
  coherence length, 4
  critical current, 4, 5, 312, 315
  processing
    annealing, 312
    in situ, 313
    oxygen loss, 316
    oxygen source, 314
    surface segregation, 313
  in radiation sources, 164
  in sources, 164
  in SQUIDS, 77, 86, 93
  in three-terminal devices, 363
  surface impedance, 6, 241, 268
Hot electron effect, 64

Hybrid device
  superconductor–semiconductor, 328
  Super-Schottky, 344

I

Insertion loss, 253, 265
Integrator, two-pole, 63
Interconnects, superconducting, 31
Intermediate frequency (IF), 174

J

Josephson junction
  critical current, 53, 87
  current-phase relation, 53
  current-voltage characteristic, 53
    hysteretic, 53
    nonhysteretic, 54
  delay-power characteristics, 103
  edge, 61
  frequency, 7
  impedance, 32
  noise, *see* Noise
  penetration depth, 11, 39, 138
  plasma frequency, 54
  power dissipation, 32, 33
  punch-through, 33
  relaxation oscillation, 17
  relaxation-oscillation driven, 17
    balanced, 19
    energy resolution, 18, 19
    noise, 18, 19
  resistivity shunted, 16, 53, 54, 56
  rise time, 105
  switching speed, 105
  thermal fluctuations, 9
  turn-on delay, 105
  voltage-frequency relation, 53

K

Kramers–Kronig transform, 172

L

Langevin equation, 55, 58
Latch, edge-triggered, 210, 211

Lithography
   electron beam, 59, 61
   photo, 59
Logic gate
   direct coupled, 111, 112
      four junction logic, 113
      Josephson Atto Weber switch (JAWS), 111
      resistor coupled, 112
   hybrid, 114
   magnetically coupled, 107
      current injection device (CID), 109
      three-junction interferometer (JIL), 107
   modified-variable-threshold-type, 114, 219
   record speed, 197
   switching delay, 117

M

Magnetometer, 73, 75, 76, 77, 91, 93
Memory
   cells, 118
      Abrikosov vortex, 123
      destructive read-out, 119
      non-destructive read-out, 118
      variable threshold, 121
   circuits, 128
      random access, 129
      read only, 131
Microprocessor, 130
Mixer
   gain
      available, 176
      coupled, 176, 185
      double sideband (DSB), 185
      single sideband (SSB), 174, 185
   Josephson, 170
   noise, see Noise temperature
   power
      LO, 181
      saturation, 171, 181, 182
   quasioptical, 188, 189
   saturation power, 23
   Schottky diode, 170
   SIN (semiconductor-insulator-normal), 171, 188
   SIS (semiconductor-insulator-semiconductor), quasiparticle, 170, 174

   waveguide, 181, 185
Molecular beam epitaxy, 273, 277
Monopole detector, 79

N

Neuromagnetism, 78, 92, 93
Noise
   $1/f$, 63, 64, 65, 87, 88, 90, 91, 92, 254
   Nyquist, 55, 56, 57, 82, 83
   phase, 253, 266
   quantum, 56
   shot, 177
   in SQUIDs, see SQUID, noise
   thermal, 55, 57, 63, 71, 87
Noise equivalent power (NEP), 171, 191
Noise temperature
   IF amplifier, 171, 186
   mixer, 171, 177, 178, 185
   quantum, 179
   receiver, 171
Non-equilibrium device, 353
   injection-controlled link, 354, 360
   quiteron, 354, 357
      switching speed, 358
   stacked-junction, 355
      gain, 356
Nyquist criterion, 207

O

OR-AND cells, 123
Oscillator
   Josephson, 170
   local, 176

P

Penetration depth, 6
Phase diagram
   of Nb-Ge, 306
   of Nb-Sn, 301
Power spectrum, Lorentzian, 64, 65
Power supply
   ac, 125
   dc latch, 127
   multiphase, 127
   punch-through, 126
Proximity effect, 336, 338
Pulse compression, 232, 245, 247, 268

Q

Quality factor, 241, 253
Quantizer, 200
   SQUAD, 205
   voltage-to-frequency, 200

R

Receivers, 22, *see also* Mixer
   photoexcitation, 23
   bolometer, 24
Refrigeration, 9
Relaxation oscillator, 66
Resistively shunted junction (RSJ) model, 137, 147, 149
Resonance
   nuclear magnetic, 83
   nuclear quadrupole, 83
Resonator, 235
   stripline, 252
Responsivity, 171, 190
RF coupling coefficent, 179

S

Sampling
   frequency, 208
   interval, 204
Schottky barrier, 330, 340
Series-resonant circuit, cooled, 62
Shift register, 219
   flux-shuttle-type, 233
   flux-transfer-type, 220, 221
   resistor-junction type, 223
Single electron logic, 40
Single-flux-quantum pulse, 25, 28, 33
   clock frequency, 36
   logic gate, 39
   NOT gate, 36
   rapid, 33
Slew rate, 63
Sources
   arrays, 136, 142, 143
      distributed, 159, 163
      impedance, 143, 145
      power, 143, 145
      radiation linewidth, 153
      two-dimensional, 145, 152, 163

phase-locking, 139, 141, 146
   arrays, 143, 149, 152
   effects of capacitance, 156
   equivalent circuit, 146, 147, 153
   to external radiation, 146, 147
   locking strength, 150, 151, 158
   mixing current, 147, 150
single junction, 136, 137
   fluxon oscillators, 140, 141, 142
   impedance, 139
   power, 139, 140, 141
   radiation linewidth, 141, 146
   small junction, 137, 144
submillimeter, 135
Specific capacitance, 183
Spectral analysis, 261
   Fourier, 234
   chirp, 234
Spin noise, 83
SQUAD, 205
Squeezed state, 178
SQUID
   dc, 15, 52, 56, 59, 60, 62, 64, 81, 85, 87, 88, 93
   flux, modulating, 61, 62, 63, 71, 75
   flux-locked loop, 61, 62, 63, 71, 75
   flux noise, 57, 63, 75, 89
      energy, 57, 58, 63, 64, 72, 75, 87, 88, 89, 90
   in geophysics, 77, 93
   inductance, 10
   input coil, 59, 60, 78
   multi-junction, 213
   noise temperature, 72, 82, 83, 85
   one-junction, 213, 214, 217
   rf, 14, 52, 66, 67, 68, 71, 73, 74, 87, 89, 93
   transducer, flux-to-voltage, 57
   transfer coefficient, 57, 58
   two-junction, 203, 213, 216, 220
   voltmeter, 81
   zero point fluctuations, 83, 85
Substrate
   dielectric constant, 240, 255
   epitaxial
      lattice mismatch, 285, 287, 296
      magnesium oxide, 293
      niobium–iridium, 306
      sapphire, 287, 290, 302

strontium titanate, 310, 315
loss tangent, 240
preparation, 285, 289
silicon, 245, 258
Superradiance, 142, 143
Surface
 analysis, 283, 284
 resistance, 6, 240
 of superconductors, 241, 259
Susceptance
 imbedding, 180
 quantum, 173, 174, 180
Susceptometer, 79, 80

T

Tank circuit, 68
Test, beat-frequency, 211
Thin film
 molybdenum–rhenium, 299
 niobium, 290
 niobium–germanium, 306, 308
 niobium–iridium, 306
 niobium nitride, 293, 294
 niobium–tin, 301
Three-terminal device, 326, see also Transistor
 fluid analogy, 326
 magnetic field controlled, 364, 366
Threshold characteristic, 208, 211
 distortion, 213, 217
Throughput, 171
Time-bandwidth product, 231
Transformer
 cooled, 62
 flux, 75, 76, 77, 91
 impedance, 247
 superconducting, 15
Transistor
 semiconducting
  bipolar, 328, 329
  field effect, 328
  hot electron, 330, 335
   ballistic transport, 330, 331
   Landauer's model, 330
   switching speed, 330
  metal base, 329
 superconducting
  current injection, 364
  field effect, 338
   critical current, 339, 347

gain, 350
Gray's, see Non-equilibrium device, stacked-junction
metal base, 332
 ballistic, 333
 Schottky barrier, 334
 semiconductor-isolated, 335
 tunneling, 359
Transmission line
 dispersion, 13
 planar, 239
 stripline, 256
TRAP, 205
Trap, electron, 64
Tunnel barrier
 artificial, 373
 barrier height, 377, 383
 deposited oxide, see specific material
deposition technique
 electron-beam, 378, 381
 ion beam, 380
 sputtering, 375, 382
 thermal evaporation, 380
dielectric constant, 380, 382
material
 Al_2O_3, 292, 300, 303, 374, 378, 382, 385
 AlF_3, 380, 385
 AlN, 298
 CaF_2, 303, 380
 Ge, 375
 MgO, 294, 296, 380, 385
 $PrBa_2Cu_3O_7$, 298
 Si, 377
 SrO, 303
 Te, 377, 385
 ZrF_4, 380, 385
native oxide, 292, 293
oxidation, 377
pinholes, 375, 377, 381, 383
surface layer technique, 373, 374
tunneling length, 385
Tunnel junction
 capacitance, 105
 current–voltage characteristic
  ideal, 275
  of Mo–Re, 300
  of NbN, 296
  of Nb_3Sn, 305
 electrode

396 INDEX

Mo–Re, 300
Nb, 291
Nb$_3$Sn, 304
NbN, 294, 380, 381
Pb, 292
Pb–Bi, 293, 296, 304
Ta, 293
negative resistance, 174
photon-assisted, 172, 173, 174, 176
power dissipation, 10
quasiparticle current, 10, 170
subgap resistance, 292, 296, 300, 304

V

Voltage standard, 24
voltage doubling, 25

W

Weak link
coherence length, 345, 346
critical current, 341
grain booundary, 8, 88, 92
S–N–S junctions, 7